TURNING STONE TO BREAD

A Diachronic Study of Millstone Making in Southern Spain

Timothy J. Anderson

SOUTHAMPTON MONOGRAPHS IN ARCHAEOLOGY NEW SERIES 5

2016

HP

SOUTHAMPTON MONOGRAPHS IN ARCHAEOLOGY NEW SERIES 5

ISBN: 978-0-9926336-5-3

A CIP record for this book is available from The British Library

To my family, Aurora Pulido, Erik and Jan Anderson,
and my parents Justice and Mary Ann Anderson

Design and layout by dboreal
http://www.dboreal.com

Published by The Highfield Press, 9 Lower Swanwick Road,
Southampton SO31 7HG, United Kingdom
http://highfieldpress.org/

This book is available from Oxbow Books Ltd, 10 Hythe Bridge Street, Oxford, OX1 2EW
or from Casemate Academic, 1950 Lawrence Rd., Havertown PA 19083, USA
www.oxbowbooks.com

Printed in Great Britain by The Short Run Press Ltd, Exeter

2016

Front cover: Moclín, Granada (GR-1). Back cover: clockwise from upper left: Cerro de Limones, Níjar, Almería (AL-1); Las Calzadillas, Seville (SE-1a); Cantera de los Frailes, Cabra, Córdoba (CO-1); Barranco de los Guillares, Padul, Granada (GR-7). All photographs by T. J. Anderson.

SOUTHAMPTON MONOGRAPHS IN ARCHAEOLOGY

The Stone of Life
Querns, Mills and Flour Production in Europe up to c. AD 500
David Peacock

Farming and Fishing in the Outer Hebrides, AD 600 to 1700
The Udal, North Uist
Dale Serjeantson

The Life and Death of Querns
The Deposition and Use-contexts of Querns in South-Western England from the Neolithic to the Iron Age
Susan R. Watts

Interpreting Shipwrecks
Maritime Archaeology Approaches
Edited by Jonathan Adams and Johan Rönnby

Turning Stone to Bread
A Diachronic Study of Millstone Making in Southern Spain
Timothy J. Anderson

Insight from Innovation
New Light on Archaeological Ceramics
Edited by Emilie Sibbesson, Ben Jervis, and Sarah Coxon

Published by The Highfield Press
Distributed by Oxbow Books

View of the millstone quarry of Moclín (GR-1, in the catalogue). This is without a doubt the largest producer of millstones in the Province of Granada in the 19th century and probably the site alluded to in the following Royal Decree.

"Don Marcelino Franco, merchant of Pontevedra, presented to our Lord the King the request to import forty-two flour millstones reputedly acquired in the kingdom of Portugal; H. M. has refused the request, because of the Law that forbids the import of this article due to the ***many millstone quarries in different provinces of Spain, chiefly in Alava, in Granada and other locations of Andalusia****, where all the needs are met by the abundant exploitation of these stones".*

From *Reales Decretos y Órdenes de S. M. que producen resolución general en su Real Hacienda expedidos en el año de 1829*. Madrid, 1830: 219 (photograph and translation by T. J. Anderson).

CONTENTS

Acknowledgements

1. Introduction 1

1.1. Millstone quarries everywhere 1

1.2. Study area 5

1.3. Previous research on Iberian millstone quarries 6

1.4. Terminology: querns, millstones and quarries 8

1.5. Chronological framework 9

1.6. The problems of identifying millstone quarries 10

1.7. Linguistic questions 12

2. Sources, resources and fieldwork 13

2.1. Sources 13

2.1.1. Conventional sources 13

2.1.1.1. Old geographical dictionaries and other chronicles (59 sites)
2.1.1.2. Historical archives and censuses (19 sites)
2.1.1.3. Molinological studies (28 sites)
2.1.1.4. Historical and archaeological research (14 sites)
2.1.1.5. Millstones in museum depositories
2.1.1.6. Geological studies and itineraries (6 sites)
2.1.1.7. Toponyms
2.1.1.8. Personal communications (11 sites)

2.1.2. Unconventional sources 18

2.1.2.1. Local history reports posted on the Internet (20 sites)
2.1.2.2. Travel itineraries on the Internet (11 sites)
2.1.2.3. Photographs, videos and postcards on the Internet (3 sites)

2.2. Resources: cartography, geology and toponymy 19

2.3. Fieldwork 19

2.3.1. Identifying sites in the field 19
2.3.2. The state of conservation of sites 22
2.3.3. Equating written sources with field finds 23

3. Millstone quarry products and milling installations 25

3.1. Introduction 25

3.2. The saddle quern 26

3.3. Generalities about the rotary quern 26

3.3.1. The Iron Age rotary quern 29
3.3.2. The Roman rotary quern 31
3.3.3. The Medieval rotary quern 34
3.3.4. The Contemporary rotary quern 36

3.4. Human and animal-driven mills 38

3.4.1. The Iberian rotary pushing mill 38
3.4.2. The Roman rotary ring-mill 40
3.4.3. The Roman Pompeian mill 42
3.4.4. The Medieval to Contemporary *tahona* 44

3.5. Watermills 46

3.5.1. Roman watermills 46
3.5.2. Medieval to Contemporary watermills 48

3.5.2.1. *Aceñas* (vertical-wheeled mills)
3.5.2.2. *Rodeznos* (horizontal-wheeled mills)

3.6. Windmills 52

4. Millstone quarry geology 55

4.1. Generalities 55

4.2. Millstone rocks 58

4.2.1. Sedimentary millstone rock 59

4.2.1.1. Limestone tufas or travertines
4.2.1.2. White limestones and dolomites
4.2.1.3. Pebble conglomerates
4.2.1.4. Biocalcarenites

4.2.1.5. *Rosso ammonitico* limestones
4.2.1.6. Other sedimentary rocks

4.2.2. Metamorphic millstone rock (schists) 61
4.2.3. Igneous rocks 61

4.2.3.1. Granitoids
4.2.3.2. Volcanic rocks

4.3. The spread of millstone quarries 62

5. Millstone quarries: terminology, topography, techniques 63

5.1. Terminology 63

5.2. The topography of millstone quarries 64

5.3. "Prospecting" quarries 66

5.4. Production techniques (extraction and fashioning) 67

5.4.1. The tools of the millstone maker 68

5.4.1.1. Direct percussion tools (picks)
5.4.1.2. Indirect percussion tools (hammer, chisel)
5.4.1.3. Leverage tools
5.4.1.4. Other tools

5.4.2. Extraction: cutting cylinders directly from bedrock 70

5.4.2.1. Horizontal extractions
5.4.2.2. Vertical extractions

5.4.3. Splitting cylinders from bedrock 76

5.4.4. Extraction: detaching angular blocks from bedrock 82

5.4.5. Quern and millstone fashioning 85

5.4.5.1. Fashioning rotary querns
5.4.5.2. Fashioning Modern and Contemporary monolithic millstones
5.4.5.3. Fashioning composite millstones

5.5. Quantifying the duration of making a millstone 96

6. Millstone quarry classification 99

6.1. Surface block millstone quarries (MQ-1) 100

6.1.1. Small surface block collection (MQ-1a) 100
6.1.2. Extracting from large surface blocks (MQ-1b) 101

6.2. Bedrock millstone quarries (MQ-2) 106

6.2.1. True extractive quarries (MQ-2a) 106
6.2.2. Block detachment quarries (MQ-2b) 108

6.3. Mixed exploitations: surface block and bedrock quarries 109

6.4. The morphology of extractive millstone quarries 110

6.4.1. Bench quarries 110
6.4.2. Pocket quarries 111
6.4.3. Edge quarries 112
6.4.4. Pit quarries 113
6.4.5. Trench quarries 114
6.4.6. Extensive contiguous shallow quarries 115
6.4.7. Extensive dispersed shallow quarries 116
6.4.8. Subterranean quarries 117

6.5. Quantifying the types of millstone quarries 118

6.5.1. Surface block and bedrock quarries 118
6.5.2. Quantifying the morphology of millstone quarries 119

7. Millstone quarry toponymy **121**

7.1. *Molares* and its derivatives 122

7.2. *Cantera* toponyms 124

7.3. *Piedra, Berrocal* and other toponyms 125

7.4. Toponyms potentially indicative of quarry infrastructure 126

7.4.1. Roads 126
7.4.2. Smithies 126

7.5. Toponyms deriving from Arabic 126

8. Millstone quarry infrastructure **127**

8.1. Tool repair and maintenance 127

8.2. Debris management 130

8.2.1. Downslope debris heaps 131

8.2.2. Backfilling finished sectors 132
8.2.3. Lateral cordons (trench quarry debris) 133
8.2.4. Retaining walls 134

8.3. Residences and shelters 135

8.3.1. Long-term residences 135
8.3.2. Short-term shelters 136

8.4. Transporting millstones 137

8.4.1. Generalities 137
8.4.2. Short-distance transport 139
8.4.3. Long-distance transport 143

8.4.3.1. Inland cart transport
8.4.3.2. Fluvial and maritime transport
8.4.3.3. Rail transport

9. The men behind the millstones 147

9.1. The millstone makers 147

9.2. Permanent or seasonal work 148

9.3. Master millstone makers and crews 149

9.4. Earnings 151

9.5. Occupational hazards 152

10. Millstone quarry ownership and control 155

10.1. Prehistory to the Iron Age 155

10.2. Antiquity 155

10.3. Islamic rule 157

10.4. Post-Islamic rule and the Modern period 157

10.5. Contemporary period 159

10.6. Millstone quarry concessions 161

11. Millstone quarry chronology 165

11.1. Introduction 165

11.2. Chronological indicators 166

11.2.1. Written sources 166
11.2.2. Morphometric indicators (millstone diameters) 170
11.2.3. Rock type indicator 176
11.2.4. Other chronological indicators 176

11.2.4.1. *Mola*-based place names
11.2.4.2. Geographical proximity to a dated feature
11.2.4.3. Relative chronology and vertical stratigraphy
11.2.4.4. Extraction techniques indicator
11.2.4.5. Oral information

11.3. The chronology of millstone quarries in southern Spain 179

11.3.1. Introduction: the chronological assemblages 179
11.3.2. Chronological assemblage 1 181

11.3.2.1. Prehistoric and Early Protohistoric quarries
11.3.2.2. Late Iron Age rotary quern and millstone quarries
11.3.2.3. Roman quern and millstone quarries
11.3.2.4. Undated rotary quern quarries
11.3.2.5. Middle Age rotary quern and millstone quarries

11.3.3. Chronological assemblages 2a and 2b 194

11.3.3.1. Assemblage 2a: quarries dated from 1481 to 1794
11.3.3.2. Assemblage 2b: quarries of the Contemporary period

11.3.4. Assemblage 3: undated quarries 198

11.4. Millstone quarry longevity 198

12. From quarry to mill: millstone distribution 203

12.1. Generalities 203

12.2. Quern distribution in Prehistory and Early Protohistory 205

12.3. Iron Age quern and millstone distribution 206

12.4. Roman millstone distribution 211

12.4.1. Roman rotary quern distribution 212

12.4.1.1. The eastern volcanic rotary querns sector
12.4.1.2. The southern biocalcarenite rotary quern sector

12.4.1.3. The western granite rotary quern sector
12.4.1.4. The distribution of rotary querns of other sedimentary rocks
12.4.1.5. Quantifying Roman rotary quern distribution

12.4.2. Roman Pompeian and ring-mill distribution 223

12.4.2.1. Roman Pompeian mill distribution
12.4.2.2. Roman ring-mill distribution

12.5. Millstone distribution in the Middle Ages 229

12.6. Millstone distribution in the Modern period 230

12.7. Millstone distribution in Contemporary times 232

12.7.1. "Long-distance" millstone distribution 233
12.7.2. "Longer-range" regional millstone distribution 234
12.7.3. "Shorter-range" regional millstone distribution 236
12.7.4. Local millstone distribution 238
12.7.5. Interpreting the distribution of Contemporary millstone quarries 240
12.7.6. The French millstone "invasion" 243

13. From stone to bread 245

13.1. Grit in the bread 245

13.2. White and dark rocks, millstones and bread 246

13.2.1. White and dark millstone rocks and quarries 246
13.2.2. "White" upper stones and "*baza*" lower stones 247
13.2.3. Dark and white bread millstone quarries 248

14. Summary, conclusions and perspectives 253

14.1. Introduction 253

14.2. Sources, resources and fieldwork 253

14.3. Millstone quarry products and milling installations 256

14.4. Millstone quarry geology 265

14.5. Millstone terminology, topography, techniques 266

14.6. Millstone quarry classification 268

14.7. Millstone quarry toponymy 273

14.8. Millstone quarry infrastructure 274

14.9. The men behind the millstones 277

14.10. Millstone quarry ownership and control 278

14.11. Millstone quarry chronology 279

14.12. From quarry to mill: millstone distribution through time 281

14.13. From stone to bread 282

14.14. Final comments and perspectives 282

Bibliography and sources 285

Annexes 311

Annex 1: List of millstone quarries
Annex 2: Spanish-English glossary of millstone quarry terminology
Annex 3: Extracts from the dictionaries of S. Miñano and P. Madoz
Annex 4: Pinilla de Jadrasitque: quarry to mill distances

Catalogue of millstone quarries in southern Spain

pdf download at

http://highfieldpress.org/

Andalusia 330

- Granada (GR) 330
- Almería (AL) 377
- Jaén (J) 402
- Córdoba (CO) 419
- Málaga (MA) 454
- Cádiz (CA) 471
- Seville (SE) 503
- Huelva (HU) 525

Murcia (MU) 537

Castilla La Mancha 547

- Ciudad Real(CR) 547
- Albacete (AB) 561
- Toledo (TO) 566
- Cuenca (CU) 570
- Guadalajara (GU) 573

Valencia 586

- Alicante (A) 587
- Valencia (V) 590
- Castellón (CA) 597

Extremadura 603

- Badajoz (BA) 604
- Cáceres (CS) 614

Madrid (M) 626

Acknowledgements

This book has benefited from the assistance of many people. I would like, in the first place, to acknowledge my family who suffered through the years of research and writing. I especially thank Aurora Pulido who had to put up with me throughout this ordeal. This is also for Erik and Jan, my children.

This book is a highly revised version of my doctoral thesis completed in December 2013 under the direction of Alain Belmont of the History Department of the Université Pierre-Mendès-France at Grenoble, France and the Laboratoire de Recherche Historique Rhône-Alpes (LARHRA - UMR 5190 CNRS). I would like to acknowledge my gratitude to Alain Belmont and the supervisory committee members Jean-Pierre Brun, Olivier Buchsenschutz, Natàlia Alonso, Michel Tarpin and David Peacock. With the recent passing away of David Peacock, the millstone quarry community has lost a scholar and wonderful person. This book is therefore also for him.

I must thank a number of friends and colleagues for contributing so much of their time, expertise, energy and encouragement. Pilar Pascual, Pedro García and Joaquín Sánchez, as millstone quarry specialists in Spain, helped me in innumerable ways. Gurli Meyer, Tor Grenne and Tom Heldal of the Norwegian Geological Survey (NGU) shared hours of fieldwork and the elation of "discovering" several ancient millstone quarry sites. Jane H. Scarrow of the Department of Geology of the University of Granada undertook petrographical analyses, as well as an updated general overview of the geology of Spain. Ignacio Montoro, Mara Pulido and Jaime Medina assisted in a number of manners, in particular in interpreting old Spanish texts. Finally I owe particular gratitude to Jeff Hodges, Mary Ann Anderson (my mother) and Lyn Cutler for their painstaking proofreading.

I also would like to recognise the following individuals for their valuable assistance:

Museum personnel

Saturnino Agüera, Museo Arqueológico de Mazarrón, Murcia
Ildefonso Alcalá, Museo de Jódar, Jaén
Irene Aranega, Museo de Carteia, San Roque, Cádiz
Salvado Bravo, Centro de Interpretación de Baelo Claudia, Cádiz
Carlos Calvo, Museo y Centro de Interpretación de Alcalá la Real, Jaén
Marcos Campos, Museo Histórico Municipal de Priego de Córdoba
María del Mar Capel, Museo Arqueológico de Úbeda, Jaén
Rafael Carmona, Museo Histórico Municipal de Priego de Córdoba
Marcelo Castro, Museo de Linares, Jaén
María Comas, Museo Municipal de Cartagena, Murcia
Åsa Dahlin Hauken, Museum of Stavanger, Norway
José Díaz Pascual, Museo Municipal Raices Conileñas, Conil, Cádiz
Ivan García, Centro de Interpretación de Baelo Claudia, Cádiz
Fátima Gimeno, Museo Arqueológico de Murcia
Esther Gurri, Museu de Badalona, Barcelona
Anna Gutiérrez García-Moreno, Institut Català d'Arqueologia Clàssica
Silverio Gutiérrez, Museo de Historia de Villanueva de Córdoba

Juama Massò, Museo Comarcal Salvador Villaseca, Reus, Barcelona
María Martínez, Museo Arqueológico de Mazarrón, Murcia
Luis de Miquel, Museo Arqueológico de Murcia
José Antonio Moreno, Museo Histórico Municipal de Cabra, Cordoba
Ángel Muñoz, Centro de Interpretación de Baelo Claudia, Cádiz
Ignacio Muñiz, Museo Histórico Arqueológico de Almedinilla, Córdoba
Ana Navarro, Museo de Almería
Encarnación Navarro, Museo Comarcal de Miguel Guirao Géo, Vélez Rubio, Almería
Ildefonso Navarro, Museo Arqueológico de Estepona, Málaga
Manuel Navarro, Museo de Historia, Etnografía y Arqueología de Melilla
Xavier Nieto, Museo Nacional de Arqueología Subacuática, Murcia
Domingo Ortiz Soler, Museo Histórico Municipal de Vera, Almería
José María Palencia, Museo de Bellas Artes de Córdoba
Soledad Pérez, Museo Nacional de Arqueología Subacuática, Murcia
Arturo del Pino, Museo de Almería
Julio Ramón, Museo de Huesca, Aragón
Manuel Ramos, Museo de Almería
Fernando Rojas, Centro de Interpretación de Baelo Claudia, Cádiz
Rubén Sabio, Museo Nacional de Arte Romano, Mérida, Badajoz
Joaquín Salmerón, Museo de Siyâsa, Cieza, Murcia
Marta Santos, Museu d'Arqueologia de Catalunya-Empúries
Ángel Saorien, Museo de Siyâsa, Cieza, Murcia
María Ángeles Sevillano, Museo de los Caminos, Astorga, Léon
José Suárez, Museo de Málaga
Isidro Toro, Museo Arqueológico de Granada

Archaeologists, historians, ethnologists

Andrés Adroher, Granada
Belén Alemán, Almería
Natalia Alonso, Lleida
Jean-Pierre Brun, Paris, France
Pablo Casado, Porcuna
Daniel Castella, Avenches, Switzerland
Javier Castro, Navarra
Aníbal Clemete Cristobal, Placencia, Cáceres
Fernando Colmenarviejo, Colmenar Viejo, Madrid
Guillermo Contrera, Granada
Juan Donoso, Ciudad Real
Pedro García, La Rioja
Ricardo Gómez Ruiz, Campillo, Huelva
Antonio Granero, Albox, Almería
Luc Jaccottey, Besançon, France
Eduardo Kavanaugh, Madrid
José Lájara, Ibi, Alicante
Stéphanie Lepareux-Couturier, Paris, France
Samuel Longepierre, France

Antonio Malpica, Granada
Ágata Marquieguie, Ibi, Alicante
Francisco Martínez López, Murcia
José Antonio Martínez, Rota
Loïc Martínez, Lyon, France
José Antonio Martínez, Rota, Cádiz
Nicolas Minvielle, France
Ignacio Montoro, Granada
José Naranjo, Guaro, Málaga
Juan Ortiz, Zagra, Granada
Juan Palomo, Villanueva de Córdoba
Pilar Pascual, La Rioja
David Peacock, Southampton, UK
Auroro Pulido, Granada
Fernando Quesada, Madrid
Helena Romero, Ciudad Real
Miguel Ángel Vargas, Almadén de la Plata, Sevilla
David Williams, Southampton, UK

Geologists

Aitor Cambeses, Granada
Lidia Catarino, Coimbra, Portugal
Antonio Daza, Belmez, Córdoba
Juan Manuel Fernádez, Granada
Alberto Gil, Almadén de la Plata, Sevilla
Joaquín Sánchez, Menorca
Jane H. Scarrow, Granada
Juan Soto, Granada

Millers, molinological amateurs and municipal personnel

Lucas Acuña Carabantes, Gerena, Seville
Prudente Arjona, Rota, Cádiz
Francisco Castillo, Fuentealbilla, Albacete
Manuel Delgado, Berruecos, Madrid
Juan Jesús Donoso, Granátula, Ciudad Real
Montaña Domínguez, Plasencia, Cáceres
Francisco Estepa, Casariche, Seville
Ana María Ferrín, Barcelona
Alonso García, Zalamea, Huelva
Juan García Cerdá, Montesa, Valencia
Helena Garrido, Guijo de Galisteo, Cáceres
Antonio Hernández, Isleta de Moro, Almería
José Herrera, Casariche, Seville
Miguel Ángel López, Guijo de Galisteo, Cáceres

José Márquez, Castillo de Locubín, Jaén
Manuel Merat, Estepa
David Molino, Castellón
Francisco Moreno, Las Canteras
Marcos Nieto, Sigüenza, Guadalajara
Carlos Osorio, Madrid
José Antonio Polop, Canals, Valencia
Antonio Quesada, Granada
Cándido Rodríguez, Guijo de Galisteo, Cáceres
Javier Pérez Ross, Bohonal de Ibor, Cáceres
José Sánchez Ruiz, Alcázar de San Juan, Ciudad Real
Ángel Serrano, Loja, Granada
Antonio Manuel de la Vega, Medina Sidonia, Cádiz

Technical and administrative assistance

Jean-Paul Collet, LARHRA, OCR recognition of Spanish texts
Rocío Liñán (Archives of Granada)
Amina Quashie, LARHRA

1. INTRODUCTION

1.1. Millstone quarries everywhere

Before the current commercial network producing bread far from where it is consumed, wheat and other cereals were ground and baked locally. Throughout history, after the phases of harvesting and storing, grains had to be ground for consumption, an activity that took place in different types of mills driven by humans, animals, water or wind. The evidence for millstones from the early days of the agricultural revolution of the Neolithic to very recent history is widespread in southern Spain, as seen through querns and millstones in museums collected on archaeological excavations or examples decorating public spaces.

The first veritable grinding tools were saddle-shaped querns, driven with a to-and-fro movement. With the advent of the Iron Age came the circular querns and millstones driven by a rotary motion, a technological advance that was developed further in Antiquity with the introduction of more sophisticated animal-driven or water-driven mills. Owing to the hydraulic engineering capabilities of Spanish Muslims, watermills spread through southern Spain in the 8th and 9th centuries AD (Imamuddin 1981: 107). Closer to our times, the geographer Muhammad al-Idrisi, toward the middle of the 12th century, describes four buildings over the Guadalquivir River in Córdoba, each containing four mills (Martínez Martínez 1987-1988: 226). Henry Swinburne, in his book of travels, quotes the translation of an Arabic text by Abi Abdaalah ben Alkalhibi Abianeni from 1378 describing the surroundings of Granada a century before the fall of Islamic domination:

"I have heard the names of above three hundred hamlets in the environs of Granada: within sight of the city walls may be reckoned fifty colleges and places of worship, and above three hundred water-mills" (Swinburne 1787: 257-258).

The *Repartimientos* of Valencia, land registers penned in the initial phases of the Christian Reconquest, record more than one hundred transfers of mills to Christian settlers (Glick 1994: 982). In light of all the evidence of the ubiquity of mills, one can speculate on where the millers acquired the thousands of hard, compact and abrasive millstones that played such a fundamental role in the daily nutrition of humans.

In the 19th century, the geographical dictionaries of Sebastian Miñano (1826-1829) and Pascual Madoz (1845-1850) specifically record the presence of mills in almost every village in Spain, not to mention the larger towns and cities (see annex 3 for specific references). Koldo Lizarralde, in his quantification of the milling installations recorded by Madoz, has itemised

22,492 flour watermills, 676 windmills and 1476 *tahonas* (animal-driven flour mills) (Lizarralde, 2010, http://murcia.mu/~molinosa/wp-content/uploads/2012/05/MADOZ-KOLDO.pdf). If we take the total number of the flour mills counted by Lizarralde and double it (based on two millstones per mill), we attain a number near 50,000 millstones for the middle of the 19th century. This, of course, is a minimal count because most milling installations possessed more than one pair of stones. Moreover, millstones at this time were not characterised for their longevity because they had to be dressed, in some cases, on a daily basis. In any event, owing to the huge demand, millstone production centres of the 19th century were omnipresent.

It must be noted that at the time that the dictionaries of Madoz and Miñano were penned, the Industrial Revolution had not yet reached Spain, and the Peninsula was still without a complete railway network. French siliceous stones from the quarry districts such as La Ferté-sous-Jouarre in the Paris Basin, deemed the best in the world because of the fine, white flour they yielded (and soon to dominate the European market of bread millstones), were yet to be introduced on a large-scale in Spain due in part to wars and social troubles (Piot 1860: 73). Spanish millers therefore supplied their mills with millstones from local and regional quarries, which also explains their ubiquity across the 19th-century landscape of Iberia.

The situation of Spanish millstone quarries changed in the second half of the 19th century with the introduction of French millstones and then the abandonment in the 20th century of traditional bread mills. As a result, these often spectacular quarry sites, with their high faces at times marked with multiple tubular extraction hollows, were abandoned. Today they are often overgrown by vegetation. Some are converted into dumping grounds, or worse, destroyed by modern construction. This is the case of the Montjuïc near Barcelona, once one of the largest and most celebrated millstone quarries in the western Mediterranean basin.

Neglect of this heritage is even reflected in the scientific community. Whereas watermills and windmills have recently attracted an immense amount of attention and rivers of ink, specialists have remained virtually mute on the quarries that provided the stones to the mills. This neglect is especially visible in the south of Spain as seen by the spread of sites identified in specialised articles before and after the initiation of this research in 2007 (Fig. 1.1- 1.2).

It is therefore my intent to undertake this unique and comprehensive balance of millstone quarries in the southern half of the Iberian Peninsula, an element of heritage mostly forgotten and now often in danger of destruction from both people and nature. It is definitely a heritage worth conserving. As a millstone quarry specialist has recently noted in the summary of his book: "Whereas ancient churches tell the story of religious beliefs, and ruins of castles evoke the bygone power of the elite, millstone quarries are monuments to the glory of work and everyday life" (Belmont 2006a).

Fig. 1.1 (Opposite page): Maps of the Iberian Peninsula indicating a) the millstone quarries identified in research previous to the current study and b) the current spread of 138 millstone quarries (or millstone production districts). The concentration of sites in the upper map in the Province of La Rioja is the work of Pilar Pascual Mayoral and Pedro García Ruíz (e.g. Pascual & Ruíz 2011). The second concentration on the Balearic Island of Minorca is the work of Joaquín Sánchez Navarro (e.g. Sánchez Navarro 2011). Only two quarry sites were published in the study area (dashed line) before or at the outset of this current research in 2007: Cantera de la Rambla Honda (AL-3) (Martínez & Granero 2005) and Fortuna (MU-4) (Matilla Seiquer 2001).

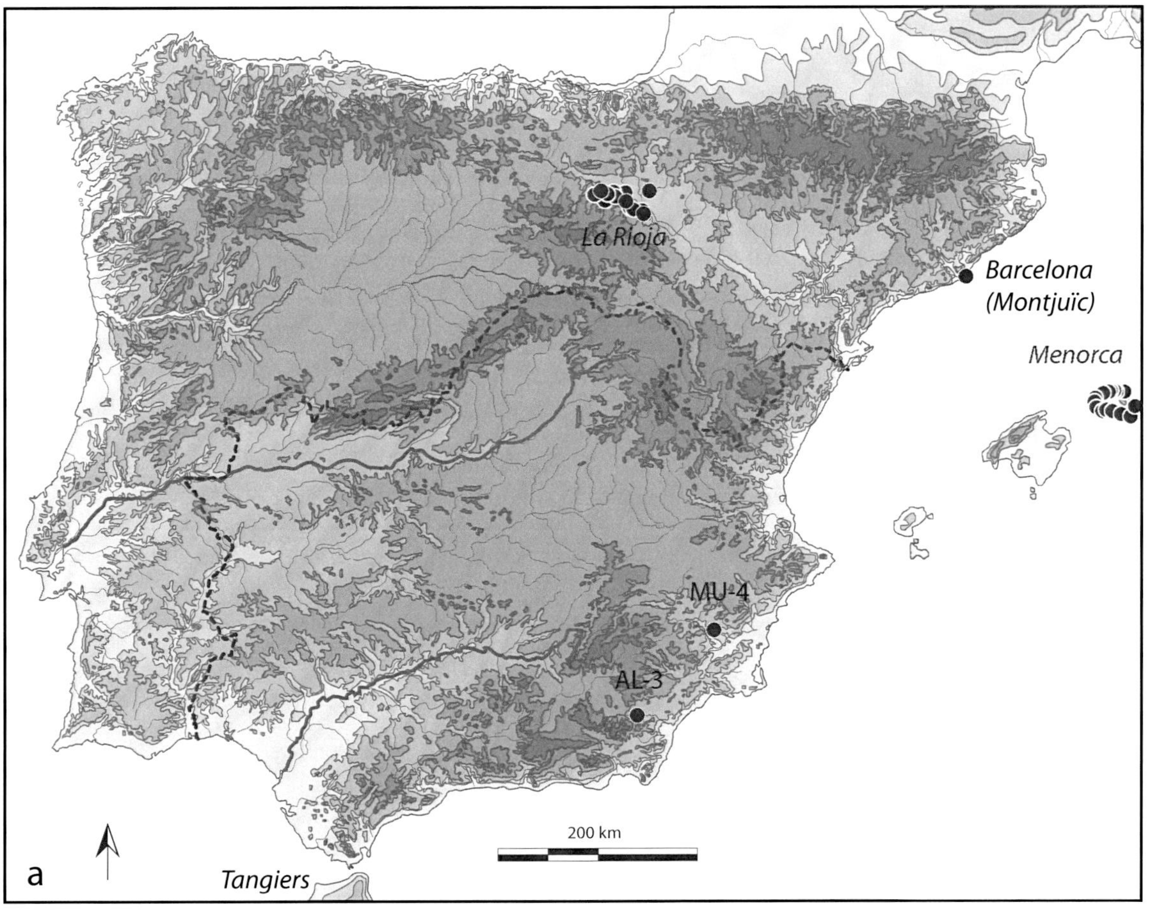
La Rioja
Barcelona
(Montjuïc)
Menorca
MU-4
AL-3
200 km
a
Tangiers

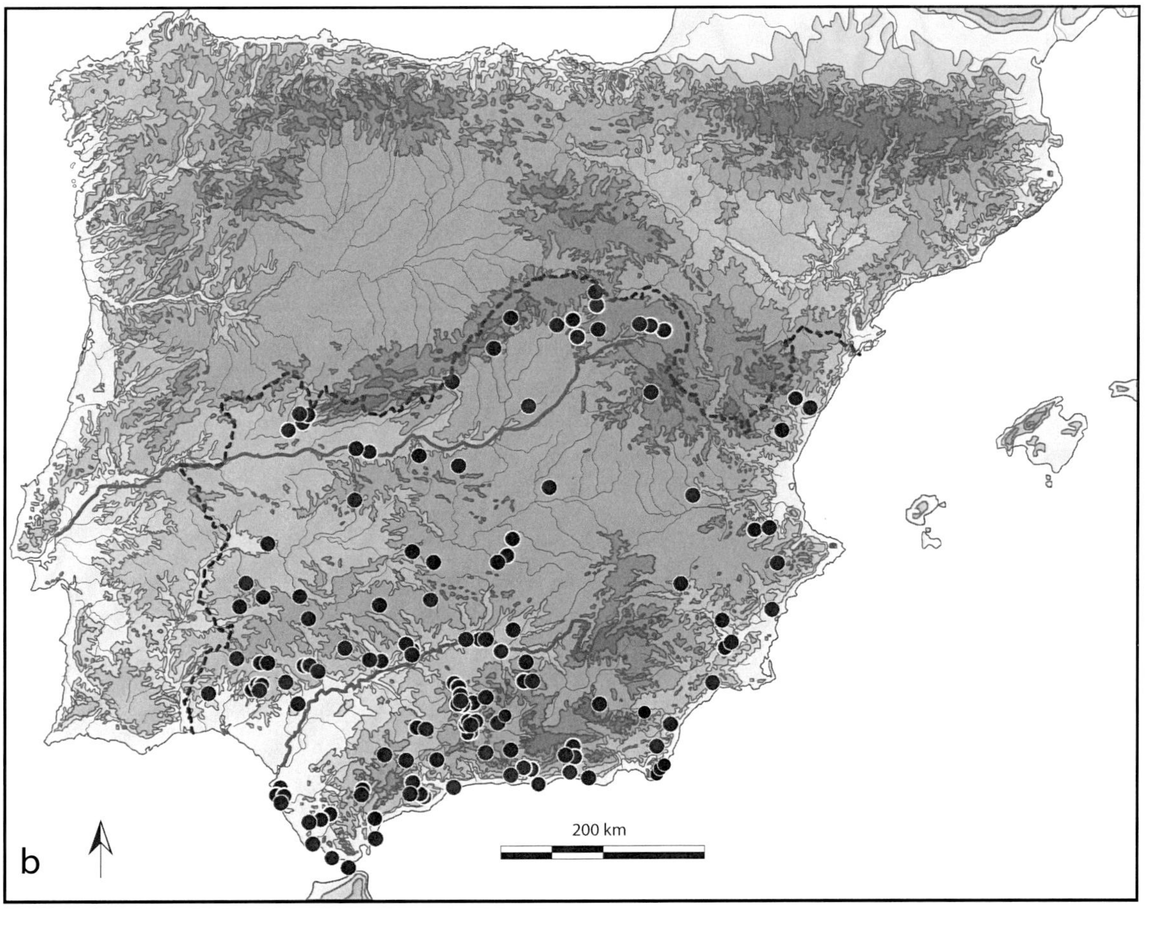
200 km
b

Fig. 1.2: Examples of millstone quarries in southern Spain: a) Los Guillares, Padul, Granada (GR-7); b) Cantera de los Frailes, Cabra (Córdoba) (CO-1); c) Cantera, Moclín (Granada) (GR-1a); d) El Lachar, Jimena (Jaén) (J-2); e) Plasencia (Cáseres) (CC-12); f) Zagra (Granada) (GR-5); g) Teba (Málaga) (MA-2); and h) Cerro Limones (Almería) (AL-1). The codes after the name of each site refer to the catalogue at the end of this work.

1.2. Study area

The area under study comprises six different autonomous communities, corresponding roughly to the southern half of Spain (Fig. 1.3). These autonomous communities are Andalusia, Murcia, Castilla La Mancha, Valencia, Extremadura and Madrid. With the exception of Madrid (M) and Murcia (MU), the communities are sub-divided into provinces: **Andalusia**: Granada (GR), Almería (AL), Jaén (J), Córdoba (CO), Málaga (MA), Cádiz (CA), Seville (SE), Huelva (HU); **Castilla La Mancha**: Ciudad Real (CR), Albacete (AB), Toledo (TO), Cuenca (CU), Guadalajara (GU); **Valencia**: Alicante (A), Valencia (V), Castellón (CS); **Extremadura**: Badajoz (BA) and Cáceres (CC). The abbreviations of these provinces, in conjunction with a number (*e.g.* GR-1), are used throughout this work and in the catalogue to identify the different millstone quarries. Hence, GR-1 is the code for site 1 in the Province of Granada (see catalogue).

From a geographical standpoint, the area comprises a large sector of the Inner Plateau, the Sierra Morena mountains, the Guadalquivir River Basin and the Betic mountains. To the west, the area reaches the Portuguese border, while to the south-west and south-east, the area is bordered by the Atlantic Ocean and the Mediterranean Sea. Its northern border corresponds roughly to an east-west line formed by the Tagus River flowing from east to west. Guadalajara and Castellón, the northernmost provinces of the autonomies of Castilla La Mancha and Valencia, as well as Madrid, are included because they comprise a number of key sites from the historical and petrographical points of view. For practical reasons, research halted at the Portuguese border. Production at the subterranean millstone quarry of Alvito in southern Portugal is therefore not included.

Fig. 1.3: Map of the study area with the borders of the six autonomies and their respective provinces (abbreviated). Indicated also are localities referred to throughout the text (Brañosera in the Palencia Mountains, Alvito in Portugal, Montjuïc by the city of Barcelona and Macael in Almería).

1.3. Previous research on Iberian millstone quarries

As mentioned before, contrary to research on Modern and Contemporary mills, in particular watermills, little work has focused specifically on the quarries that produced the millstones. An early study of the Cerro Redondo quarry in Zaragoza (Cisneros *et al.* 1985) has the merit of including geological analyses. The authors, however, did not consider the possibility that the porphyric granite extractions, ranging between 60 and 80 cm in diameter, could be anything else but column drums. Since the Cerro Redondo publication, millstone quarry research has increased significantly in the last decade. Pilar Pascual and Pedro García in the Provinces of La Rioja and Soria (Pascual & García 2001, 2001-2002, 2002, 2003a-d) and Joaquín Sánchez on the Balearic Island of Menorca (2001, 2005, 2006) are the veritable pioneers of millstone quarry research in Spain. Articles summarising their work appeared recently in the proceedings of a colloquium held in Rome (Pascual & García 2011; Sánchez 2011) organised by David Williams and David Peacock.

Pilar Pascual and Pedro García, recently allied with Javier Castro, have also identified millstone workings in Navarra and the Basque country (Pascual *et al.* 2011; Castro 2012), whereas Joaquín Sánchez has begun surveying the neighbouring Balearic Island of Majorca. Most of their finds are recorded in the Atlas of Millstone Quarries available on the Internet (eu.millstonequarries: http://meuliere.ish-lyon.cnrs.fr).

For Iron Age millstone procurement, there are two articles that delve, although only briefly, into the question of millstone procurement. The first is that of Checa *et al.* (1999) on the querns of *Numantia* (Soria). The second is the work of the team led by Alonso *et al.* (2011) on the mills of the settlement of Els Vilars (Catalonia). For more recent periods in Catalonia, M. Sancho (2008) has briefly noted the quarry of Vilamolar (Lleida) in her study of the exploitation of natural resources in the Middle Ages, and Gutiérrez García-Moreno (2009), in her study of ancient building block quarries, has alluded to the millstone workings at the Montjuïc Mountain at Barcelona. This last site is by far the most celebrated millstone quarry on the Iberian Peninsula (Fig. 1.4-1.5).

Far from the Spanish mainland, but not devoid of interest, is the research by the team of Amelia Rodríguez on the volcanic quern quarries of the pre-colonial Guanche inhabitants of Tenerife in the Canary Islands. This indigenous population, unaware of iron technology, scored rotary querns from bedrock with stone tools (Rodríguez & Cruz 2001).

Fig. 1.4: Photograph from c. 1915 of the sandstone quarry of Montjuïc (Barcelona) with an arrangement of millstones (and other products, notably conical oil rollers). The scene, although certainly staged, reveals the predominance of the pick as a cutting tool (from the blog of Ana María Ferrin, http://amf2010blog.blogspot.com.es/2012/03/gaudi-y-la-piedra-un-magico-acuerdo.html).

Fig. 1.5: Undated photograph of the Montjuïc quarry. To the right, in the foreground, is a half-fashioned millstone. A second millstone can be seen in the background to the left of the compressor. Source unknown.

One of the more complete works dedicated specifically to the millstone industry is the short monograph by Maestro Hernández (2011). His research is based in part on fieldwork, historical archives and interviews with the descendants of the *moleros* (millstone makers) from the mountains of Palencia.

In the study area, the following articles stand out: Matilla (2001) on the oil roller and millstone quarry of Fortuna (MU-4), Murcia; Montero (2008) on the spectacular *rosso ammonitico* quarry of Los Frailes near Cabra (CO-1), Córdoba; Martínez López and Granero (2005) and its revised version (Martínez López *et al.* 2011) on a conglomerate quarry near Albox (AL-3) Almería; Altamirano and Antón (2012) on the Patriarca site (CO-12) on the outskirts of the city of Córdoba; and Molina and Cultrone (2012) on the provenance of the millstones from archaeological contexts in Baza, Granada. None of this research goes beyond the domain of the case study.

Indirectly related to millstone quarries are a number of quern and millstone studies that will be cited throughout this study. Those that stand out are the following: Borges 1978; Py 1992; Checa *et al.* 1999; Alonso 1999; Asensio *et al.* 2001; and Berrocal Rángel 2007. This last study is the most extensive typological work at the national level based on an incomplete catalogue of 406 millstones compiled in the 1940s by Augusto Fernández de Avilés, director of the National Museum of Archaeology of Madrid.

The backbone of this study, nonetheless, is the research that I have undertaken since 2007. This includes data from a series of unpublished reports I have penned based on millstones stored in provincial and municipal museum depositories such as Priego de Córdoba and Almedinilla (Córdoba); *Baelo Claudia* (Cádiz); Vélez Rubio and Almería (Almería); Málaga (Málaga); Baza (Granada); Murcia; and Linares, Úbeda and Alcalá la Real (Jaén), and Mérida (Extremadura).

Along with the unpublished reports, I have published several papers on the subject of Iberian millstone quarries from Protohistory to the Middle Ages. The first paper appeared in the proceedings of the conference of Saint Julien sur Garonne in France (Anderson 2011a). This was followed by two articles in the proceedings of the Rome colloquium, the first being a collaboration with the geologists Tor Grenne of the Norwegian Geological Survey and Juan Manuel Fernández Soler of the University of Granada, on the subject of quern and millstone production in Antiquity in the volcanic provinces of the Cabo de Gata (Almería) and Campo de Calatrava (Ciudad Real) (Anderson *et al.* 2011a). The second was a general review of millstone quarries in southern Spain, a collaboration with the geologist Jane Scarrow of the University of Granada, with emphasis on the Modern and Contemporary productions identified through the Internet and by consulting old texts (Anderson & Scarrow 2011). The third, a diachronic study of rotary querns and their quarries in southern Spain, is a collaboration with J Scarrow and Aitor Cambeses that was recently published to the proceedings of the colloquium held in Bergen, Norway (Anderson *et al.* 2014). A fourth article, published in the proceedings of a millstone meeting held in Arbeca, Catalonia, published in Spanish, focused the classification of millstone quarries (Anderson 2014). Other articles are currently pending publication notably a collaboration with the archaeologist Luc Jaccottey (INRAP, France) and the geologists Jane Scarrow and Aitor Cambeses of the University of Granada submitted to the proceedings of the conference held in 2011 in Lons-le-Saunier, France examining the question of Pompeian mills in the Iberian Peninsula (Anderson *et al.* in press).

This study also benefits from unpublished data stemming from my work on the excavation of three millstone quarries in France, Claix (Charente) and Mont Vouan (Haute-Savoie) under the direction of Alain Belmont (University of Grenoble, LARHRA), as well as two projects, notably a monograph, in collaboration with Åsa Dahlin Hauken (2014a), related to the ancient rotary querns stored in the Museum of Stavanger, Norway (Dahlin Hauken & Anderson 2014a and 2014b). These projects were undertaken in the framework of the Norwegian Millstone Landscapes research project directed by Gurli Meyer, Tom Heldal and Tor Grenne.

1.4. Terminology: querns, millstones and quarries

The term "millstone" in this study, depending on the context, is used in both its broad and narrow sense. "Millstone" in its broad sense includes any type of grinding stone, from the early quern driven with a to-and-fro motion, to the rotary handmill, to the large and heavy stones driven by watermills. In its more narrow sense "millstone" refers specifically to the large stones, most often at least a metre in diameter, driven by water and wind. The equivalent terms in Spanish (*muela*) and French (*meule*) also refer to all types of stones, from the Neolithic to Contemporary times.

All of these different terms are at times ambiguous, leaving one wondering if the reference is indicative of a handmill or a larger stone powered by a more sophisticated mechanism. The English language, contrary to Spanish and French, benefits from the term "quern" to specify smaller hand-driven mills. Thus, in this work, "quern" is exclusively used to designate the small mills driven by hand, whereas "millstone" is most often reserved for the larger circular stones with more complex driving mechanisms such as human or animal pushed mills, watermills or windmills.

The term "quarry" is defined simply as a place where stone is extracted. In English, the term "quarry" applied to millstone extraction is, in a certain sense, a misnomer. According to the *Oxford English Dictionary*, the word has an origin in Middle English and is a variant of the Old French *quarriere,* based on Latin *quadrum* "a square", a shape far from that of a circular millstone. In spite of the irony of the origin of the term, it is adopted in its general sense as a place where millstones were extracted. The Spanish and French languages, contrary to English, benefit from specific terms to designate millstone quarries (*molera* and *meulière).*

For our purposes, the term "quarry" is stretched in the sense that it is not only the place where stone is extracted directly from bedrock by cutting or by detachment, but also specifies natural features such as taluses or screes at the base of a slope, or river beds or ravines where rocks were repeatedly gathered to be carved into millstones.

Concerning the personnel related to millstone quarries, in Spanish and French, contrary to English, there are specific terms (*molero* and *meulier)* to designate the millstone maker. One of the rare images of *moleros* "at work" in Spain is that of the quarry of Brañosera in the mountains of Palencia (Fig. 1.6). In Spanish the term *molero* is also at times associated with a millstone merchant.

*Fig. 1.6: Photograph dating to March 1933 of millstone makers (*moleros*) at the quarry of Brañosera in the mountains of Palencia (from Cuevas Ruiz 2006).*

1.5. Chronological framework

Due to the embryonic state of research on both millstone quarries and millstones themselves in the study area, the production sites can only be placed, at best, in broad chronological niches. A large group of sites defy a chronological classification and can fall into any of various periods. For the study I have adopted the Spanish dating system.

The oldest two quarries producing saddle querns date from the Late Neolithic, towards 3000 BC. A third saddle quern quarry dates from the transition of the Chalcolithic and the Bronze Age, roughly from 2000 and 1500 BC. The Iron Age is only represented by one site dating to the mid-2nd century BC and associated with the Iberian Culture. A series of Roman sites span several centuries from the 2nd century BC until about the 4th-5th century AD. The transition from Late Antiquity to the Early Middle Ages, marked by the Germanic invasion, is hazy and some productions attributed to the Roman period might actually be later, from the 6th and 7th centuries. The Middle Ages is particularly poorly known from the point of view of millstone production, in spite of the major political and social changes that took place with the onset of the Islamic domination in 711. The expulsion of the last Islamic ruler from Granada in 1492 marks the end of the Middle Ages. The Christian Reconquest of all of the territory in the study area, however, was a long gradual process which explains the series of historical documents in Castilian that predate 1492. The few centuries of the Modern period, dated between 1492 and 1789, sees a slight increase in millstone quarries documented by written sources. It is the advent of the Contemporary period in 1789, date of the French Revolution, that marks a peak in millstone production. These are often recorded in geographical dictionaries of the first half of the 18th century. Beyond the 20th century there are very few records of millstone workings.

1.6. The problems of identifying millstone quarries

Differentiating quarries that produce millstones for flour mills from quarries with other types of cylindrical extractions (*i.e.* column drums, edge rollers, sharpening stones, troughs) is not necessarily a simple task. Cylindrical roughouts, extraction hollows and tool marks for all of these types of products are practically identical.

Cerro Bellido (see catalogue, SE-4) in the Province of Seville is an example. The site is reputed to be a Roman exploitation of column drums (Fig. 1.7). The rock is a coarse and porous limestone, similar to rocks often chosen for millstones and the parallel diagonal linear tool marks on the quarry faces are identical to those seen in millstone quarries. The main difference is that the drums, although of a diameter (1.00 m), like many millstones, are roughly twice as thick as millstone blanks.

In this particular case, abandoned millstones found by old watermills in the nearby riverbed of the Yeguas River, has led local historians to speculate that unused drums from Cerro Bellido were recycled in the Middle Ages as millstones (Francisco Estepa, pers. comm.).

Fig. 1.7: View of the Roman column drum quarry of the Cerro Bellido (SE-4, Casariche, Seville) and detail of an abandoned drum.

The last generation of olive rollers for *almazaras* (oil mills), often seen decorating public places throughout southern Spain, are granite cones usually more than one metre in diameter at their base. Due to their shape, there is no possible confusion with cylindrical millstones. There are, however, older edge rollers, for example at the 15th-century mill of Nigüelas (Granada) that are used with upright cylindrical or slightly trunco-conical stones (Fig. 1.8). Depending on the type of rock and their shape, the quarries producing these *almazara* edge rollers, of which at least a dozen are identified in southern Spain in the Geographical Dictionary of Pascual Madoz (1845-50), could be difficult to distinguish from those making flour millstones.

Quarries producing cylindrical sharpening stones, a common tool as seen in a painting by Goya (Fig. 1.9), are also difficult to distinguish from cereal millstone workings. These workings called *amoladeras* were plentiful from old geographical texts. The most famous in southern Spain, cited in ancient texts, exploited reddish sandstones around Alhambra (Ciudad Real) (Benítez de Lugo 2001: 13). Montoro (Córdoba) is also known for its sharpening stones scored from fine reddish *Bundsandstein* (*piedra molinaza*) (Clementson 2012: 3-5). To complicate matters, historical archives also indicate that reddish millstones for grain mills were scored at Montoro (CO-14) (Córdoba de la Llave1988: 843, footnote 22).

Finally, other industries such as the sugarcane workings in the 19th century along the coast of Granada, also required millstones. The edge rollers on display in the Castle of Almuñecar (Fig. 1.10) were imported, according to their distinctive rock type (biocalcarenite), from quarries in the Cádiz area a few hundred kilometres away. These medium-sized cylinders could easily be confused with medieval grain grinders.

Differentiating extractive hollows from features totally unrelated to millstone production is also a problem. Circular extractions have been interpreted as features as varied as brine evaporation cavities for salt, animal drinking troughs, prehistoric sun symbols or marks to delimit the boundaries of property (Fig. 1.11).

Finally, production sites devoid of extractions hollows, where angular blocks were scored and subsequently hewn into segments for composite millstones (banded with plaster and secured by iron bands), leave little evidence in the field that permit them to be identified as grain millstone workings.

Fig. 1.8: Reconstruction of a Moorish donkey-driven olive oil mill dating to the 15th century (Nigüelas, Granada). Outside of its context, the vertical edge roller could easily be confused with a cereal millstone (photograph by Eve Andersson).

Fig. 1.9: Rotary sharpening stone (1808-1812). Painting by Francisco de Goya. (Budapest Museum of Fine Arts). The stone appears to be a fine-grained, rose sandstone.

Fig. 1.10: Millstones of shell-rich ostionera *stone on display at the Castle of San Miguel of Almuñecar (Granada). These medium-sized edge rollers (65 cm in diameter) are reputed to have been used in the recent sugar cane industry.*

Fig. 1.11: Example of the difficulty of Interpreting extraction sites. Quern extraction or property boundary? (Ávila) (from the blog of M. Serna Martínez; http://terraeantiqvae.com/profiles/blogs/el-medievo-a-las-puertas-de-vila#.UV0wl45FfuE).

1.7. Linguistic questions

In this work, I have resorted to the use of English names for well-known localities in Spain. For example, the autonomous community of *Andalucía* is "Andalusia" and its capital *Sevilla* is "Seville." Most locations, however, do not have English counterparts so the original Spanish name, including accents, is retained.

Spanish personal names consist of both patronymic and matronymic names received at birth. This complicates bibliographical references because a number of different authors share the same name. In these cases, so as to avoid confusion, both the patronymic and matronymic are noted, following the Spanish tradition.

2. SOURCES, RESOURCES AND FIELDWORK

Conventional sources, in particular, historical archives, old geographical dictionaries and molinological studies, have provided the bulk of the inventory of 138 millstone quarries (or millstone quarry districts) that appear in the catalogue of this study. Newer, unconventional sources, for the most part historical accounts or hiking itineraries on the Internet, represent about a third of the sites. A series of tools related to cartography, geology, toponymy, available over the Internet, have also been extremely useful.

At times, sites are identified in more than one source. Unfortunately, most of the information collected from both conventional and unconventional sources is extremely perfunctory. Most of the quarries are only mentioned in passing - rare are the texts that offer more than a few lines describing the site. Even scarcer are those with illustrations and descriptions of the rock type. The following is a detailed description of the different sources.

2.1. Sources

2.1.1. Conventional sources

2.1.1.1. Old geographical dictionaries and other chronicles (59 sites)

The geographical dictionary compiled by Pascual Madoz and his collaborators (*Diccionario Geográfico-Estadístico-Histórico de España y sus Posesiones de Ultramar*, in 16 volumes (1845-1850) (Fig. 2.1a), is by far the most valuable written source for identifying millstone quarries. Of the 138 sites identified in this work, 43 were drawn from Madoz. The earlier, less complete, geographical dictionary of Sebastián Miñano (1826-1829) accounts for about a half dozen sites in the study area (Fig. 2.1b).

The series *Crónica General de España*, organised by Province (Granada, Málaga, Cádiz, etc.) and published two decades after Madoz' work, singles out some sites not recorded by either Miñano or Madoz. Other travel narratives (*i.e.* Cruz y Bahamonde 1813), local histories (*i.e.* Martínez y Delgado 1875) and exposition catalogues (Paris Exposition 1867) also pinpoint millstone workings. The drawback of all these sources, including Madoz, is that they rarely offer more than the basic information about a quarry. Since some municipalities are quite large, a number of the sites have yet to be pinpointed in the field.

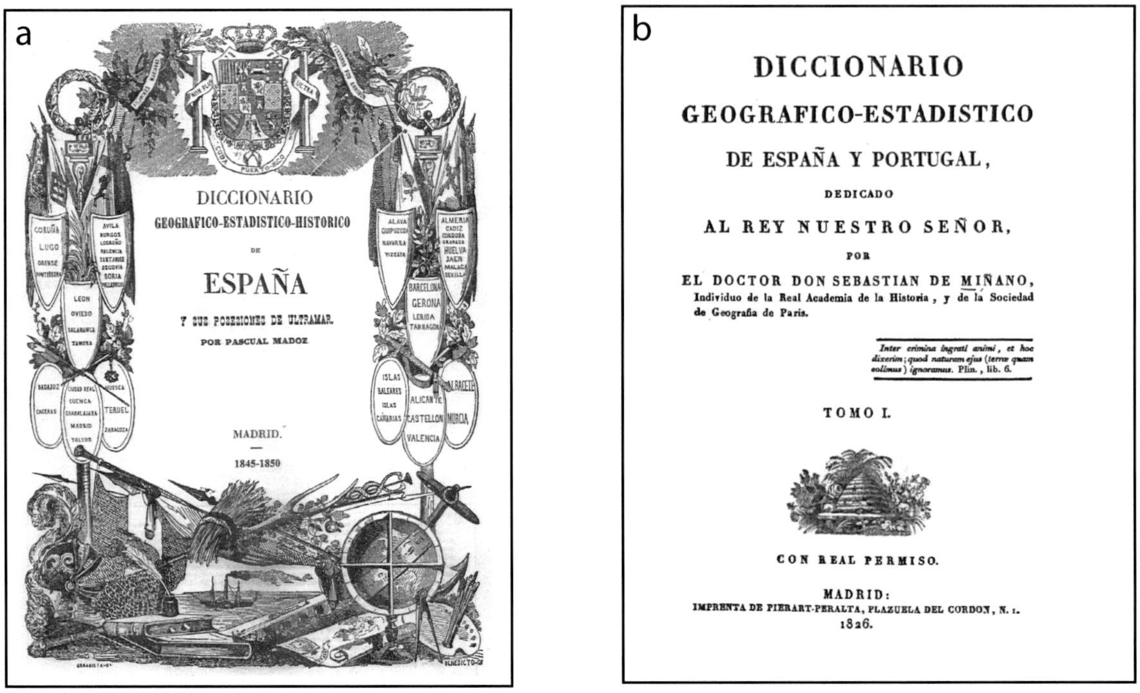

Fig. 2.1: Cover pages of the geographical and historical dictionaries known as "the Madoz" and "the Miñano": a) Diccionario Geográfico-Estadístico-Histórico de España y sus Posesiones de Ultramar, 16 Vol., 1845-1850; b) Diccionario Geográfico y Estadístico de España y Portugal, 11 Vol., 1826-1829.

2.1.1.2. Historical archives and censuses (19 sites)

A series of old documents, in particular censuses, ordinances and commercial transactions from historical archives, are a vital source of information, in particular for the time span between the end of the Islamic domination and the 19th century, before the advent of geographical studies. These documents not only identify about 20 millstone exploitations, but throw light on questions related to ownership and commerce.

A series of codices, codes, ordinances and other legal documents from the Middle Ages stored in national, provincial and municipal archives reveal interesting data about millstone quarries, in particular from the Middle Ages during the struggle between the Islamic and Castilian domination (Table 1). The early *Fueros* (Legal Codes) of Sepúlveda (Segovia) (1076), Cuenca (1190) and Béjar (Salamanca) (end of the 13th, beginning of the 14th centuries) provide directives about the ownership and conditions of the leasing of millstone quarries. A legal dispute over millstone extraction in 1501 between the townspeople of Málaga and Antequera (Málaga) even reached the court of the Catholic monarchs Ferdinand and Isabel in Granada. A municipal ordinance in Loja (Granada) in 1502, delivered by the *Pregonero* (town crier), warned of the consequences of not declaring millstone extraction. A similar ordinance in Llerena (Extremadura) dating to 1566 warns "foreigners" or "outsiders" of the consequences of extracting millstones without official authorisation.

Notarial protocols also record commercial agreements and, at times, cite the location of quarries and throw light on when they were exploited. This is the case for the sites of Santa Ana de Albaida (CO-7), Hornachuelos (CO-13), Montoro (CO-14) (Córdoba) cited in protocols

Table 1: List of millstone quarries cited in national, provincial and municipal archives.

Site code	Site	Year	Archive Location	Archive reference number	Bibliography
CO-07	Albadia	1486	Córdoba	1486.04.02, AHPC, PNCo, 14-21, 1, 44r	Córdoba de la Llave 2003:306
CO-13	Hornachuelos	1481	Córdoba	1481.02.18, AHPC, PNCo, 14-17,3, 73r	Córdoba de la Llave 2003:306
J-8	Villargorda	1499	Jaén	1499.11.26, AHPJ, PNJa, 9, 415r	Córdoba de la Llave 2003:306
MA-1	Torcal (?)	1501	Granada	A.G.S. R.G.5., XII-1500 [sic, date 1501]	Fernández López 1982
GR-2	Loja	1502	Loja	AML, Leg. 49, p. 9. 1 cuartilla + 1.	Municipality of Loja
CO-10	Bélmez	1606	Pozo-blanco	not specified	González Peralbo blog (see cat.)
CO-10	Bélmez	1616	Pozo-blanco	not specified	González Peralbo blog (see cat.)
CO-6	Iscar	1557	Córdoba	1557.01.30, AHPC, PNBa, 4P, 70v	Córdoba de la Llave & Varela 2011: 106
CO-14	Montoro	1481	Córdoba	1481.26, APC, 14-5, 3, 38 r.	Córdoba de la Llave 1988
CA-3	Rota	1719	Madrid	A. H. N., Nobleza, Osuna, libro 19, p. 104.	Martínez, J. A. (pers. comm.)
CA-3	Rota	1745	-	not specified	Martínez, J. A. (pers. comm.)
TO-2	Ventas con Peña Aguilera	1587	Madrid	A. H. N., Nobleza, Frias, leg. 313/2 a	Baltanas 1998: 35-36

dating from 1481 and 1486. The sites of Iscar (CO-5) and Molino de la Piedra (CO-6) in the Municipality of Baena (Córdoba) are likewise cited in protocols from 1557. Knowledge of the *molero* Juan de Bargas and the quarry of Bélmez (CO-10) (Córdoba) was obtained from the study of notarial protocols dating from 1600 and 1616.

Several censuses, in the form of questionnaires, compiling data about the geography, demography, production, natural resources and commerce of Spain, furnish information about a half dozen sites. The oldest, known as the *Relaciones topográficas de los pueblos de España* (1574-1578), ordered by King Philip II, comprised 24 queries. This census identifies millstone workings at Ventas con Peñas Aguilar (TO-2) (Toledo) and Montarrón (GUA-8) (Guadalajara). A second major census, two centuries later, is known as the *Catastro del Marqués de Ensenada* (1750-54). The 40 questions of this survey were answered by representatives from 15,000 localities throughout Spain. Curiously, although this work mentions and illustrates watermills throughout the Spanish landscape (Fig. 2.2), it only sheds light on two millstone quarries: Fuentealbilla (AB-1) in Albacete and El Berrueco (CA-8) in Cádiz. A third census was undertaken exclusively in the Province of Extremadura (*Interrogatorio de la Real Audiencia de Extremadura de 1791*) and identifies sites at Alconera (B-1), Llerena (B-2) and Llera (B-3). Finally, the geographical and historical dictionary of Tomás López de Vargas Machuca from the late 18th century, a project that remains unpublished, provides data about a few sites, notably Colmenar de Oreja (Madrid), (M-2), Sierra de Utrera (Ma-8) by Casares (Málaga) and Otívar (GR-13) (Granada).

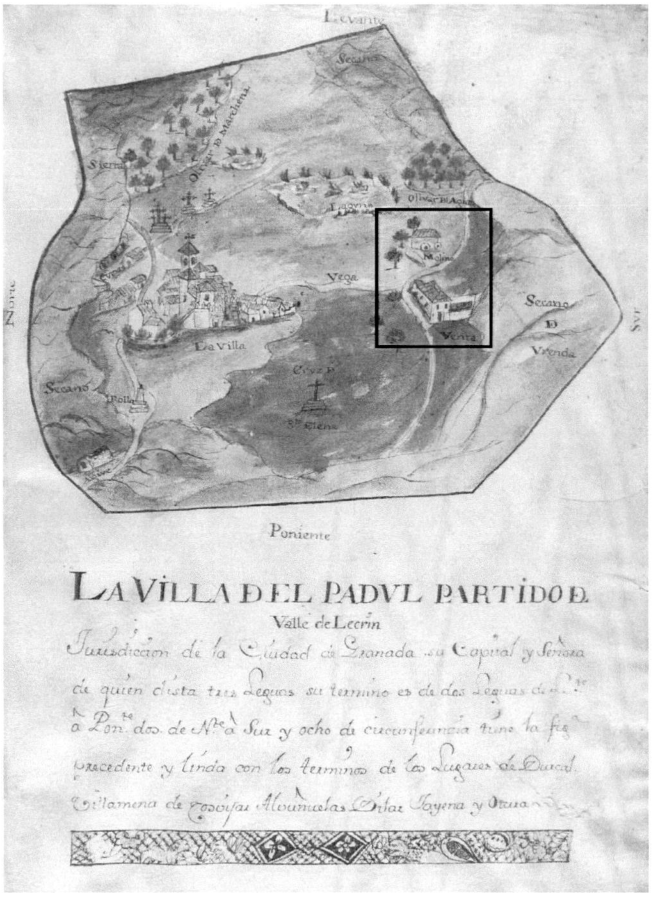

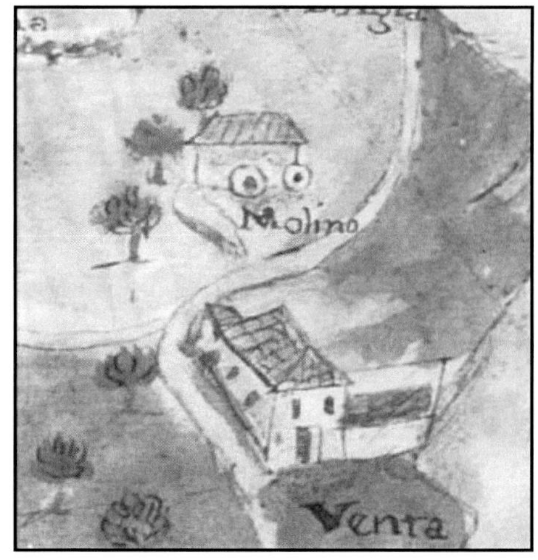

Fig. 2.2: Extract and detail of the illustration of Padul (Granada) of the entry to the Census of the Marqués de Ensenada. The watermill was erected along the irrigation canal. Although almost every town illustrated in the census shows at least one mill, millstone quarry workings, despite sometimes being cited in the responses to the queries, are never illustrated (from El Catastro del Marqués de Ensenada en el Antiguo Reino de Granada, Instrumentos de Descripción, Archivos 16, A.H.P.G, CD, no date).

2.1.1.3. Molinological studies (28 sites)

In the last decades, in southern Spain, there has been growing interest in the question of ancient mills, especially watermills. The organisation ACEM (*Asociación para la Conservación y Estudio de los Molinos*), for example, has been actively studying and conserving a number of mills and has organised a major international conferences every two to three years. Nonetheless, in the proceedings of these conferences only two studies specifically focus on the subject of millstone quarries (Rambla Honda: Martínez López & Granero 2005; Cabra: Montero 2008).

Other molinological research in the form of monographs or extended articles, such as those of Gómez Ruiz (2003) on the watermills of the Odiel River of Huelva, Cara Barronuevo *et al.* (1999) on the mills of the Alpujarra mountains of Almería, Córdoba de la Llave & Varela (2011) on the mills of the Guadalquivir River Valley and Córdoba de la Llave (1988, 2003) on medieval watermills in Córdoba, provide valuable information about millstone workings. For example, an article by Baltanás (1998), related to the renovation of a 16th-century watermill in Colmenar de Orejo (Madrid), indicates in passing, that the newest millstones for the refurbished mill were acquired from the quarry at Ventas con Peña Aguilera (Toledo), a granite exploitation about 100 kilometres away.

But once again, the focus of the vast majority of these articles is the mill and not the millstone quarry. Hence the information about the source, petrography and dimensions of the millstones that equipped the mills is usually perfunctory.

2.1.1.4. Historical and archaeological research (14 sites)

Other historical and archaeological research unrelated to molinology has also provided direct or indirect information about millstone production. Quern workings are identified in a study of Roman settlements in the Minateda-Agramón region at the volcanic outcrop of Cancarix (Albacete) (AB-2) (Jordan *et al.* 1984). Recent excavations of the settlements in the Municipality of Pinos Puente (Granada) (Anderson 2010, unpublished) and Cerro de la Cruz (CO-15) (Almedinilla, Córdoba) (Quesada *et al.* 2010; Quesada *et al.* 2014) have also indirectly provided information about potential local sources of limestone tufa millstones. Archaeological surveys such as that of the Sierra de Molar (Alicante) (A-1) (Gutiérrez Lloret *et al.* 1998-1999) and provincial site inventories (Almonaster la Real, Huelva, HU-8; Mazarrón, Murcia, MU-2; Vega de los Morales, Córdoba, CO-4) also provide valuable information as to the location of millstone quarries.

2.1.1.5. Millstones in museum depositories

Petrographic analyses of querns and millstones in museums can also yield interesting clues as to the presence of local or regional quarries. The large assemblage in the Museo National de Arte Romana in Mérida (Extremadura), for example, dominated by granitoids, certainly reflects work from local and regional granite outcrops. In similar fashion, the biocalcarenite querns in the collections of *Baelo Claudia* (Cádiz) and volcanic rocks (notably lamproites) in Murcia point to local and regional workings.

The assemblages of Roman quern roughouts from Mazarrón in the museum of Murcia and on display at the archaeological site of Oreto y Zuqueca in Ciudad Real are types of artefacts that usually do not travel long distances from their source (especially when fragmented) and most likely point to nearby quern workings.

2.1.1.6. Geological studies and itineraries (6 sites)

Several sites are identified in both old and new geological studies, as well as in geological itineraries. Production is cited, for example, in geological papers of the sites of Berrueco (Madrid) (M-1); Sierra Utrera (Málaga) (Ma-8); Linares (Jaén) (J-7); Arenales, Trassierra (Córdoba) (CO-9); Pinilla de Jadraque (Guadajara) (GU-2) and Cobeta (Guadajara) (GU-9).

2.1.1.7. Toponyms

Toponymic literature has not resulted in the identification of any millstone quarry in southern Spain. Yet the place name search engine for Andalusia, available on the Internet (*Buscador de nombres geográfico, Junta de Andalucia*: http://www.ideandalucia.es/nomenclator/), has helped pinpoint the location of several quarries and has identified a series of other potential production sites that have yet to be confirmed. Moreover, this branch of study has been of great assistance in interpreting the nature of sites identified by other methods. The most significant example is that of *Piedras molares* near Villanueva de Córdoba as the name literally means "*millstone stones*". Silverio Gutiérrez, director of the local Historical Museum of Villanueva de Córdoba, has confirmed the existence of granite workings in the surroundings of this place name. Yet millstone workings, although probable, have yet to be identified.

2.1.1.8. Personal communications (11 sites)

Personal communications have been essential for identifying sites in the field cited by old written sources. For example, Ángel. Serrano of Loja, Granada, a former miller and son and grandson of millers, led me to the workings at Loja (GR-2/3) mentioned in old archives and texts. Other sites such as Alhama de Granada (GR-6) and Los Guillares (GR7), among others, would never have been found without help from the local Tourist Information Office or from Municipal Officers.

Personal information was also particularly important to identify sites that were never, to my knowledge, recorded by written sources. La Merced (GR-4a/b/c), for example, was brought to my attention by the historian Juan Ortiz of Algarinejo. Similarly, the Roman site of Cerro de Limones (AL-1) and the coastal site of Calahonda (GR-9) were communicated respectively by the geographer Loïc Martínez and the archaeologist Isidro Toro, Director of the Archaeological Museum of Granada.

2.1.2. Unconventional sources

2.1.2.1. Local history reports posted on the Internet (20 sites)

The new Internet technology has opened up the doors to a large number of local historians and amateurs to disseminate, by means of blogs and other web sites, local information that would otherwise remain silent. This is the case, for example, for the spectacular Guillares (GR-7) (Granada) millstone quarry, recorded in the web site of the Municipality of Padul and unknown (at least for the moment) in conventional written sources and archives.

Although this type of information must be considered with caution, it is an invaluable source to identify millstone quarries, especially when the web site is accompanied by photographs that confirm the nature of the site. About two dozen quarries can be attributed to Internet sources, notably Rota (CA-3/4/5), Ubrique and Benaocaz (CA-11/12) in Cádiz and Plasencia (CC-1) and Guijo de Galisteo (CC-5) in Cáceres.

Establishing contact with the authors of historical reports has also been extremely beneficial. The contact with Prudente Arjona and José Antonio Martínez of Rota, Montaña Domínguez of Plasencia (Extremadura), Alonso Alonso (Huelva), Juan Ortíz (Zagra and Loja), has led to the identification of other unidentified sites, as well as to photographic and archival records.

2.1.2.2. Travel itineraries on the Internet (11)

About a dozen quarries were identified through hiking or biking itineraries. The presence of these exploitations along these itineraries is probably not a coincidence. Modern tourist routes often coincide with paths and roads that were trudged in past times by millstone transporters. Some sites, such as Soneja (CS-1) (Castellón) and El Láchar (J-2) (Jaén), benefit from simple explanatory panels. El Torcal (MA-1) (Málaga) is along a secondary path in the protected karstic national reserve of El Torcal. Probably the most unique site identified by this means is Isla de la Paloma (CA-7) (Cádiz), recorded in a scuba diving itinerary.

2.1.2.3. Photographs, videos and postcards on the Internet (3 sites)

Photographic stills and videos posted on the Internet are also a new, unconventional means of identifying millstone quarries. Photographs account for Guijo de Galistea (Cáceres) (CC-5) and Albardado, Córdoba (CO-10), whereas Almaden de la Plata (Seville) (SE-1), Zagra (Granada) (GR-5) and Granátula de Calatrava (Ciudad Real) (CR-5) came to my knowledge through videos on *YouTube*.

Old photographs or postcards available on the Internet also offer valuable insight into the old quarry work, products and transport. This is the case for the sites of Gerena (SE-7) (Seville) and Colmenar Viejo (M-3) (Madrid).

2.2. Resources: cartography, geology and toponymy

The resources that figure heavily in this work are the following:

SIGPAC: Sistema de Información Geográfica de parcelas agrícolas. This online mapping service provided maps and divers topographical information, general data about place names, as well as orthophotos (http://sigpac.mapa.es/fega/visor/).

Google Earth, *Google Satellite View* and *Google Street View* provided maps and photographs.

IGME: Mapa Geológico de España. This online service provided geological maps at the scale of 1:50,000, as well as explanatory booklets (http://www.igme.es/internet/cartografia/cartografia/magna50.asp).

SEC: Sede Electrónica del Catastro (https://www1.sedecatastro.gob.es/OVCInicio.aspx). This online service provided detailed names of the different parcels revealing at times toponyms related to millstone workings.

Nomenclátor Geográfico de Andalucía. This online service provided a database of place names, as well as a locater, for the Autonomy of Andalusia (http://www.ideandalucia.es/nomenclator/).

2.3. Fieldwork

2.3.1. Identifying sites in the field

Field surveys are essential to this study. Although written sources are essential for identifying and dating, it is the examination of a quarry in the field that provides the data on the specificity of the rock, extraction techniques, extraction dimensions, and other related features such as debris heaps, roads, smithies and dwellings. Fieldwork is also essential to determine additional phases of production not recorded in written sources.

Of the total 138 sites identified through old texts, the Internet, etc., more than a third (53) were surveyed in the field. Most of this work was done alone or with local guides. In a few cases, these visits benefited from the company of geologists from the Norwegian Geological Survey or from the University of Granada. Visits to a dozen sites, after long-planned outings, were unsuccessful because they were either not accessible (fenced off on private property, in hunting grounds, protected by fierce animals) or simply not found. Only a few, such as Los Guillares (GR-7), Moclín (GR-1) and Alhama de Granada (GR-76) were visited more than once.

Because of the lack of time and the long distances, visits consisted, for the most part, of brief, superficial inspections. It was not possible to undertake a proper systematic topographical surveys producing scaled drawings, much less proper modern excavations. The fieldwork resulted for the most part in a general portrayal of the site (an approximate counting and description of the extraction hollows). Because of my previous experience excavating the complex of Châbles in Switzerland (Anderson *et al.* 2003), I was particularly sensitive to the potential of features of infrastructure such as paths and roads, smithies and dwellings. On account of the limited amount of time spent for each survey, an emphasis was placed on photographing as much as possible to establish a record for further study in the office.

Access to most of the millstone quarries in the study area was simple, in particular to the sites of Carchuna (GR-9), Trafalgar (CA-1) (Fig. 2.3), Guardias Viejas (AL-4) (Fig. 2.4) or Hoya del Paraíso (AL-2) that are on or near the coastline. Other workings such as Zagra (GR-5) or La Merced (GR-4) are beside roads and could be attained directly by vehicle. Sites such as Moclín (GR-1) or Plasencia (CC-1) are in or near towns. Some exploitations, however, were in less accessible, mountainous terrain and moderately difficult to attain. The Cerro de Limones (AL-1), for example, required an hour's hike up a steep slope (Fig. 2.5).

Fig. 2.3: View of the easily accessible quern quarry on the coast of the Bay of Trafalgar (CA-1) (Cádiz).

Fig. 2.4: View of the easily accessible millstone quarry of Guardias Viejas (AL-4) (Almería).

Fig. 2.5: View from the south-west of the Cerro Limones mountain. Access to the volcanic quern quarry, located at the top of the mountain, is moderately difficult.

Fig. 2.6: View of the millstone quarry of Castillo de Locubín (Jaén). Access to the site on a steep slope is moderately difficult.

The drawback of written sources is that they do not necessarily provide precise information as to the whereabouts of the quarries. Certain sources were misleading and lead to hours and hours of unfruitful field work. The reference of Madoz to Alcalá la Real (Jaén) (Madoz, Vol. 1: 382), for example, includes both a place name (*Las Canteras*) and the name of the mountain *Acamuña* (today *Camuña*), data that helped narrow down its location. The problem was that this site was not, as reported, in the Municipality of Alcalá la Real but in the neighbouring Municipality of Castillón de Locubín.

At times, the only information of locality was the name of the municipality. This is tantamount to searching for a needle in a haystack .It was pointless, for example, to even attempt to find the Otívar quarry (GR-13) in 57 km^2 of rough, mountainous terrain. Other sites that were not visited were those of Vera (Almería) (AL-9) in an area of 58 km^2, Linares (Jaén) (J-7) in 197 km^2, Zalamea la Real (Huelva) (HU-6) in 238 km^2 and Chillon (Ciudad Real) (CR-3) in 208 km^2. These sites were, nonetheless, retained in the catalogue, and we retain the hope that someday they will be found.

Millstone workings even more difficult to pinpoint are those only identified at the level of their Judicial District, that is, vast territorial units comprising a number of municipalities. Madoz (1847: Vol. 9, 42) mentions such workings in the *Partido Judicial* of Guadix (Granada), equivalent to a surface measuring nine by seven leagues (about 500 km^2). Another example is the "spongy" *toba* rock quarry (or quarries) exploiting flour millstones in the Judicial District of Zafra (Extremadura) (Madoz 1850, Vol.16: 443), a surface corresponding to about 600 km^2. Obviously I did not attempt to locate these sites in the field.

In any case, without the assistance of local information and guidance, many millstone workings would never have been visited in the field. The names of the people who have helped me to find the different sites are recorded in the individual entries of the catalogue.

2.3.2. The state of conservation of sites

The state of conservation of millstone quarries in the study area is variable from site to site. Some, due to subsequent rock exploitation, are no longer recognisable as millstone quarries. This is the case of El Berrueco (CA-8) and La Pila (CA-9) in Cádiz, or Sierra del Molar (A-1) in Alicante where vast modern workings for construction material continued until recent times and have obliterated the older extraction features.

Millstones and querns have clearly been ransacked by pillagers at many sites. The most visible example is the Cerro de Limones (AL-1) most of the better conserved querns appear to have been gathered for removal (Fig. 2.7a). In fact, many blanks and roughouts from both Cerro de Limones and Hoya del Paraíso (AL-2) (Níjar) now decorate the nearby towns of Presillas Bajas and Rodalquilar (Fig. 2.7c-d). Pillaging is not restricted to the smaller querns. Large millstones from Lora de Estepa (SE-2) now decorate a number of private gardens. The transport of these heavy stones required a crane. This type of destruction of sites, unfortunately, continues with total impunity. Recently (2015) a sector of the Barrio del Pilar quarry (CC-1) in the city of Plasencia, Caceres, was destroyed during the construction of a playground for children.

Fig. 2.7: Example of the pillaging of a millstone quarry. The plunder of sites in the volcanic district of the Cabo de Gata (Almería) is a severe problem for future millstone quarry research. It is, however, due to pillaging that knowledge was gained of the existence of certain sites.

Human intervention is not the only source of harm to these sites. Many quarry faces or abandoned cylinders are heavily weathered by agents of nature resulting in the disappearance of the old tool marks. Certain exposed sectors of Castillo de Locubín (J-1), for example, are smoothed over, greyish and devoid of tool marks while other protected faces and roughouts are fresh and bright yellow (Fig. 2.8).

2.3.3. Equating written sources with field finds

The question at times arose if a site identified in the field could be securely equated with a site described in an old written source. When the information from the text is explicit and offers a place name or other topographical data, then we can be confident that the find in the field is the same as that of the text. At times, however, when there is little or no written information about the site's whereabouts, there is the possibility that the sites do not coincide. For the purposes of this study I have therefore assumed that the sites identified in the field correspond to those in the written records. Future research will validate or negate these connections.

Fig. 2.8: Views of two abandoned millstones at the millstone quarry of Castillo de Locubín (J-1). The first cylinder is in an open area, exposed to weathering and covered with moss. The second example, near the base of a rock cliff, is protected from weathering and shows the original colour of the stone.

3. MILLSTONE QUARRY PRODUCTS AND MILLING INSTALLATIONS

3.1. Introduction

Before examining the millstone quarries themselves, I will briefly review the wide variety of millstones produced in these quarries from the earliest prehistoric saddle querns to the large millstones that equipped mills as late as the 20th century. In the centuries that have transpired since the earliest to-and-fro querns, a number of factors, including the introduction of rotary motion and the invention of special means of assembling and tentering millstones (spindle, rynd and other fittings), as well complex means of traction (gearing), have had a profound effect on milling technology, resulting in a wide variety of mills driven by either humans, animals, water or wind (Fig. 3.1). These changes in technology are also reflected, to a certain extent, in the techniques of extraction and the choice of rocks.

In the study area, with the introduction of the rotary motion in the Iron Age, appear simultaneously small hand querns and larger mills pushed from a standing position. In Roman times, later versions of these larger mills will take the form of Pompeian or ring-mills driven by humans or animals. They will continue to evolve throughout the Middle Ages until Contemporary times in the form of *tahonas* (animal-driven mills).

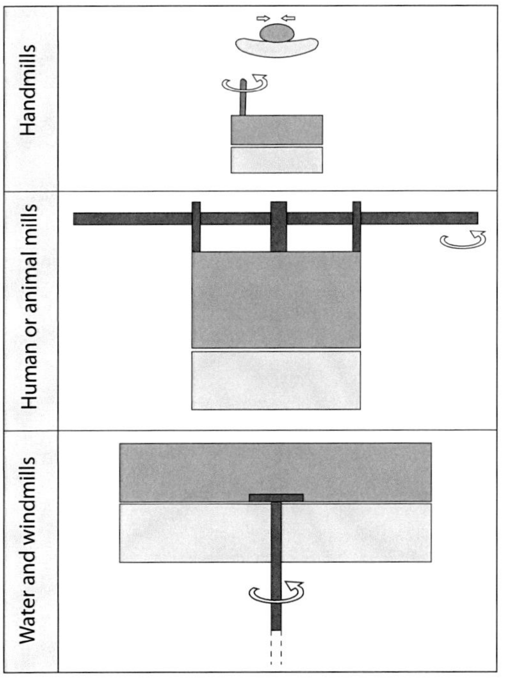

As to the watermill, in spite of mounting evidence of its presence elsewhere in Europe in Roman times, there is practically no evidence of its presence in *Hispania*. The watermill will, nonetheless, proliferate under later Islamic rule and will remain the main type of flour mill until the 20th century. Windmills will play a minor role in grinding grain and like the more modern versions of the animal-driven mill, will be restricted for the most part to regions with limited access to running water.

Fig. 3.1: Schematic representation of the three major types of grain mills produced in quarries according to their motive forces: handmills driven by the arm; mills driven by humans or animals; water and windmills driven by harnessing natural forces (drawing by T. Anderson).

3.2. The saddle quern

The earliest grinding mechanism known in the study area is the saddle quern (Fig. 3.2). This mill is made up of an elongated concave or flat nether stone and a smaller rubber or *mano*. Known since the Neolithic, this type of mill endured for centuries in a variety of forms, sizes and rock types before giving way, towards the middle of the first millennium BC, to the rotary quern and larger pushing mill (Alonso *et al.* 2011: 56). There are potentially three saddle quern quarries in the study area: Zujaira (GR-12) (Anderson, unpublished), El Barronal (AL-10) (Haro *et al.* 2006) and Cabezo de Oliva (MU-2) (Agüera *et al.* 1999).

It is worth noting that a sophisticated version of this early non-rotary mill, the hopper rubber (or Olynthan mill), with a distribution concentrated for the most part in the eastern Mediterranean, has not been identified in the study area. The closest examples are at Lattes in southern France (Py 1992: 185) and at the Sec shipwreck off the Balearic Island of Majorca (Arribas 1987).

Fig. 3.2: Examples of saddle querns of various lithologies. a) Late Neolithic calcareous sandstone quern from the site of Los Pensadores (Granada), excavation T. Anderson 2010, unpublished; b) Chalcolithic garnet-mica schist from Las Rajas (Granada) excavation T. Anderson 2010, unpublished; c) Early Bronze Age garnet-mica schist querns from Fuente Álamo (Murcia); d) Iron Age sandstone and conglomerate querns from Numantia (Soria); e) granite querns from the Early Iron Age site of Cancho Roano (Badajoz) (photographs a, b, and d by T. Anderson; photograph c from http://www.elargar.com/caracterizacion/Artefactos/Instrumentos/; photograph e from blog of Jesús López, http://extremosdelduero.blogspot.com.es/2011/12/edificio-protohistorico-de-la-mata.html).

3.3. Generalities about the rotary quern

The rotary quern comprises two small circular stones assembled by means of different types of fittings. The upper stone (runner or *catillus*), mounted on a stationary lower stone (nether, *meta*), was driven by hand with a steady, continuous circular motion. It was driven from a stationary position, probably at floor level as often seen in ethnographical studies, by one person with one leg outstretched and the other bent.

The product that was ground in these mills (usually cereal) was gradually fed manually through the eye of the upper stone and exited in the form of meal or powder along the periphery of the stones and was gathered in a 'flour catcher' in the form of a cloth, skin or wooden box below the mill.

It is becoming more and more apparent that these rotary querns served to grind a number of different products. In fact, two recent Spanish studies of macroscopic remains show that they served to grind rye and acorns in pre-Roman times (Checa *et al.* 1999: 66-67), whereas fish for fish flour and fish oil, as well as cinnabar for pigments, were ground by these handmills in Late Roman contexts (Domínguez & Bernal 2011: 457). Furthermore, rotary handmills are known to have served in Roman metal-working contexts for grinding ore (García Romero 2002: 620-621; Gutiérrez Soler & Corpas 2011: 25-26) and in Contemporary times, to grind animal fodder.

In terms of establishing quern typology, upper stones are much more susceptible to change than lower stones, due to the variety of means of traction and assemblage (handle and rynd cuttings). Basically, upper stones present two different types of profiles or sections that, at times, can reflect if they were scored from rounded surface blocks or from a quarry by means of direct extraction.

The first type of upper stone section is the hemispherical or domed upper stone, labelled "beehive" in the British Isles. Although it could be argued that this shape reveals the millstone maker's desire to reduce weight to facilitate manipulation and transport, it is more probable that it reflects the fashioning of small, rounded surface boulders. When conditions of conservation are favourable, the boulder's original cortex can be observed on unworked sectors of the surface (Anderson *et al.* 2003: 42; Green 2011: 125; Pascual & García 2011: 287-288). Upper stones with 'doughnut' profiles is a variation of this type that also reflects the carving of rounded surface boulders. Neither of these types, however, are common in the collections of southern Spain.

The second, and by far the most common quern section type, is cylindrical. Both upper and lower quernstones from true extractive quarries, scored directly from bedrock by means of cutting trenches, bear, almost invariably, vertical or slightly inwardly tilted edges (Anderson *et al.* 2003: 42). This shape is also characteristic of querns hewn from angular blocks previously pried from bedrock. Cylindrical querns therefore reflect a more complex, organised (potentially large-scale) model of work that contrasts with the simple collection and carving of surface blocks. Cylindrical querns also are easier to stack for storage and transport than hemispherical upper stones.

In spite of the introduction of other more sophisticated and efficient milling installations, notably the watermill, rotary querns did not disappear at the end of Antiquity. In the following pages I will demonstrate that they endured through time and that their morphology and fittings evolved, culminated in their use for grinding animal fodder in Contemporary times.

In figure 3.3, I present an overview of the major typological changes of rotary querns in southern Spain from the Late Iron Age to Contemporary times (the Modern period between 1492 and 1789 is excluded due to lack of data). These features, in particular that of diameter and rock type, described individually below in their chronological context, make up the backbone of the criteria for dating quern quarries.

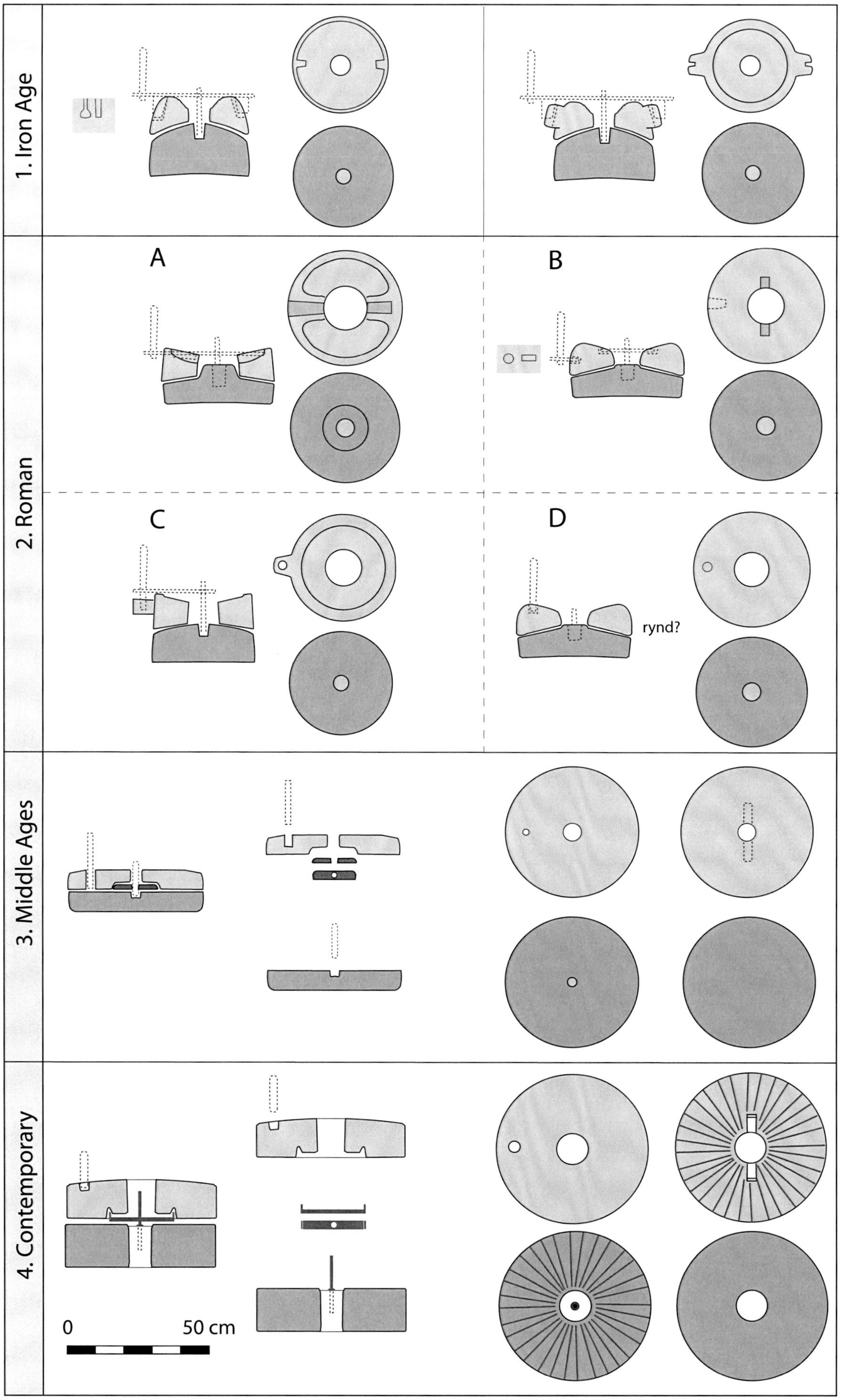
1. Iron Age
2. Roman
A
B
C
D
rynd?
3. Middle Ages
4. Contemporary
0
50 cm

3.3.1. The Iron Age rotary quern

Among the hundreds of rotary querns deposited in museums throughout southern Spain, less than half dozen can be ascribed with a high degree of certainty to the Iron Age Iberian Culture.

Based on studies undertaken elsewhere in Spain, for the most part in Catalonia (Burés *et al.* 1993; Checa *et al.* 1999; Asensio *et al.* 2001; Guérin 2003; Portillo 2006; Alonso *et al.* 2011), as well as the few reasonably dated examples from the study area, in particular the assemblage from the Late Iberian settlement of Cerro de la Cruz, Almedinilla (Quesada *et al.* 2014), certain typological tendencies emerge for Iron Age rotary querns (Anderson *et al.* 2014) (*cf.* Fig. 3.3).

Pre-Roman querns (Fig. 3.4), to begin with, tend to be smaller than their Roman, Medieval and Contemporary counterparts. They most often measure between 30 and 35 cm in diameter. As elsewhere in Europe, they also tend to be proportionally thicker.

Other characteristic features include cuttings in the form of opposite vertical straight slits, "L" shaped slots and "inverted keyhole" slots carved into the upper stone's edge that served to lodge the wooden fittings to assemble, tenter, lift and drive the upper stone. 'Inverted keyhole' cuttings, for example, are present among the querns of Numantia (Soria) and diverse sites in Catalonia (*e.g.* Burés *et al.* 1993: 134, fig. 14; Asensio *et al.* 2001: 68, Plate 6, M-7 and M-17). In the study area this feature is visible on the examples at Ventorros de San José (Granada) (Fig. 3.4 a), in a private collection at Zujaira (Granada) and Municipal Museum of Jódar (Jaén) (Fig. 3.4 b).

Opposite ear-like appendices or lugs projecting from the edges (fashioned directly from the rock) with vertical cuttings, also meant to lodge wooden driving rigs, are also known on Late Iron Age querns. An example is the quern on display in the Municipal Museum of Alcalá la Real (Fig. 3.4 d). These lugs can at times bear "inverted keyhole" cuttings like those mentioned above, as seen at Cerro de la Cruz (Almedinilla, Córdoba) (Quesada *et al.* 2014) (Fig. 3.4 e).

Yet the presence of lugs on upper stones is not a conclusive criterion for dating rotary querns to the Iron Age. These appendices are present, although presumably less common, on upper stones from Roman contexts, as seen in several examples at the Roman city of *Baelo Claudia* (see below) and other examples distributed throughout museums in southern Spain. The impression is that quern lugs in later contexts are single (rarely double) and no longer have vertical slits or "inverted keyhole" cuttings but vertical sockets serving as handle holes. In any case, these lugs reveal the continuity of the "lugged" handmill tradition from the Late Iron Age Iberian Culture to the Roman times.

As in the case of the earlier saddle querns, no certified Iron Age quern quarry has, to date, been positively identified in the field.

Fig. 3.3: Opposite page. Schematic representation of the major rotary quern types in southern Spain from the Iron Age to the Contemporary period. In general, querns progressively increased in diameter while grinding surface angles decreased becoming flat. Handle cuttings shifted from opposite lateral slots to a single hole on the upper surface and rynd fittings relocated from the top to the bottom (from Anderson et al. *2014).*

Fig. 3.4: Examples of rotary querns attributed to the Iron Age. a-b) Upper stone (ø: 32 cm) and lower stone (ø: 34 cm) from Ventorros de San José (Granada), possibly a pair; c) upper stone (ø: 38 cm) with inverted "keyhole" handle cuttings (Jódar, Jaén) d) upper stone with lugs with vertical slits (Alcalá la Real, Jaén); and e) top and side view of an upper stone with a lug bearing an inverted "keyhole" cutting (Cerro de la Cruz, Almedinilla, Córdoba). Petrography: a-b, d) porous limestone; c) probably conglomerate; e) granite (from Quesada, et al. *2014, nº cat. 7, fig. 6).*

3.3.2. The Roman rotary quern

Roman rotary querns (*cf.* Fig. 3.3) are typical finds at excavations of Roman rural settlements, towns and cities. They are, at times, found in conjunction with other more complex, industrial, milling equipment such as cylindrical mills and Pompeian mills (see below). The number of Roman rotary handmills in the study area surpasses by far that of rotary querns from all the other chronological periods together. This high number can be attributed, in part, to the disproportional focus of fieldwork on the Roman period. It certainly also reflects, when compared with the preceding Iron Age Iberian Culture, an ever-growing population in Antiquity.

The most common type of rotary quern upper stone (*catillus)* (Fig. 3.3a and Fig. 3.5) in southern *Hispania* is cylindrical. Its upper surface has a rim (a) around its edge. The central eye is bordered by two receptacles (b) for grain (shaped like bat wings) separated by opposite (c) radial handle slots. The grain was fed progressively from these hopper-like receptacles into the mill, through the eye, by means of vibration and gravity. The radial slots served to lodge (d) an iron bar or crosspiece (never conserved), about 45-50 cm long and about three centimetres wide. The hole in the centre of the crosspiece (like the hole of a rynd), coupled with (e) the spindle of the lower stone, served to assemble and tenter the mill to keep it centred during rotation. Its projection beyond the mill to one side through a gap in the rim served also as a base for a vertical handle (f) (see also Fig. 3.6a; Fig. 3.7). Two examples in the study area indicate that the iron crosspiece was secured by (g) molten lead (Palomo & Fernández 2007: 357, fig. 8; Linares museum, inv. no. CE02430).

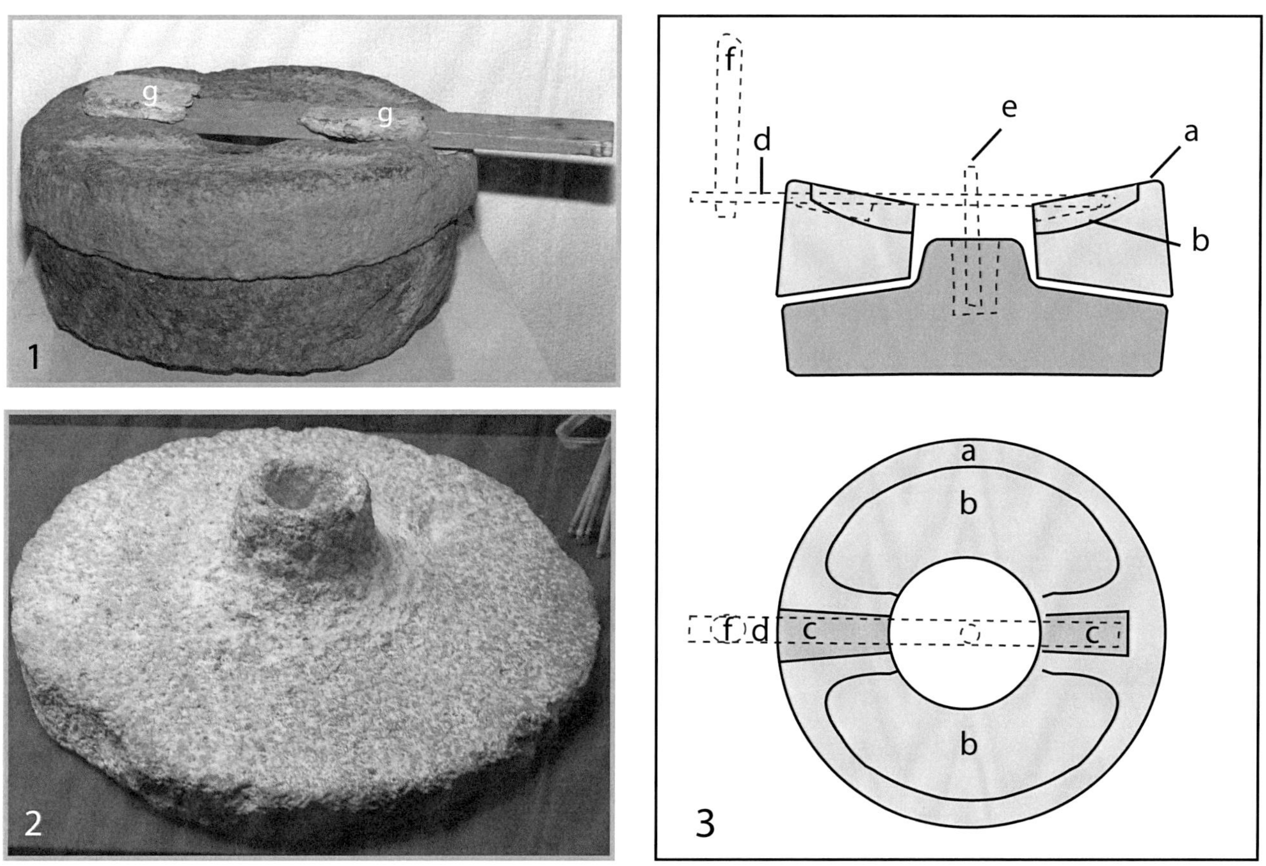

Fig. 3.5: 1) Roman quern upper stone (ø: 39 cm) (photograph by the Museo Arqueológico de Linares, Jaén); 2) and "sombrero" lower stone (ø: 36 cm), Museo Municipal de Úbeda. Both are of volcanic rock. 3) Schema illustrating the assemblage and driving fittings (drawing by T. Anderson).

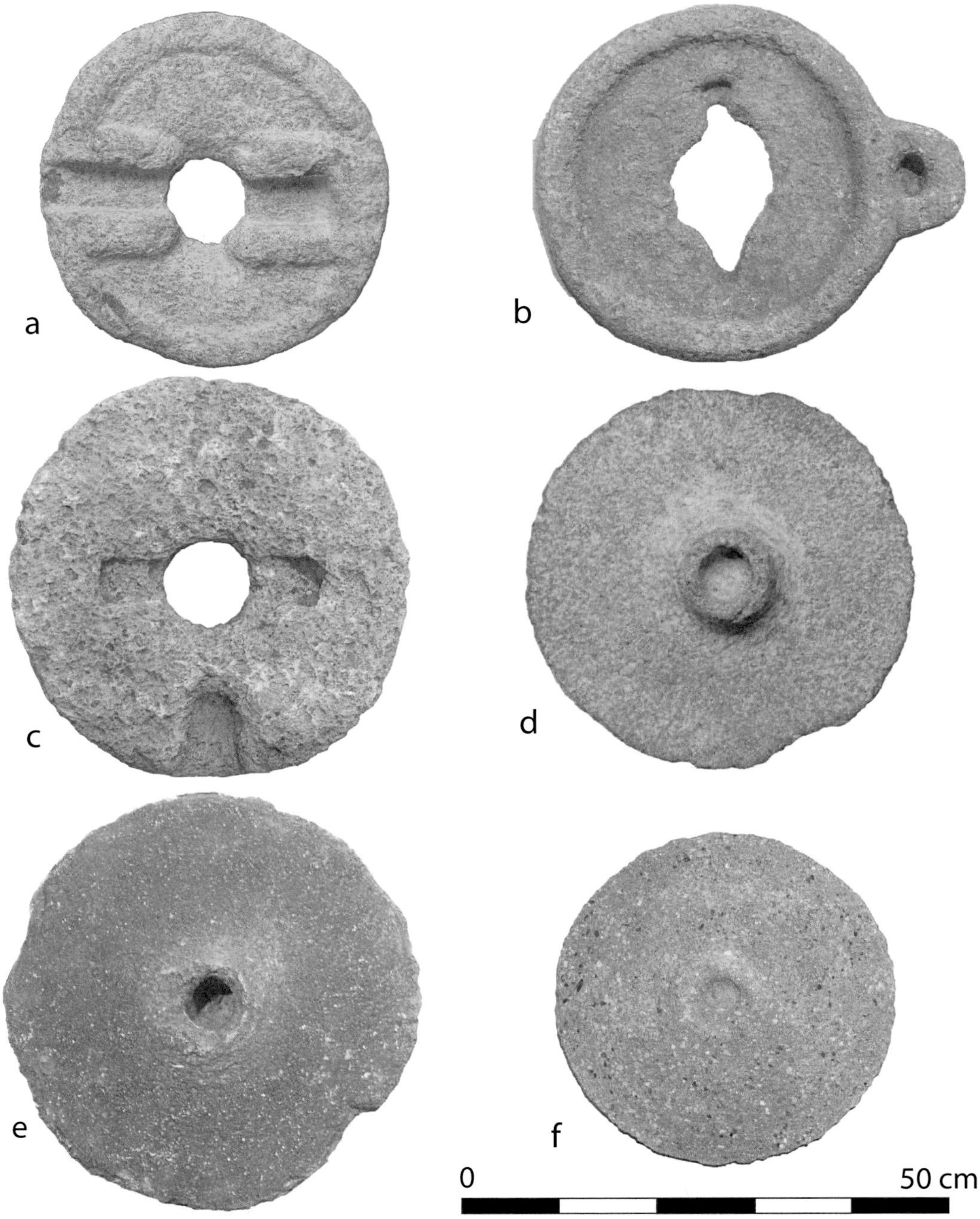

Fig. 3.6: Examples of the typological and lithological variety of Roman rotary handquerns a) upper stone rim and radial cuttings (Priego de Córdoba); b) upper stone with rim, handle lug and rynd cuttings (Murcia); c) upper stone with no rim, single radial cutting and rynd cuttings (Baelo Claudia, Cádiz); d) lower stone with prominent collar ("sombrero type") (Úbeda, Jaén); and e-f) conical lower stones (Murcia). Petrography: a, b, d, e) volcanic rock; c) biocalcarenite; f) conglomerate.

The other types of Roman rotary quern *catilli* are less frequent. A second example is characterised by a horizontal socket cut into the edge of the upper stone destined to lodge a lateral handle (*cf.* fig. 3.3b). This hole is most often circular, rarely rectangular. It is worth noting that although this is probably the most common type of Roman quern handle fitting elsewhere in Europe, it was not adopted on a large-scale in southern Spain.

A third type of Roman rotary quern *catillus* was driven by means of one (rarely two) lateral lugs bearing a vertical handle socket (*cf.* Fig. 3.3c). The handle is independent from the rynd that was probably made of iron, and lodged in two small rectangular cuttings in the upper surface on opposite sides of the eye.

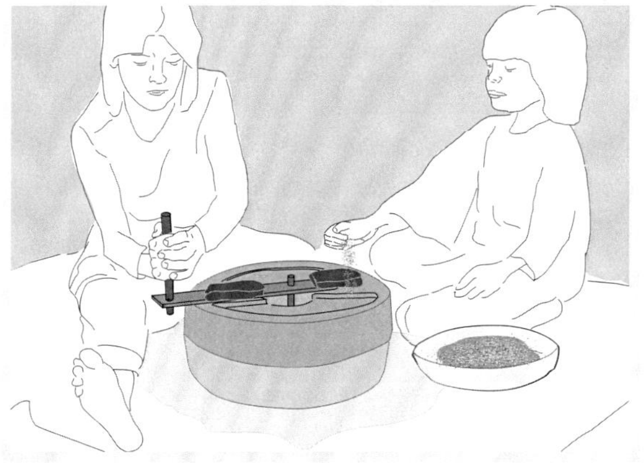

Fig. 3.7: Reconstruction of a Roman domestic milling scene in the south of Spain. The catillus *bears a horizontal crosspiece (handle and rynd rig) fastened with lead in radial slots (drawing by T. Anderson, from Anderson* et al. *2014: 112, fig. 1).*

A fourth type, also not widespread, had a vertical handle socket cut into the upper surface. This was the most simple of handle fittings, and it dates, presumably, to the transition from Late Antiquity to the Early Middle Ages (*cf.* Fig. 3.3d). In this model, it is still unclear if the rynd slots are on the top or the bottom of the upper stone. In any case, the vertical handle hole on the upper surface became predominant in the Middle Ages and has endured until recent times.

It is noteworthy that no example of "elbow" handle hole cuttings from the edge to the upper grinding surface (to lodge a metal sleeve for a wooden handle), known among the volcanic productions of Catalonia (Vivar 2004: 104, fig. 1) and the Eifel in Germany (Anderson *et al.* 2003: 42, fig. 32,2), is present in the study area.

Lower stones in general show fewer typological features and hence, when devoid of context, are more difficult to date. The Roman upper stone cited above with the radial handle and rynd cuttings is associated with two types of lower stones (*metae*). The first consists of the "normal" conical model with or without a small lip around the eye (resulting from wear). This is the most common type of *meta* known throughout the Roman world.

A second type of lower stone presents a unique feature: a prominent collar around the eye (Fig. 3.3a and Fig. 3.6d) that resembles a hat (hence *sombrero*). This collar is a premeditated feature carved at the quarry and not a result of wear. Although analogous collars to facilitate feeding the grain are known on upper stones elsewhere in Europe, notably in southern France in the rhyolite productions of the Esterel (Désirat 1996: 133, fig. 59; Longepierre 2011: fig. 101), this morphological feature seems to be unique to Roman lower stones of querns in south-eastern Spain. This feature is neither documented in Catalonia (Portillo 2006), nor in southern France (Py 1992; Longpierre 2011). The closest example to south-eastern Spain is a lower stone from *Conimbriga* (Borges 1978: 127, no. 52; Plate VII, 52) in Portugal. However, the Portuguese model does not present a cylindrical section. Quarries producing these *sombrero* lower stones are Cerro de Limones (AL-1) and Hoya del Paraíso (AL-2) (*cf.* Fig. 3.5.2).

Although most often associated with domestic activities, querns are also well known to have formed part of the equipment of the Roman army. All said, no examples incised with military inscriptions along their edge, known elsewhere in Europe, are known in Roman *Hispania*.

A number of Roman rotary quern quarries have been identified in the study area, notably at the volcanic sites of *Sisapo* in Ciudad Real (CR-1) and Cerro de Limones and Hoya del Paraíso in Almería (AL-1, AL-2), as well as the *ostionera* sites of Trafalgar (CA-1) and Rota (CA-3) in Cádiz.

3.3.3. The Medieval rotary quern

If progress has been made on the study of rotary querns and their quarries dating to the Roman period, research on rotary querns and quarries of the Middle Ages is still in a nascent stage. There is a huge gap corresponding roughly to the thousand years of Visigothic and Islamic domination. The number of medieval rotary querns in museum depositories is limited due to the scarcity of settlement excavations and few are described or illustrated in publications. The notable exception is the lot from the settlement of El Castillejo, Montefrío, Granada (Motos Guirao 1987: 465-467, 473-474, 479-480, PL. I-IV). Based on this work, as well as on observations in museums (in particular in the Province of Murcia), medieval rotary querns can be differentiated from their earlier Roman counterparts (*cf.* Fig. 3.3; Fig. 3.8; Fig. 3.9).

The first difference is their diameter. Medieval models, although at times still small (between 30 and 40 cm and sometimes as small as 20 cm, Fig. 3.9), usually measure between 40 and 50 cm, well above the average diameter of the Roman models. A second major difference is that the sections of medieval querns are flatter (discoidal), with practically horizontal grinding surfaces. Upper stones are more simple, devoid of rims and receptacles for grain, and they were driven by means of a vertical handle placed in a hole cut towards the edge of the stone. This system is a world apart from the older, more complex Roman model with radial handle cuttings. In the Middle Ages, the diameter of eyes also decreases and rynd cuttings appear to relocate from the top to the base of the upper stones.

The evolution from a pronounced to a practically flat grinding surface angle suggests that cereal grain (or other products) were fed directly through the eye and not loaded into a hopper-like receptacle. The systematic change of the handle fitting from a radial groove to a vertical socket could also indicate that the mill was driven by a long vertical rod connected, for example, to the ceiling of the house. Another option is that it was driven from the side by a horizontal connecting rod. All said, these more complex driving systems using torque, known elsewhere in Medieval Europe, are hypothetical.

Several quarries of this mill type have been identified in the study area, notably at Zagra (GR-5) and Almadén de la Plata (SE-1).

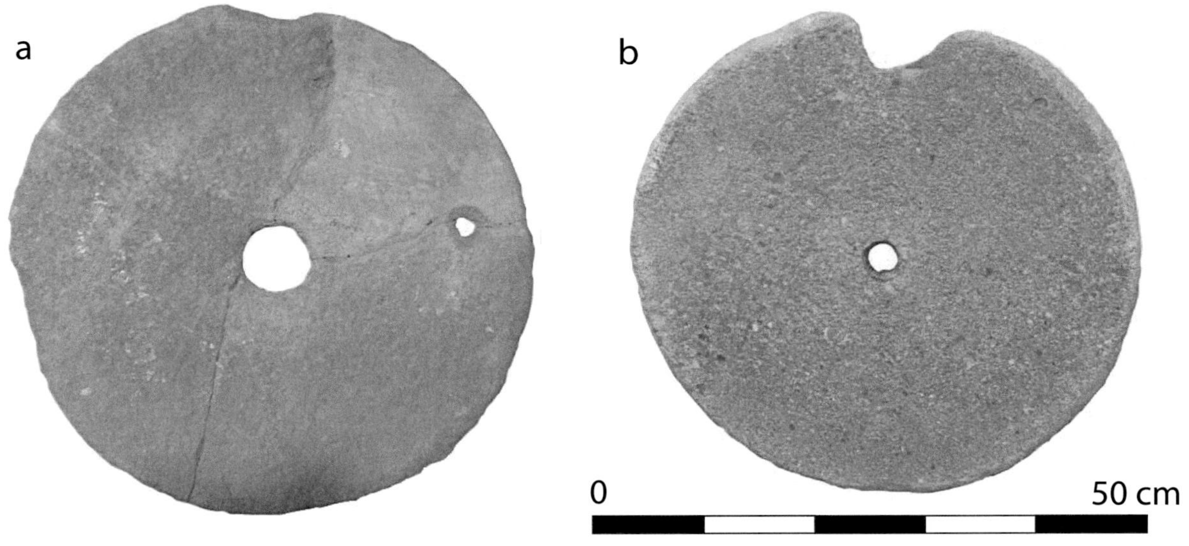

Fig. 3.8: Medieval discoidal rotary querns: a) Upper stone ø: 48 cm (Murcia); b) lower stone ø: 47 cm (Siyâsa, Murcia). The eye of the lower stone shows traces of iron oxide indicating an iron spindle.

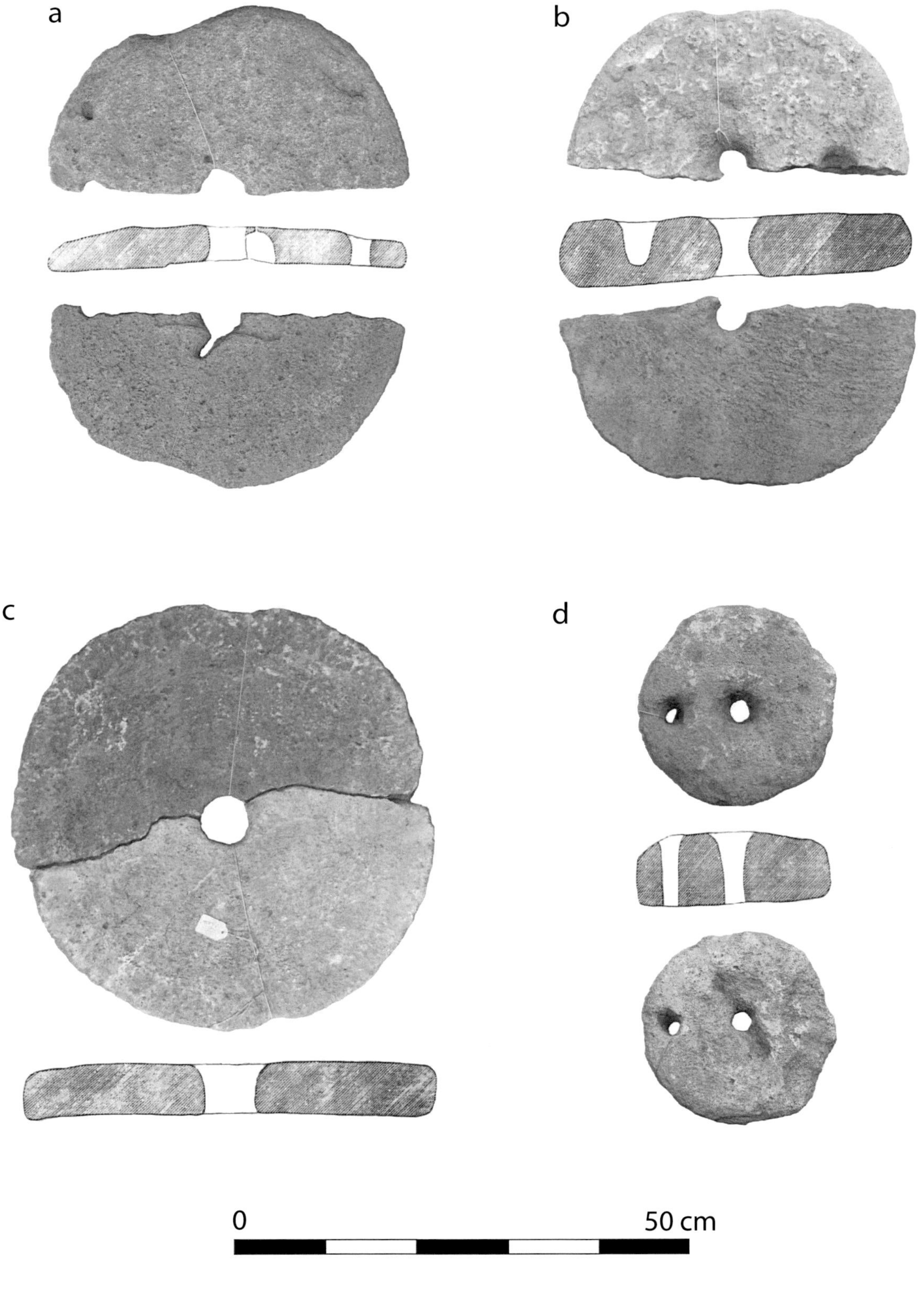

Fig. 3.9: Rotary querns from the 1981 excavation of the medieval settlement of El Castillejo, Montefrío (Granada) dating from the 9th to the 12th century (Motos Guirao 1987: 480). a) Upper stone with handle hole and lower rynd cutting, ø c. 38 cm; b) upper stone with no rynd cutting; c) lower stone; and d) upper stone with handle and rynd cutting of a very small hand quern (photographs by T. Anderson; the sections from the original drawings from Motos Guirao 1987, plates I-III).

3.3.4. The Contemporary rotary quern

Contemporary rotary querns are spread throughout the rural landscape of southern Spain. They differ from earlier querns in that their principal use was to grind animal fodder (*molinos de cebo*) (Fig. 3.10). All the same, it is known that some of these mills ground foodstuffs for human consumption in times of crisis, notably in the dark years after the Spanish Civil War. In Priego de Córdoba (Córdoba), M. Campos has collected a dozen of these handmills (both upper and lower stones) from abandoned farms. This lot forms the backbone of my study of this type of handmill.

These querns distinguish themselves from their older counterparts, first of all, by their larger diameter (often surpassing 50 cm) and weight. Their driving fittings are standardised. They comprise a vertical handle socket carved into the upper surface near the edge and

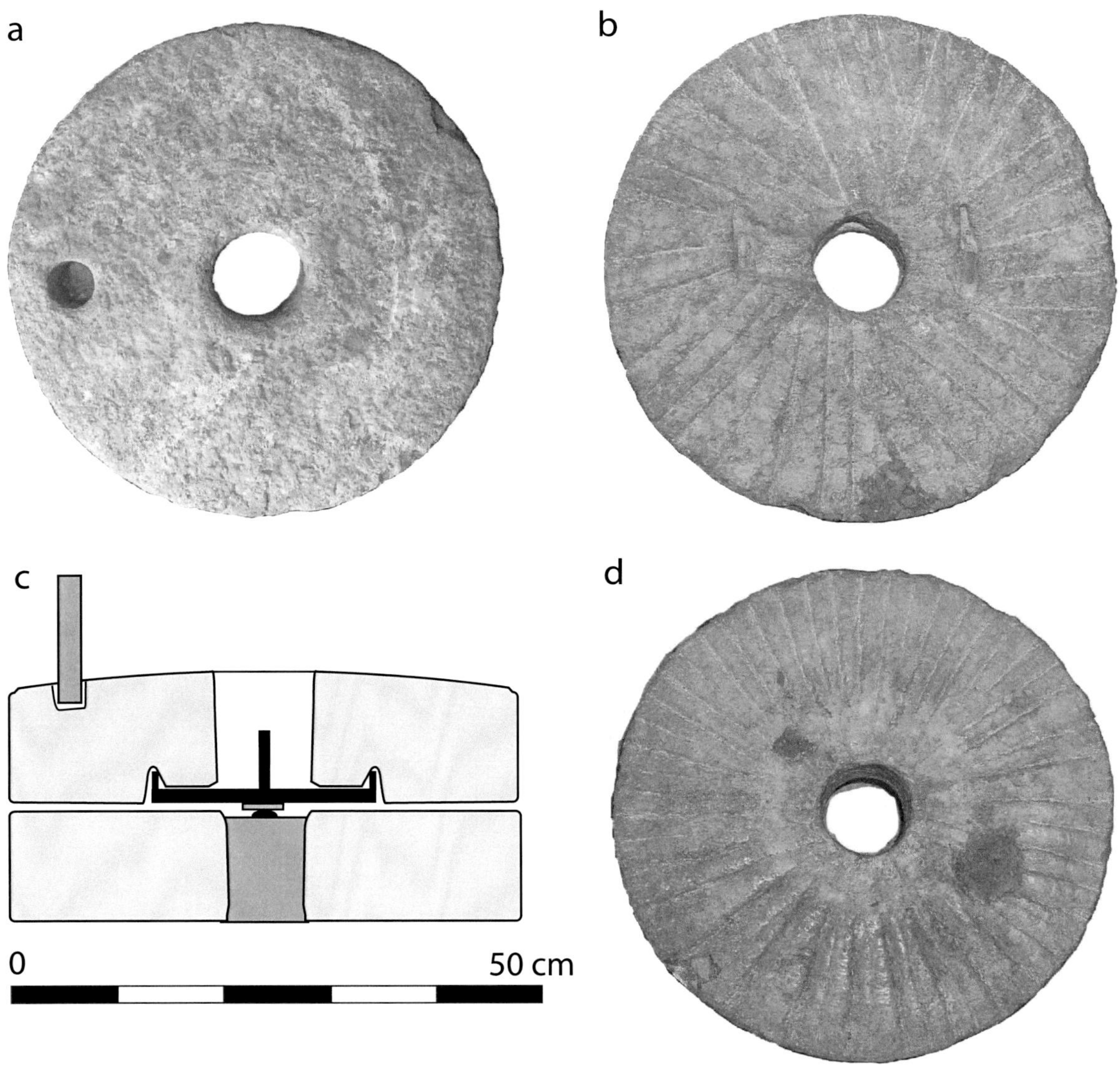

Fig. 3.10: Example of a Contemporary animal fodder handmill measuring 47 cm in diameter from the area of Priego de Córdoba with reconstructed handle and rynd (Anderson 2011b). These mills were also presumably served to grind foodstuffs for human consumption during dire times. a) upper stone, top view b) upper stone working surface c) section d) lower stone working surface.

rynd cuttings on the lower surface. These cuttings are variable in form, from the single opposite rectangular slots to complex cruciform slots (possible repairs). The eye of the lower stones is nearly always totally perforated. Their iron spindle is plugged with wood or plaster (Anderson 2011b). A specific characteristic that differentiates them from earlier hand querns is the systematic presence of radial or harp-shaped furrows on the grinding surfaces of both upper and lower stones.

Most of these handmills, according to Marcos Campos, were still in their working position on ledges at table height near the stables of the farms.

Other similar animal fodder mills are known elsewhere in Spain (*e.g.* Pascual *et al.* 2010). Their driving mechanisms differ greatly, however, in that instead of being driven from above with a handle, they were driven from below by means of connecting rods or cranks (Fig. 3.11). This explains the absence of handle holes on these upper stones.

The precise location of the production centres of these mills is not known. It is possible that they were carved from detached angular blocks, a technique that leaves little characteristic evidence in the field.

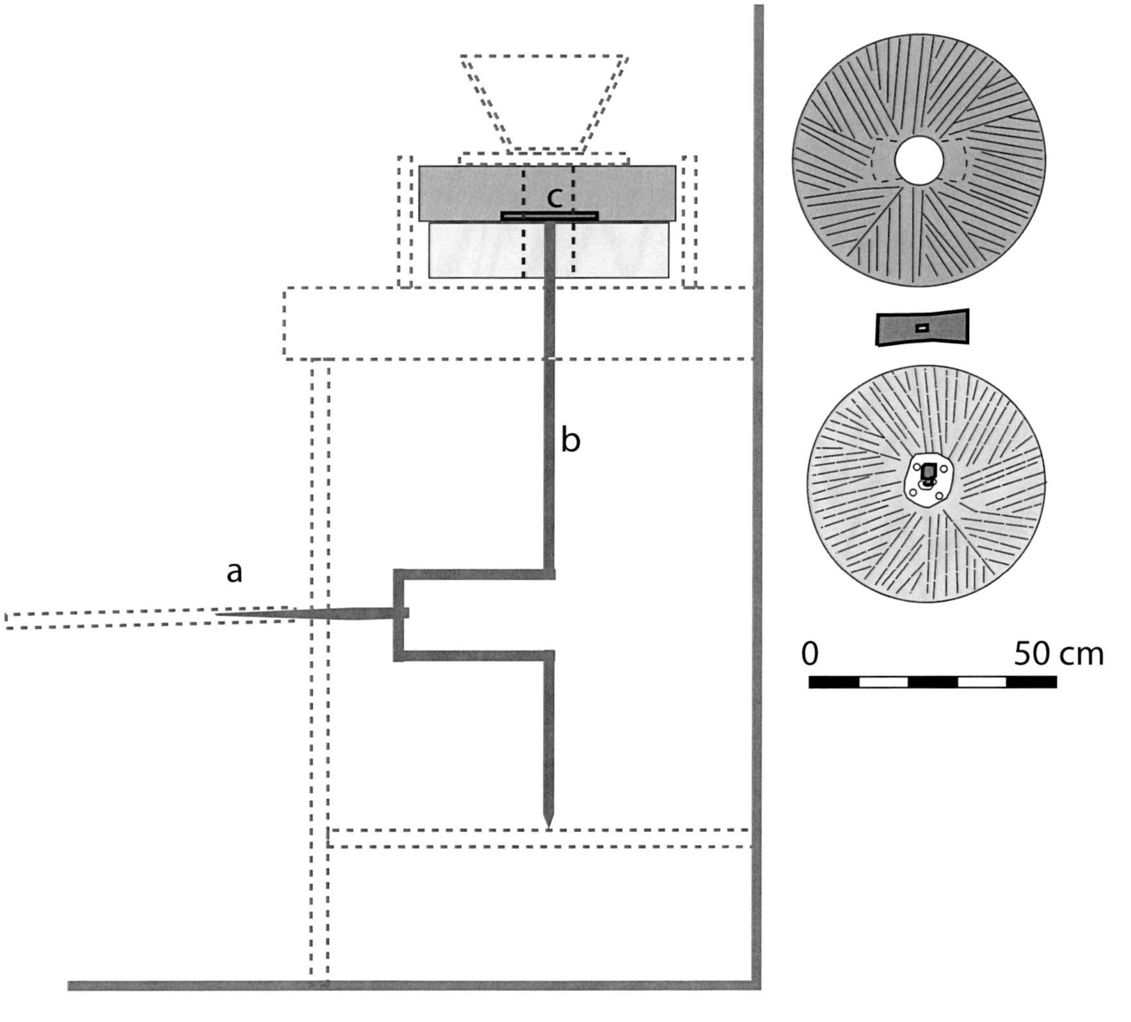

Fig. 3.11: Schema of the Contemporary (20th century) crank mill from Cubo de la Sierra (Soria). Since the driving force is transmitted from below by means of the a) connecting rod, b) drive shaft and c) rynd, the upper stone has no handle socket on its upper sufarce (drawing by T. Anderson, from Pascual et al. *2010, 10, fig. 5).*

3.4. Human- and animal-powered mills

Human- and animal-powered mills differ from rotary querns in that they are larger and thicker and are always driven from a standing position. They therefore require a peripheral ambulatory space. Some are known to have been set on stone and masonry podiums that not only raised the mill, but served as flour catchers. The origin of the driving force is in the legs and body and not in the arm. Since the driving force is not always evident (human or animal), these are known as "blood mills" in Spanish and French (*molino de sangre* and *moulin à sang*).

These mills are assembled by means of a wooden rig. This is topped by a horizontal lever, lodged in the lateral cuttings of the upper stone and into the spindle of the lower stone. They are turned by pushing the horizontal lever, like a capstan. The lateral cuttings are usually larger at their base so as to fasten the wooden frame and allow it, by lifting the lever, to be removed for maintenance. The earliest examples in the study area, known as Iberian rotary pushing mills, are found in Iron Age settlements.

Models similar to the Iron Age mill remained in use in Roman *Hispania* in the form of cylindrical mills with "ring-*catilli*". The other typical Roman human- or animal- driven mill is the Pompeian hourglass mill. It is worth noting that the Morgantina mill (the precursor of the Pompeian mill), is not recorded on the Iberian Peninsula.

3.4.1. The Iberian rotary pushing mill

The Iron Age Iberian rotary pushing mill (term from N. Alonso, pers. comm.) (Fig. 3.12) comprises a cylindrical, ring-shaped upper stone with a "D" section and a lower stone with a steep grinding surface topped by a square spindle hole. After extended wear, the lower stone diminishes in width, resulting in an elongated neck with an inwardly inflected grinding surface.

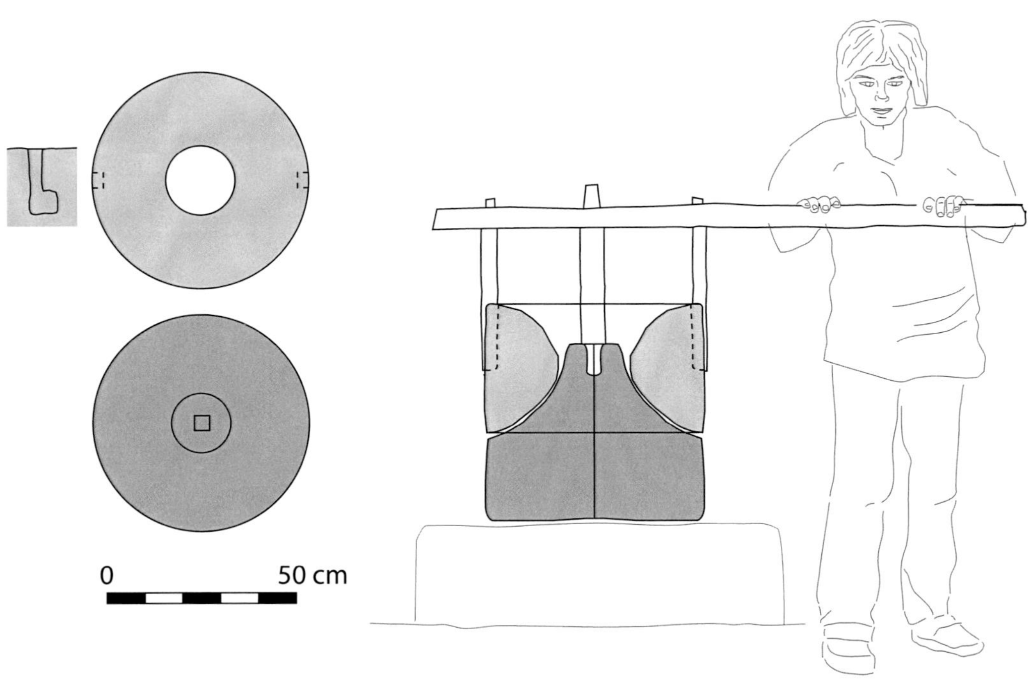

Fig. 3.12: Schema and reconstruction of an Iberian rotary pushing mill (drawing by T. Anderson).

The wooden frame is usually lodged in vertical slits, "L-shaped" or "inverted keyhole" lateral cuttings (Berrocal Rángel 2007: 292). The last two also permit the lever to serve as a means to lift and remove the upper stone for maintenance. The eye of the lower stone is either square or rectangular. Currently, there is no evidence that these Iron Age mills were driven by animals.

These mills are known throughout Late Iron Age Iberian Culture settlements in Catalonia and Valencia. In southern Spain they are common in museum collections such as Baza (Granada) and Jódar (Jaén). The only examples from secure chronological contexts were excavated at the hilltop settlements of Cerro de la Cruz, Almedinilla (Jaén) (Quesada *et al.* 2010: 88-89) (fig 3.13-3.14) and Castrejón de Capote, Higuera la Real (Berrocal Rángel 1989: 258; 284, Fig. 26.1).

Only one district producing this type of Iron Age mill has been identified. It is the limestone tufa that outcrops around the town of Almedinilla, very near the settlement of Cerro de la Cruz. The exact location of the millstone workings, however, remains unknown.

Fig. 3.13: Views of an Iberian rotary pushing mill (diameter 60 cm) from the Iberian settlement of Cerro de la Cruz (Museo Histórico–Arqueológico de Almedinilla). The stone is a local "travertine". The L-shaped lateral cuttings served both to lodge the wooden driving fittings and to lift the stone for maintenance (photographs by T. Anderson).

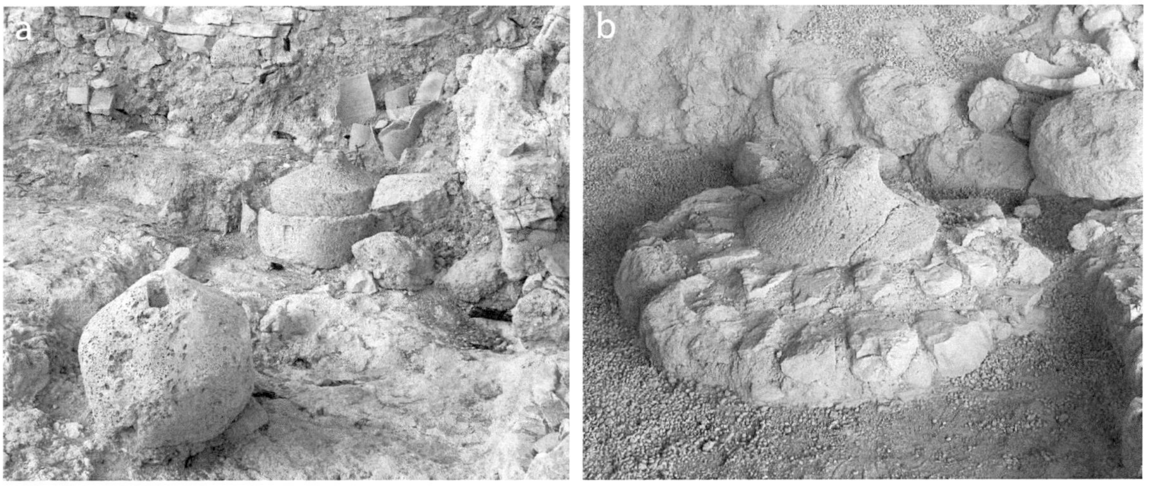

Fig. 3.14: Iberian rotary pushing mills in situ *at the Cerro de la Cruz made of "travertine" a) The broken upper stone of the background mill has slipped down to the base of the lower stone, diameter 60 cm; (photograph from http://www.rutasconhistoria.es/loc/poblado-ibero-el-cerro-de-la-cruz). b) an extremely worn "travertine" lower stone* in situ *on a rock-built podium at the Cerro de la Cruz, diameter: 59 cm (photograph by T. Anderson).*

3.4.2. The Roman rotary ring-mill

Roman rotary ring-*catilli* share the same shape, D-section, lateral cuttings and conical or bell-shaped lower stone (Fig. 3.15) with their Iron Age forbearers. For this reason, when devoid of context, it is at times difficult to differentiate Roman ring-mills from the Iberian pushing models. They differ in that the latter are usually larger, their cuttings are most often of dove-tail shape and their eyes are much more prominent. The ring-mill is also known as the *Volubilis* type because of its abundance in the Roman city of that name in northern Morocco (Luquet 1966: 306, fig. 3B; Akerraz & Lenoir 2002; Longepierre 2011: 1000; Leduc 2011). A similar mill, the Haltern-Rheingönheim type, is known in the Rhine Valley in Germany with examples as far south as Basel and Avenches in Switzerland (Anderson *et al.* 2004: 11).

It is becoming more and more clear that the ring-mill was related to both the bread and olive oil industry. As Peña Cervantes has noted, of the 35 excavations of oil presses, ten are associated with ring-mills (2010: 66). An *in situ* example is at Munigua, Seville, in the same room (two steps away) from an oil press (Peña Cervantes 2010: 146, fig. 52). These *meules annulaires*, as they are known in French, are also characteristic of the *pistrinas* of *Volubilis*. The 39 examples at Volubilis outnumber by far the three Pompeian (Leduc 2008: 482). The ring-mill may, in fact, have been "multifunctional" (a term borrowed from J.-P. Brun), used both for crushing olives and for grinding cereals (Akerraz & Lenoir 2002; Leduc 2008: 483; Peña Cervantes 2010: 574, 576, 750). Much research remains to be done on this question from both the point of view of typology and analyses of vegetable remains, in particular phytoliths.

There are at least a dozen ring-*catilli* (*anneaux*) reported in the survey of Michel Ponsich of the Guadalquivir River Valley (Ponsich 1974; 1979; 1987; 1991). A granite model is also recorded at *Conimbriga* in Portugal (Borges 1978: 125, no. 31) and a volcanic model is on display in the Museum of *Castulo* in Linares (Jaén) (inv. CEO1429) (Fig. 3.16-3.17). There are other examples of varying lithology throughout museums of southern Spain such as Murcia, Osuna (Seville), Mérida (Badajoz), Granada and Almería.

As has been pointed out (Anderson *et al.* 2011a: 154), the morphometric similarity of ring and Pompeian lower stones has engendered confusion as to the identification of the mill type (Fig. 3.18). The tendency in archaeological literature is to link all conical lower stones directly with

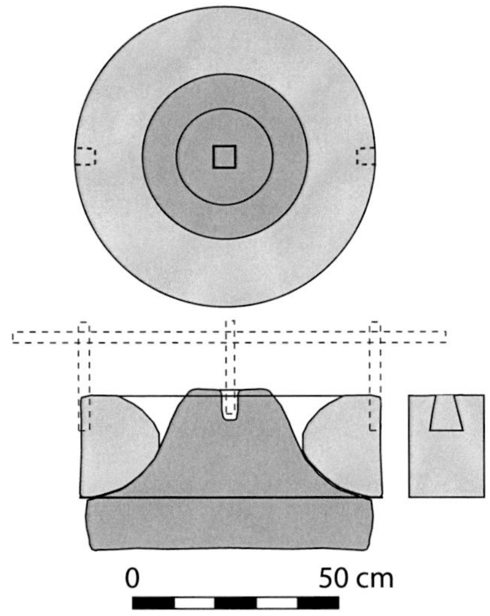

Fig. 3.15: Schema of a Roman human- or animal-driven ring-mill (drawing by T. Anderson).

*Fig. 3.16: Volcanic ring-*catillus *from the Roman city of* Castulo *(Jaén). The type of rock of the lower stone (not the original pair) has not been determined. ø: 64 cm (photograph from Linares museum, inv. no. CE01430_R).*

Pompeian mills without considering that they could in fact be coupled with ring-*catilli*. This confusion is heightened by the relatively low spatial spread of ring-*catilli*, due probably to their fragile nature rendering them difficult to recognise in a fragmented state.

The sites of *Sisapo* (CR-1) and Bolaños (CR-2) are the best examples of quarries producing ring-mills. There is also strong evidence, from examples in museum collections, of production at still unidentified volcanic lamproite outcrops in the Province of Murcia.

*Fig. 3.17: Examples of granite Roman ring-mill upper stones (*catilli*) measuring about 70 cm in diameter in the National Museum of Roman Art, Mérida (Badajoz).*

*Fig. 3.18: Roman human-driven lower stones (*metae*). When the lower stone is isolated from its upper stone, it is difficult to determine the mill type. However, due to the rarity of Pompeian mills in southern Spain, these lower stones were probably most often coupled with ring-type *catilli*: a-b) Almedinilla Museum (Córdoba), ø base: 62 cm, sedimentary rock; ø: 68 cm, volcanic rock; c) Albolote (Granada), inv. no. CE14119, ø: 83 cm, sandstone (photograph http://www.juntadeandalucia.es/culturaydeporte/WEBDomus/fichaCompleta.do?acron=MAEGR&musid=9&ninv=CE14119&volver=portal&k=CE14119&tipoBusqueda=simple); d-e) Oreto-Zuqueca (Ciudad Real) ø: 55 cm, ø: 73 cm, volcanic rocks; f) Fuente Álamo (Murcia) ø: 53 cm, volcanic rock.*

3.4.3. The Roman Pompeian mill

The Pompeian mill is the best known of the different mills and it is known throughout the Roman Empire (Fig. 3.19). With its typical hourglass-shaped upper stone (*catillus*) with lateral lugs or sockets for the driving rig and conical or bell-shaped lower stone (*meta*), this mill was either driven by a donkey (it is often referred to as the "donkey mill"), as seen in Antique iconography, or by humans.

Although the most celebrated examples are found in Italy in the ancient bakeries of the cities of Pompeii and Ostia (Peacock 1986; 1989), models are known as far as the Eifel in Germany (Hörter 1994: 32-33), the British Isles (Williams & Peacock 2011b) and the Middle East (Peacock & Williams 2014). Recent work in France reveals a number were distributed throughout the Narbonese province (Jaccottey & Longepierre 2011). While most of these mills found in France are imports of volcanic rock from the volcanic district of Orvieto (Italy), there are also copies, especially in the north of France, hewn from local sandstone (Jaccottey & Longepierre 2011).

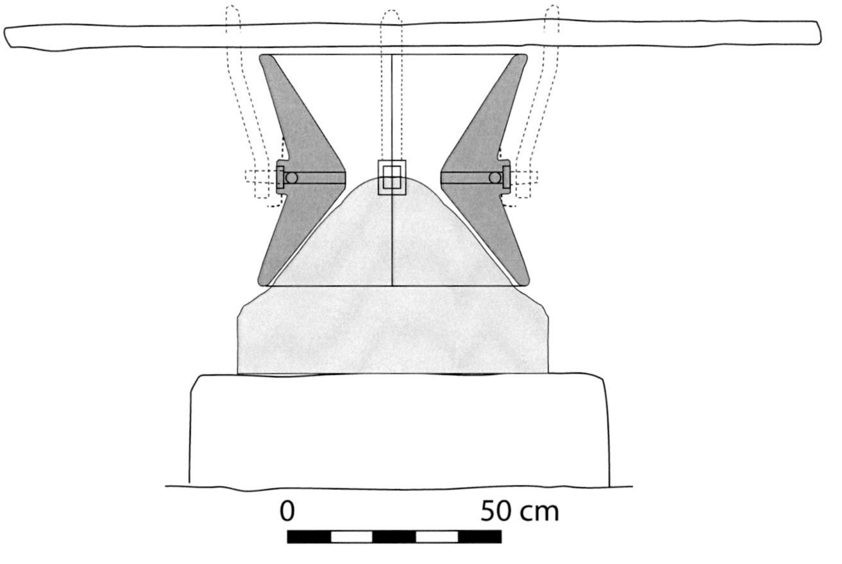

Fig. 3.19: Schema of a Pompeian mill (drawing by T. Anderson).

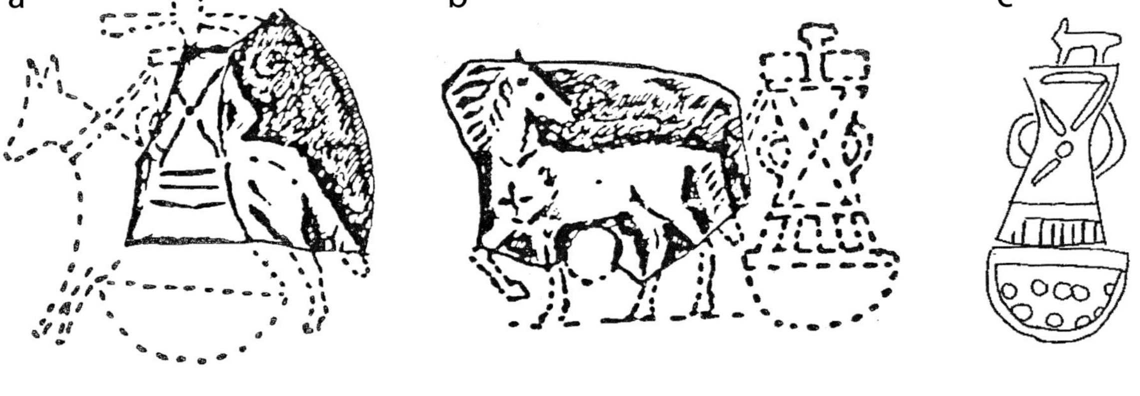

Fig. 3.20: Representations of Pompeian millstones on Roman lamps. a-b) Bilbilis, *Zaragoza (from Amaré 1988, p. 83-84, fig. 152-153); c) Murcia (from Amante 1988, p. 227, pl. IV, 72).*

Contrary to Italy and France, there are no monumental iconographic representations of Pompeian style mills on Iberian soil. The only confirmed scenes of this mill type are sketches on Roman oil lamps from Zaragoza (Amaré 1988) and Murcia (Amante 1988) (Fig. 3.20). It is not known if these scenes on locally produced lamps are copies of representations from imported models or if they were inspired from Pompeian mills in use in *Hispania's* urban centres (Amaré 1988; Amante 1988).

In any case, at present, only slightly more than a dozen Pompeian upper stones are known for the whole of the Peninsula, a number that differs with the total absence when Moritz penned his mill masterwork in 1958. Most are of volcanic material and, like those of France, are probably long-distance imports from elsewhere in the Mediterranean (Fig. 3.21).

There is nonetheless evidence, albeit modest, of Pompeian "production" in *Hispania*. A *catillus* roughout of local *ostionera* rock is in the *lapidarium* of the Roman city of *Baelo Claudia* (Cádiz) (see CA-2a), while a second *catillus* from ancient *Malaca*, hewn presumably from a porous local limestone, is in the Museum of Málaga (Fig. 3.22). A small model in the ARQVA Museum of Cartagena could, potentially, have been scored at a nearby Murcia volcanic outcrop (Anderson *et al.* in press). But these are exceptions. There is no hard evidence of Pompeian workings in southern Spain. In fact, the nearest Pompeian volcanic production, signalled by Williams-Thorpe (1988), is across the Mediterranean in northern Morocco at Farkhana. Two hourglass-shaped roughouts in the archaeological Museum of Melilla, a short distance from Farkhana, confirm this production centre.

Thus, although the number of Pompeian style mills in Spain has grown since Moritz' research (1958), this mill type, as is the case in the Roman city of Volubilis in northern Morrocco (Leduc 2008: 483), was not widely adopted in Roman *Hispania* (Anderson *et al.* in press).

Fig. 3.21: Pompeian mill from Mazarrón (Murcia). Recent analyses indicate this volcanic mill is probably an import from the volcanic district of Orvieto, Italy. The lower stone and podium are reconstructed (photograph Museo Arqueológico de Murcia).

Fig. 3.22: Pompeian upper stone from the Alcazaba de Málaga. This porous limestone (?) is possibly a local copy of an imported volcanic millstone.

3.4.4. The Medieval to Contemporary tahona

The *tahona* (sometimes written *taona* or *atahona)* (Fig. 3.23), according to the dictionary of the *Real Academia Española*, is a flour mill driven by *caballería* (*i.e.* horse, mule or donkey). This mill was the natural heir of the Roman Pompeian or "donkey" mill. The term derives from the Arabic *attahuna* meaning "mill". An alternate definition of the word *tahona* is that of a bakery. These different definitions, at times, lead to confusion as to whether the term refers to a mill or a bakery.

The *tahona* was an alternative to watermills and windmills. The early 17th-century *Covarrubias* dictionary explains that these "dry" mills were found in fortifications and other features devoid of sources of running water (Covarrubias 1611: 99). The medieval *tahona* in the Fortress of Calatrava la Nueva (Aldea del Rey, Ciudad Real), for example, is perched on the top of a mountain (Fig. 3.24).

It is beyond the scope of this study to describe in detail the technical features of this mill. All the same, in the study area, two types of medieval *tahonas* are described among the mills of the region of Córdoba. The first was driven by direct horizontal traction from the animal, whereas the second possessed a more complex gearing system similar to that of watermills (Cordóba de la Llave 1988: 839-840). Cordóba de la Llave, however, specifies that no mills of either type are identified in the urban Córdoba, dominated by watermills.

Accordingly, recent research has not identified any *tahonas* in the Alpujarra mountains of the Province of Almería or in the neighbouring areas of Murcia and Granada (Cara Barrionuevo *et al.* 1999: 148). Koldo Lizarralde, in his quantification of mid-19th century mills gleaned from the Madoz dictionary, only itemised 1,476 *tahonas* of the total of 25,000, representing less than 6% of the total (Lizarralde 2010). It is therefore apparent that *tahonas*, from the Middle Ages to Contemporary times, took on a secondary role to that of watermills.

Millstones for *tahonas* are known to have been fashioned at certain quarries in southern Spain, for example at the white limestone exploitations of Colmenar de Oreja (M-2) and El Berrueco (CA-8). Madoz states that the production of Colmenar de Oreja targeted the *tahonas* of the city of Madrid (Madoz 1847, Vol. 6: 525). In the case of El Berrueco, he specifies that, besides the yearly production of 64 millstones for watermills or windmills, the quarry made 480 "*granos*", that is, millstones for *tahonas* (Madoz 1846, Vol. 4: 290).

The use of the term *grano* (literally meaning "grain") by Madoz raises a problem. Does this term suggest a morphological difference between the production destined for watermills and windmills? Or is it simply a different name for the same product? My search in dictionaries of the 18th and 19th centuries did not shed light on the question suggesting the term was not widespread. Furthermore, from old representations (such as figure 3.23) these stones do not appear to be proportionally smaller than millstones for hydraulic installations.

In any case, the extraction process of millstones for animal-driven *tahonas* would have been identical to that of stones for other types of mills. Hence any millstone quarry could technically deliver an order of a millstone for a *tahona*.

Fig. 3.23: Extract from the Codex of pseudo Juanelo Turriano (c. 1595) depicting a tahona *(animal-driven grain mill). The scene is of a sophisticated mechanism divided into a mill house (upper level with the millstones) and wheel house (lower level with the animal) (Madrid, Biblioteca Nacional, digital image from Archimedes Project, http://dmd.mpiwg-berlin.mpg.de/author/dmd/database/).*

*Fig. 3.24: Views of the animal-driven flour mill (*tahona*) in the castle of Calatrava la Nueva (Aldea del Rey, Ciudad Real). In spite of being in the heart of the Calatrava volcanic district, the millstones are of conglomerate (photograph left http://patrindustrialquitectonico.blogspot.com.es/2011/08/un-molino-del-siglo-xii-es-las-nueva.html; photograph right by T. Anderson).*

3.5. Watermills

3.5.1. Roman watermills

The identification of Roman watermills through excavations and research in museum depositories has become more and more commonplace in the last years, in particular in central Europe. In France, besides the complex of Barbegal near Arles (Leveau 2006), watermill installations have been pinpointed in the *villae* of Mesclans and Saint-Pierre (Brun & Borréani 1998), and at Longvic near Dijon in eastern France (Jaccottey & Labaune 2010). In Switzerland, four Roman watermills have been excavated: Hagendorn, Zug (Gähwiler & Speck 1991); Rödersdorf-Klein Bühl, Solothurn (Harb 2002), Avenches/En Chaplix (Castella 1994); and Avenches/Les Tourbières (D. Castella, pers. comm.). A systematic study of ancient millstones stored in Swiss collections clearly substantiates the idea that hydraulic mills were widely distributed throughout the Roman countryside of the Swiss Plateau (Anderson *et al.* 2004).

Currently, the discoveries in France and Switzerland point to the use of the vertical waterwheel. The millstones at these sites are often bell-shaped, long-distance imports of volcanic rock measuring between 55 and 90 cm in diameter (Fig. 3.25). Two systems of transmitting motive power from the vertical spindle to the upper stone have been recognised. The first combines holes and grooves cut into the upper surface for iron cramps (in French *anille-crampons*; Castella 1994: 47, fig. 31). The second is the traditional system by means of a rynd inserted into cuttings at the base of the stone.

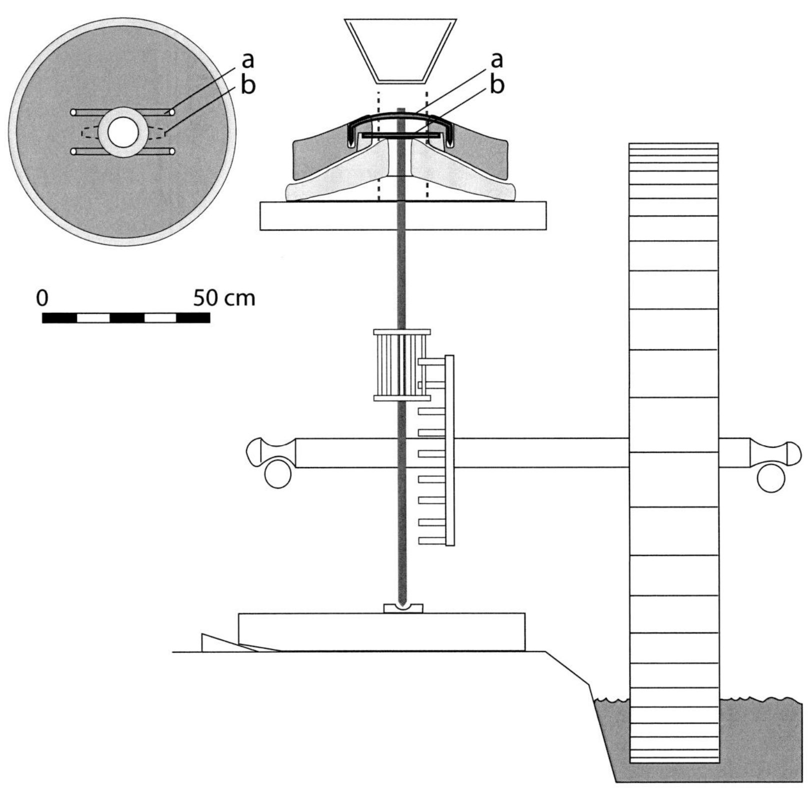

Fig. 3.25: Schematic representation of the Vitruvian vertical waterwheel. The two systems of transmission of the motive force to the upper stone are represented: a) iron cramps on the upper surface and b) rynd on the lower surface (drawing by T. Anderson based on Castella 1994).

The situation of watermills in Roman provinces of *Hispania* is far from settled. With the exception of *Conimbriga* in Portugal (Brun 1997a), no Roman watermill installation or millstone has been unequivocally identified in the Iberian Peninsula.

With the growing number of Roman watermill finds elsewhere in Europe, it seems unlikely that this type of mill was absent in *Hispania,* well known for its hydraulic works (aqueducts, dams). To date, however, only one Spanish millstone, a highly vesicular, black volcanic *meta* measuring 65 cm in diameter, (Fig. 3.26), conforms typologically to those of Roman watermills brought to light in France and Switzerland. This surface find, exposed at the entrance of the archaeological site of Oreto y Zuqueca (Granátula de Calatrava) near the Roman centre of *Oretum*, presumably equipped a watermill along the nearby Jabalón River.

Likewise, no Roman quarry producing watermills has been identified. It would be reasonable to assume, nonetheless, that the millstone exposed at Oreto y Zuqueca, in the heart of the Calatrava volcanic province, was produced in a local or regional quarry. It could come from the nearby volcanic outcrops of Cerro Columba (CR-6) or at Las Herrerías (CR-2), by Bolaños de Calatrava, only 15 km to the north-east. The source of this stone remains to be confirmed.

Fig. 3.26: Photograph and drawing of the lower stone at the entrance to the archaeological site of Oreto y Zuqueca (Ciudad Real). The size, morphology and lithology of this lower stone, a surface find, conforms to that of millstones found in Roman watermills (photograph and drawing by T. Anderson).

3.5.2. Medieval to Contemporary watermills

Cereal grinding watermills throughout these periods were built along rivers, canals and *acequias* (irrigation ditches) (Fig. 3.27). They were raised for the most part with stone and with their mechanisms made of wood. The buildings comprised two stories: a lower level (wheel house), reserved for the waterwheel and gearing machinery, and the upper level (mill house), with the millstones, grains reserves and hopper, where the grinding took place.

Throughout these periods, millstones evolved little from a morphological standpoint. The bell-shaped model from Roman times (often of volcanic rock) gave way to flat, discoidal models that varied in diameter from about 0.80 to 1.30 m. These were hewn for the most part from sedimentary stones.

There are two main types of hydraulic installations in the study area. *Aceñas* were driven with vertical waterwheels, whereas *rodeznos*, the more common type in the south of Spain, were powered by horizontal waterwheels. Other hydraulic mills, such as tide mills and boat mills, were much less common in the study area and will not be described in this study.

The development of the hydraulic milling technology in southern Spain has been the subject of much molinological research (Reyes Mesa 2000a-b; 2006; Córdoba de la Llave 1988, among others). The following descriptions, therefore, are no more than superficial portraits of the more common types of hydraulic milling installations.

Fig. 3.27: Engraving from 1842 entitled Molinos Árabes de la Mina en Alcalá de Guadaira by Genaro Pérez de Villaamil (Museo del Romanticismo, Seville). The scene corresponds to a subterranean watermill below the town of Alcalá de Guadaira (Seville). Two sets of millstones with their respective hoppers are powered by wheels harnessing water from natural wells channelled through chutes seen in the background. A millstone is propped against the wall in the shaded area to the left of the grinders. Under the vault to the left are two more millstones propped against the wall. The scene seems to reflect more the situation of the middle of the 19th century than a scene dating to the Islamic domination as indicated by the title of the engraving (from http://sevillalegendaria.blogspot.com.es/2013/05/los-canos-de-carmona-agua-e-historia.html).

3.5.2.1. *Aceñas* (vertical-wheeled mills)

The term *aceña* derives from the Arabic term for "elevator", and in the context of grain mills, refers to the vertical waterwheel set directly in the flow of a river or canal to drive one or more millstones. The driving force was transmitted by means of gears, converting the vertical motion of the waterwheel into a horizontal millstone rotation (Fig. 3.28). *Aceñas* are the direct heirs of the Roman "Vitruvian" watermill.

In the south of Spain, these mills were restricted to certain wider and deeper watercourses. In the Province of Córdoba, for example, *aceñas* were known during the Islamic domination to have been constructed along the wide Guadalquivir and Guadajoz Rivers (Córdoba de la Llave 1988: 835).

In fact, the most celebrated group of *aceñas* in the study area is the chain of eleven mills in the city of Córdoba.These mills, with early phases dating from the Middle Ages, have undergone many transformations over the years (Montero 2008). Some are represented in an engraving

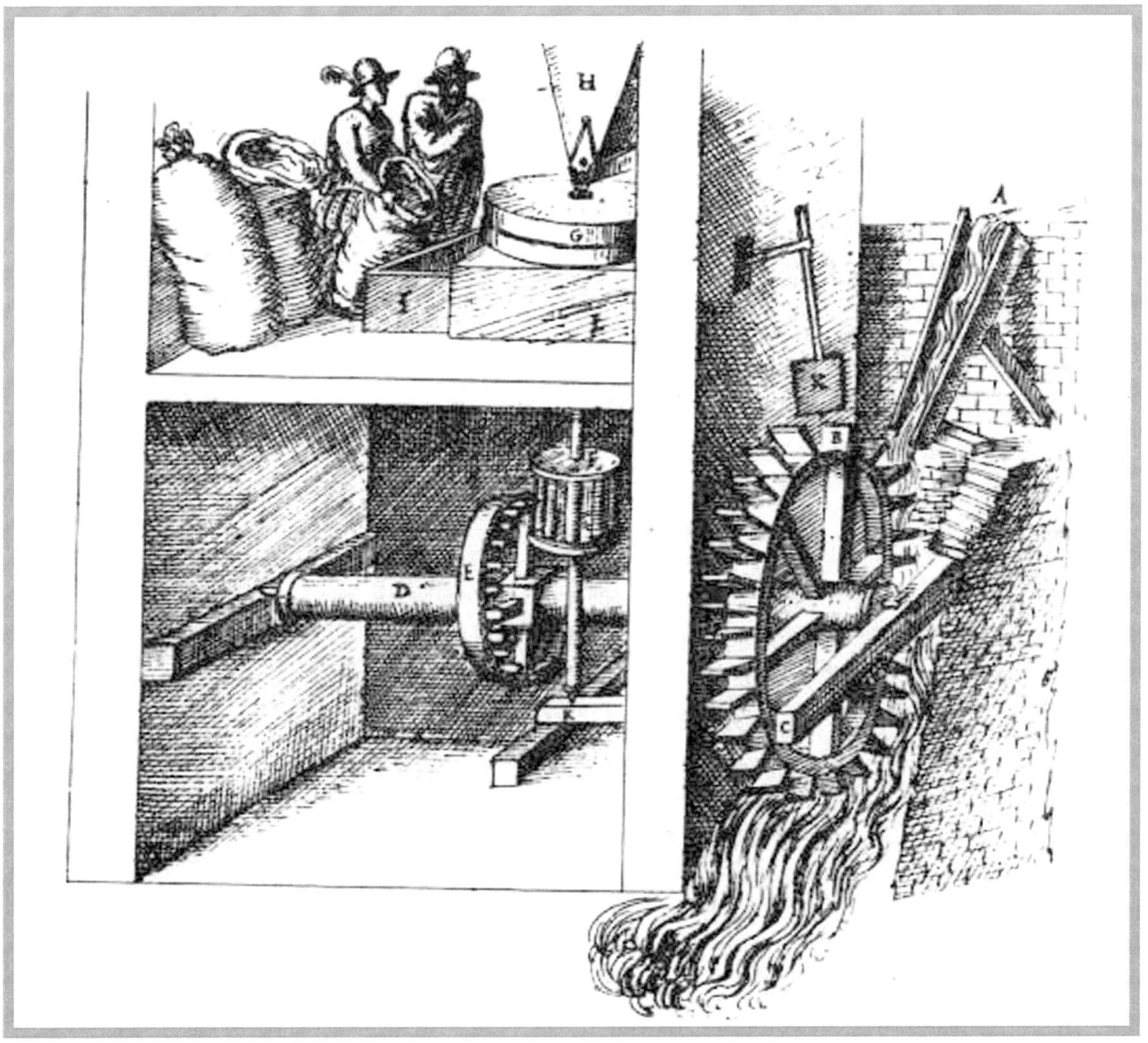

Fig. 3.28: Extract from the Codex of pseudo Juanelo Turriano (c. 1595) depicting an aceña *(grain mill with a vertical waterwheel). The structure is divided into a mill house (upper level with the millstones) and wheel house (lower level) (Madrid, Biblioteca Nacional, digital image from Archimedes Project, http://dmd.mpiwg-berlin.mpg.de/author/dmd/database/).*

by the Flemish artist Anton Van den Wyngaerde dating to 1567 (Fig. 3.29). In earlier times they were equipped with sandstones from local quarries, such as Albaida (CO-7) (Córdoba de la Llave 2003: 305). At a certain point in the last few centuries, some began to import the higher quality *rosso ammonitico* millstones from workings in the Cabra area (CO-1; CO-2), about 85 kilometres to the south-east (Montero 2008).

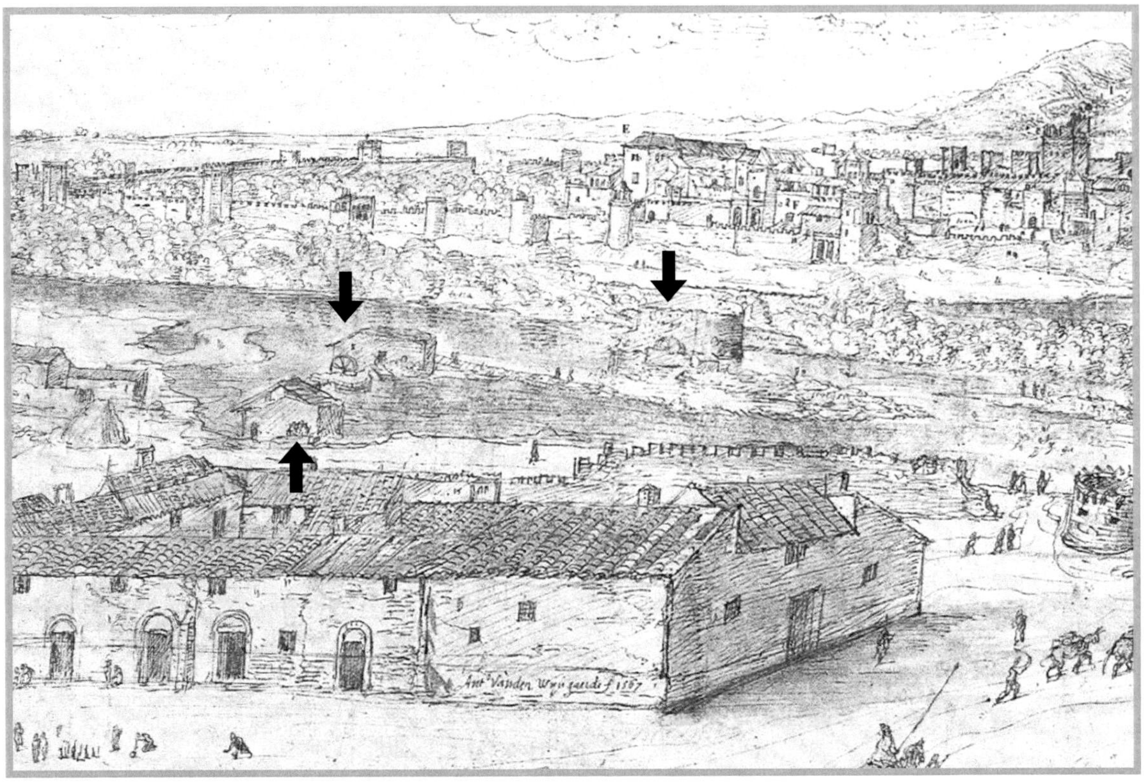

Fig. 3.29: Extract from an engraving by the Flemish artist Van den Wyngaerde (1567) with a view of the city of Córdoba. The arrows point to the waterwheels of the aceñas *in the Guadalquivir River (original illustration in the Victoria and Albert Museum of London).*

3.5.2.2. *Rodeznos* (horizontal-wheeled mills)

The *rodezno* (from the Latin *rota,* meaning "wheel"), equipped with a horizontal waterwheel (Fig. 3.30), was better adapted to the shallow or fluctuating sources of water common to the south of Spain (Reyes Mesa 2006: 213). The transmission of the driving force from the wheel to the stones was either direct, through the shaft (meaning the millstones turned at the same rate as the turbine), or indirect, by means of a gearing system (similar to that of the vertical waterwheel mills) allowing the miller to adjust the velocity of the millstones. Water was directed to the wheel either by means of a channel (*molino de rampa*), inclined at an angle of about 45°, or by a more complex feature that guided the water through a penstock (*molino de cubo*), usually a vertical cylindrical feature (like a well) several metres high and about a metre in diameter. This system permitted the jet of water it released at its base to gain strength and, by impacting the paddles, drive the wheel (Reyes Mesa 2006: 193-201).

The *rodezno*, in the study area, is by far the most common type of grain mill. As the city of Córdoba with the deep and wide Guadalquivir River epitomizes the use of mills with vertical waterwheels, the city of Granada, with its shallow rivers at the foot of the Sierra Nevada mountains, typifies the use of the *rodezno* (Fig. 3.31). In a recent study of more than 700 mills of Granada and its province, all but two applied the *rodezno* technology (Reyes Mesa 2006: 213).

The British writer Gerald Brenan, who sojourned in the Granada mountains in the 1920s wrote:

"... water-mills of the Andalusian mountain villages are of a peculiar and very primitive sort, for their wheels are set horizontally instead of vertically and are turned by the oblique percussion of the water against a number of upright vanes, arranged like spokes. The grindstones run on the same axis in a lower chamber, which greatly simplifies the mechanism and makes them economical both of space and cost. The entire mill is contained in a small two-storied building, set up against the steep hillside and more than twelve feet in height" (Brenan 1998: 72-73) .

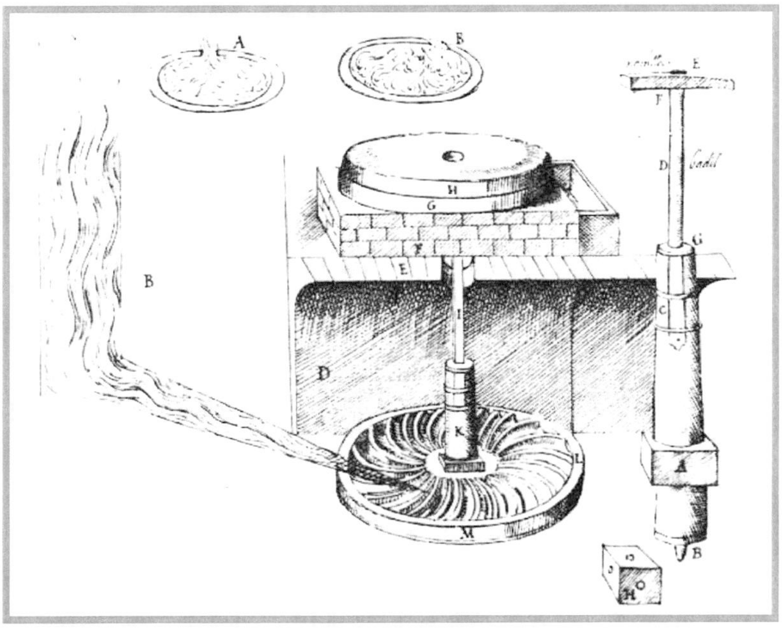

Fig. 3.30: Extract from the Codex of pseudo Juanelo Turriano (c. 1595) depicting a rodezno *(grain mill with a horizontal waterwheel) (Madrid, Biblioteca Nacional, digital image from Archimedes Project, http://dmd.mpiwg-berlin.mpg.de/author/dmd/database/).*

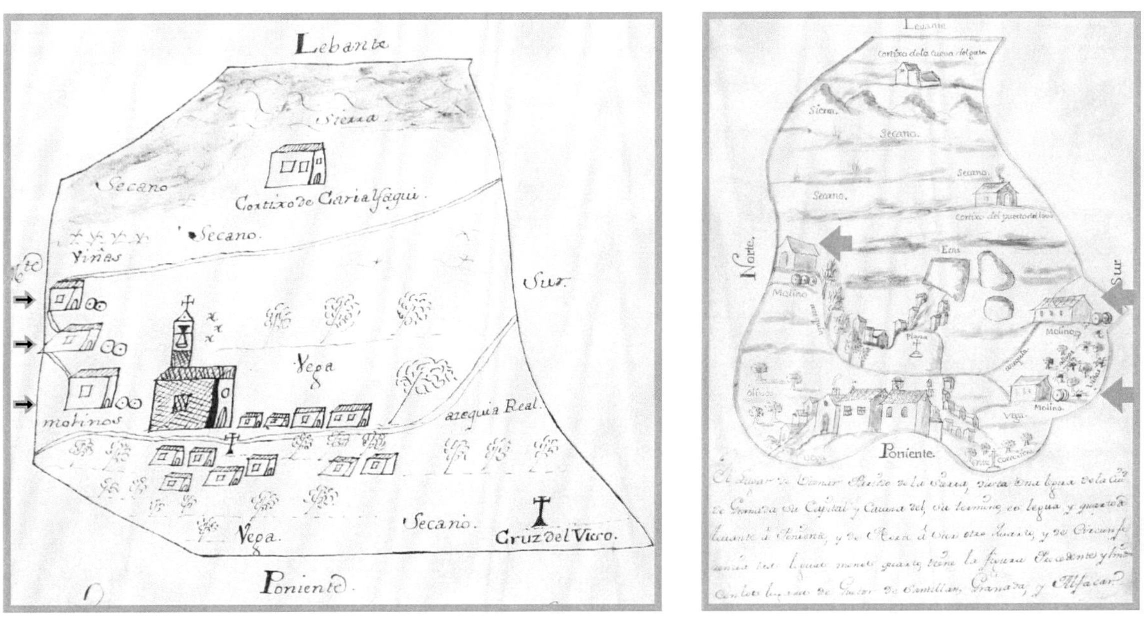

Fig. 3.31: Drawings from the Catastro del Marqués de Ensenada (1750-1754) of the towns of Nivar and Viznar (Province of Granada) showing the location of the rodezno *watermills. These drawings supplement the answers of a questionnaire and illustrate the principal features of the towns such as mills, churches, houses, surrounding farms, irrigation ditches* (acequias) *and landmarks (crosses) (from El Catastro del Marqués de la Ensenada en el Antiguo Reino de Granada, Instrumentos de Descripción, Archivos 16, Consejería de Cultura, Junta de Andalucía (CD format), no date.*

3.6. Windmills

Contrary to the impression drawn from the tale of Don Quixote, windmills represent only a modest fraction of the grain mills in Spain. Koldo Lizarralde only records 676 (about 3%) of a total of 25,000 mills that are cited in the Madoz dictionary (Lizarralde 2010). (Curiously, according to Lizaralde's numbers, the lesser-known animal-driven *tahonas*, at 1,476, nearly doubles that of windmills.) A second quantification of old mills undertaken in the province of Granada registers only two windmills in a total of 542 (Reyes Mesa 2006: 213).

Windmills in Spain, for the most part, were grouped in certain areas where water was in short supply or could not easily be harnessed. Don Quixote's La Mancha region, spread over several provinces in central Spain, is the location of the most celebrated group. These "giants" with whitewashed bodies capped with black cones can still be seen in a number of towns: Alcázar de San Juan (Fig. 3.32); Campo de Criptana and Valdepeñas (Ciudad Real); Consuegra (Toledo), Villarobledo (Albacete); and Mota del Cuervo (Cuenca). The Flemish 16th-century artist Anton Van den Wyngaerde, at the service of King Phillip II, depicted a group on the horizon outside the Belmonte (Cuenca) (Fig. 3.33). The vanes of the "Manchego" type, normally four, are rectangular, made of wooden slats covered with canvas.

The second concentration of windmills in Spain is in the south-east of the Peninsula, from the south of Alicante (Torrevieja) (Fig. 3.34), through the Campos de Cartagena of Murcia and the Campos of Níjar in Almería (Fig. 3.35). These are some of the driest regions in Spain. The vanes of these windmills, double in number, differ from their Manchego counterparts in that they have triangular sails (Rojas & Amezcua-Ogáyar 2005).

Fig. 3.32: View of a group of windmills outside the city of Alcázar de San Juan (Ciudad Real) (Photograph by T. Anderson).

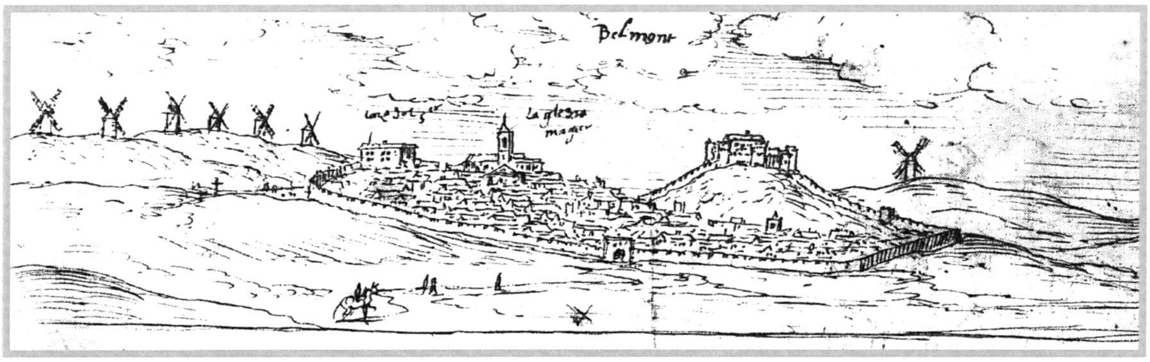

Fig. 3.33: Drawing of the windmills of Belmonte (Cuenca) by Van de Wyngaerde (1563). The Flemish artist was at the service of King Phillip II. The work is in the Victoria and Albert Museum, London (from Ibáñez 2003: 75).

Windmills of both the Mancha and the south-east share a cylindrical tower built of stone topped by a cap of wood, holding the gearing system, that could be pivoted to face the wind by means of a tail pole. The origin and the precise date of the introduction of this technology (sometime during the Islamic domination) is still a question of debate (Rojas & Amezcua-Ogáyar 2005) and beyond the scope of this work.

There is no evidence in southern Spain that millstones destined for windmills differed in any way from those for watermills (Fig. 3.36). Madoz, in his comments about El Berrueco (CA-8), lumped production of millstones for watermills and windmills in the same lot (albeit different from *tahonas*) (Madoz 1847, Vol. 6: 525) suggesting that millstones for water and windmills were identical. There is only one quarry site, Pedrizas de Piédrola (CR-4), on the outskirts of the town of Alcázar de San Juan, the location of a large cluster of windmills (*cf.* Fig. 3.32), that can be directly related to the production of millstones for windmills. These products, although not studied in detail, do not appear to differ from that of other millstone workings.

Fig. 3.34: Photograph by Jean Dableda of the town of Torrevieja, Alicante in the late 19th century. Four windmills can be distinguished on the horizon (photograph from the blog of Franciso Rebollo Ortega, http://franciscorebollo.blogspot.com.es/2011_09_01_archive.html).

Fig. 3.35: Examples of windmills in south-eastern Spain driven by vanes with triangular sails. a) Molino Viejo de Zabala, Cartagena, Murcia, photograph left by Torres Ros, http://www.forocartagena.com/t800-molinos-de-viento-y-arquitectura-del-campo-de-cartagena; b) Cabo de Gata, Nijar, Almería by T. Anderson).

Fig. 3.36: Examples of millstones in the Cabo de Gata of the province of Almería (Isleta del Moro and Las Negras) from nearby windmills. a) Coarse sandstone b) porous limestone. From the petrographical and morphological point of view, these millstones show no differences with those produced for watermills.

4. MILLSTONE QUARRY GEOLOGY

4.1. Generalities

The following text (in italics) is a sketch of the geology of the southern half of the Iberian Peninsula originally penned by Jane H. Scarrow of the Geology Department of the University of Granada for an article about millstone quarries in the south of Spain published in the proceedings of the *Bread for the People* colloquium held in Rome in 2008 (Anderson & Scarrow 2011). The text was modified and extended from the introductory chapter of *La Geología de España* (Vera 2004). In the original article, the illustrations appeared in black and white. A second, slightly modified version of this text was published in colour recently to the proceedings of the colloquium *Seen Through a Millstone* held in Bergen, Norway (2011) in an article about rotary querns and their production sites in the south of Spain (Anderson *et al.* 2014).

The geology of the Iberian Peninsula is dominated by the high central Iberian Massif. After considerable debate, this Massif, that covers most of the western half of the Peninsula, was divided into six main zones. Broadly from north to south these zones are as follows: Cantabrian (Fig. 4.1, sector A); West Asturian-Leonese (Fig. 4.1 sector B); Galicia Tras-Os-Montes (Fig. 4.1, sector C); Central Iberian (Fig. 4.1, sector D); Ossa-Morena (Fig. 4.1, sector E) and South Portuguese (Fig. 4.1, sector F). The central region has an average elevation of 660 m and is bordered to the north by the basin of the Duero River, to the north-east by the Ebro River and, broadly, to the south by the Guadalquivir Valley. Lying beyond these hydrographic confines are the Cantabrian Cordillera in the north, the Pyrenees in the east and the Baetic Range in the south.

This study of millstone quarries basically concerns the region to the south of the River Tagus that flows east to west, cutting the Peninsula into two roughly equal parts. Thus, rocks exploited for the millstone production considered here include: the Precambrian to, predominantly, Palaeozoic granite-gneiss-schist igneous and metamorphic crystalline rocks of the three most southerly zones of the Iberian Massif (Central Iberian, Ossa-Morena, and South Portuguese); the Mesozoic-to-Cenozoic limestone-sandstone-conglomerate sedimentary rocks of the southerly basins and the Baetic Range; and the Cenozoic volcanic rocks of the recent meridional magmatic activity.

The Central Iberian Zone of the Iberian Massif, encompassing large areas of Extremadura, Castilla-La Mancha and Madrid (Fig. 4.1, sector D), comprises Proterozoic and early Palaeozoic metasediments and orthogneisses, extensive Lower Ordovician 'Armorican' quartzites and

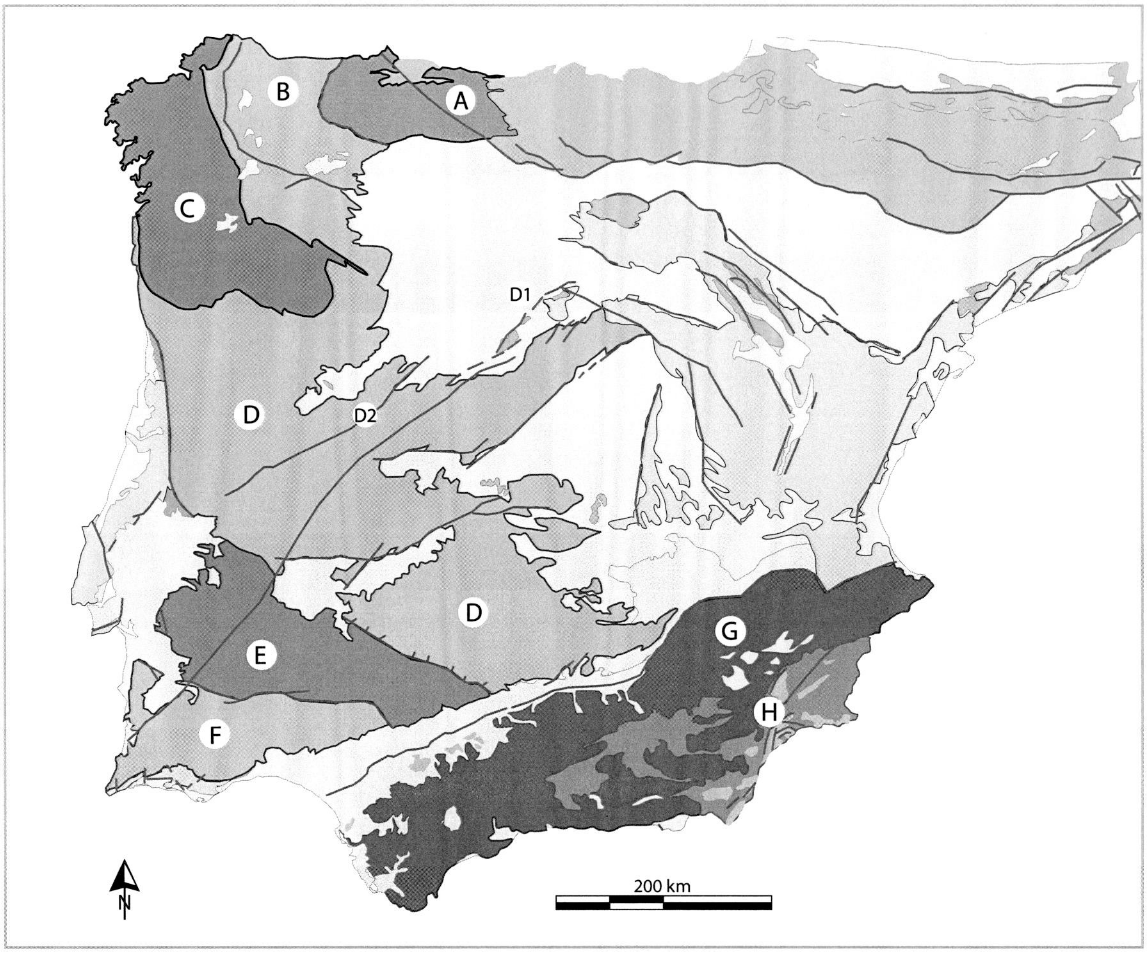

Fig. 4.1: Map of the principal geological zones of Iberia. Letters, also mentioned in the text, refer to the following regions. A. Cantabrian Zone; B. Western Asturian-Leonian Zone; C. Galicia Tras-Os-Montes Zone; D. Central Iberian Zone, D1. 'Ollo de Sapo' domain, D2. Greywacke-Schist Complex domain; E. Ossa-Morena Zone; F. South Portuguese Zone; G. Baetic range External Zone; H. Baetic range Internal Zone (from Anderson & Scarrow 2011: 261, after Farias et al. *1987). A version of this map in colour was published in Anderson* et al. *2014: 113, fig. 2, cf. http://am.uis.no/getfile.php/Arkeologisk%20museum/publikasjoner/timothy-j-anderson-etal.pdf.*

widespread Carboniferous granitoids. The Zone is divided into two domains mainly on the basis of the rocks underlying the quartzite. In the north, the 'Ollo de Sapo' domain pre-Ordovician rocks are potash feldspar augen orthogneisses, high grade regional metamorphic rocks and Variscan syn-tectonic granites (Fig. 4.1, sector D1). In the south, the Greywacke-Schist Complex domain pre-Ordovician rocks are schists and greywackes, the metamorphic rocks are low grade and the granitoids are syn- and post-orogenic (Fig. 4.1, sector D2).

The Ossa-Morena Zone occupying parts of the provinces of Huelva and Seville (Fig. 4.1, sector E) is made up of Upper Proterozoic to Carboniferous rocks. Oceanic basic igneous rocks crop out at the northern contact with the Central Iberian Zone and the southern contact with the South Portuguese Zone. The main lithologies in this region are low-grade metamorphic rocks and granitoids of varying ages including Vendian, Cambrian and Carboniferous syn- and post-orogenic intrusions.

The South Portuguese Zone (Fig. 4.1, sector F), the most meridional division of the Iberian Massif (in the Huelva province), is composed of Devonian to Carboniferous volcano-sedimentary, igneous and low-grade metamorphic rocks that, in places, are overlain by Permian sediments.

The Baetic range in the south and south-east of the Iberian Peninsula (Fig. 4.2), from Huelva all the way to Alicante, forms the most westerly sector of the Alpine orogen that borders the northern and southern margins of the Mediterranean. The Baetic cordillera is divided into two main zones, which were two separate microplates through the Mesozoic and Cenozoic: the External Zone (South Iberian paleomargin) and the Internal Zone (Alborán Domain fragment of the Mesomediterranean plate). The External Zone, to the north (Fig. 4.1, sector G), comprises Triassic to Miocene continental margin sedimentary rocks. This zone is further subdivided into the northerly Pre-baetic relatively undeformed shallow marine sediments and the more strongly deformed southerly Subbaetic lower Jurassic pelagic sediments and submarine volcanic rocks. The Internal Zone, to the south (cf. Fig. 4.1, sector H), is made up of a stack of tectonic units, from base to top: the Nevado-Filábride Complex, the Alpujárride Complex and the Málaguide Complex. In contrast to the External Zone, the pre-Mesozoic basement of this region was displaced together with the Triassic to lower Miocene cover rocks. The basement part of the two lower units preserves traces of pre-Alpine deformation, magmatism and metamorphism, which in the Alpujárride Complex includes continental crust and subcontinental upper mantle. Both the basement and cover of all three units were affected by complex Alpine metamorphism. Throughout part of the Mesozoic through to the mid-Paleogene, turbidites were deposited in a narrow ocean basin that existed between the External Iberian plate and the Internal zone Mesomediterranean plate. These marine sediments are preserved in the Campo de Gibraltar Complex.

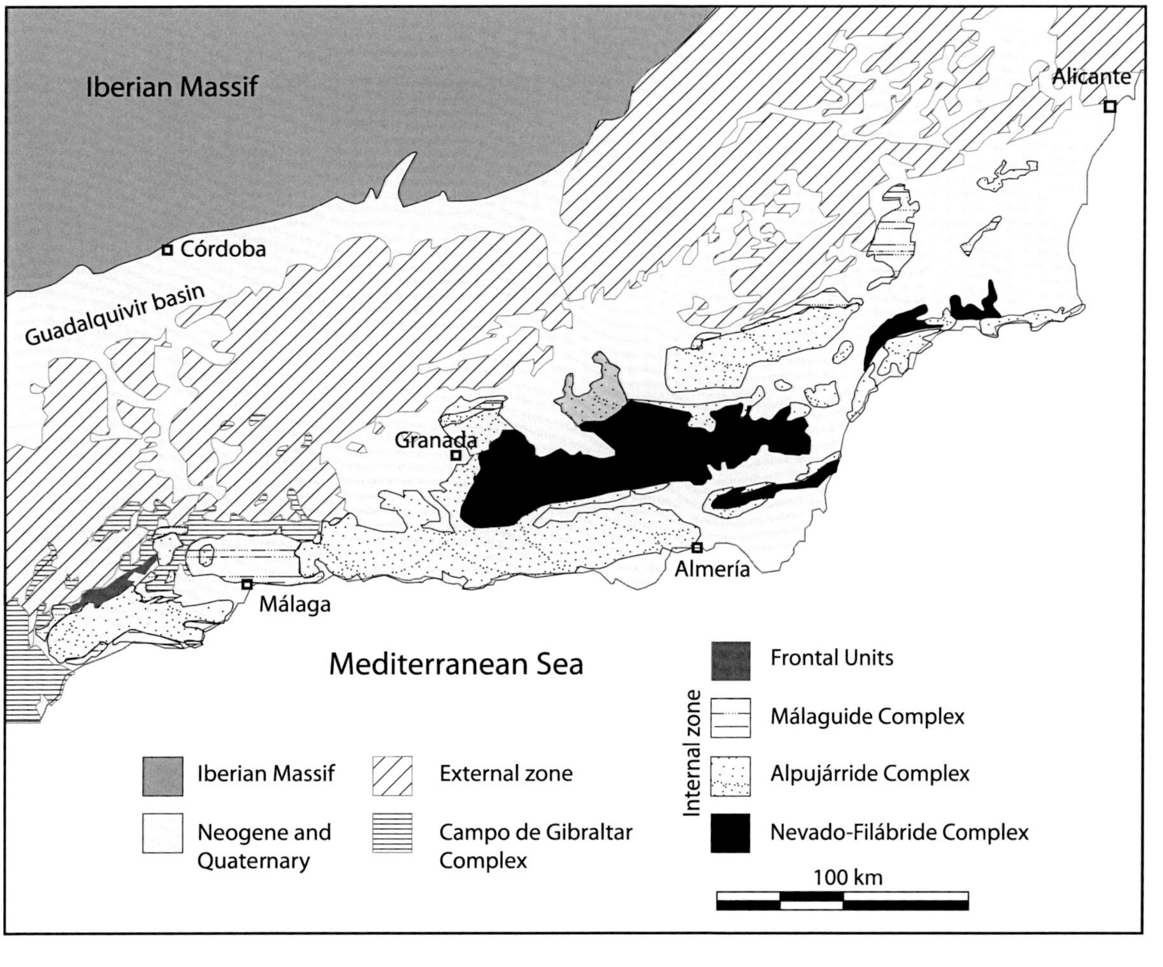

Fig. 4.2: Detailed map of the principal geological formations of the Baetic range (after Martín Algarra 2004). A black and white version of this map appeared in Anderson & Scarrow 2011: 261.

Cenozoic volcanic rocks crop out in four regions throughout the Peninsula: Gerona, Gulf of Valencia, Calatrava and Almería-Murcia. The last two of these are of interest in the current work and include the oldest volcanism, Aquitanian, in Cabo de Gata (Almería) and some of the youngest, Holocene, in the Campos de Calatrava (Fig. 4.3). The Calatrava rocks are basic alkaline and carbonatites. In the Almería-Murcia region, the igneous rocks are more varied including calc-alkaline to alkaline compositions.

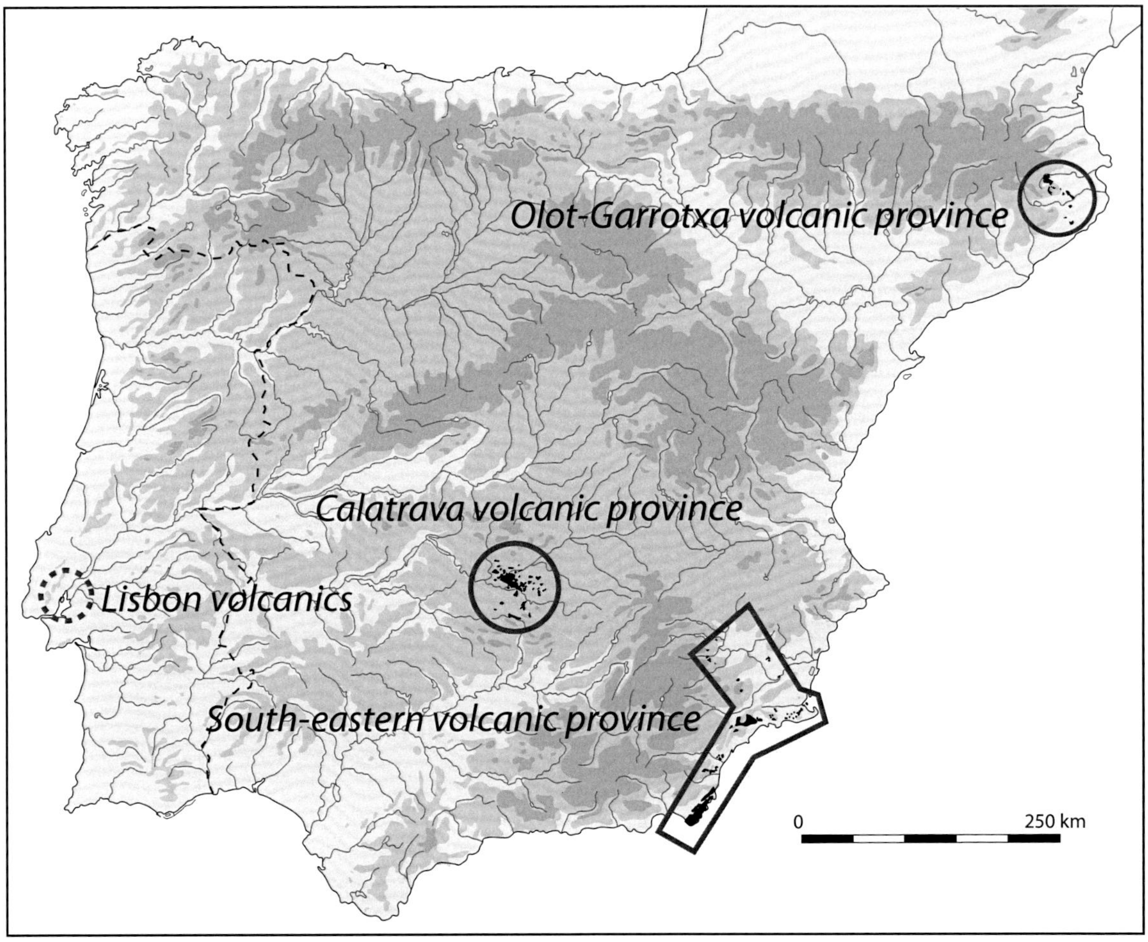

Fig. 4.3: Map of the volcanic districts in the Iberian Peninsula. The red areas correspond to Neogene Quaternary volcanism (based on the Geological Map of Spain, 1.1.000.000, IGME 1994). The Olot-Garrotxa volcanic district in Girona (Catalonia) has been certified as a millstone production centre since the 1980s (Williams-Thorpe & Thorpe 1987; Williams-Thorpe 1988). Millstone workings have since been identified in recent years in the Calatrava and south-eastern Spanish volcanic province by this author. The drawings of the volcanic districts are based on geological maps (IGME). The south-eastern volcanic province is updated from Cambeses 2010.

4.2. Millstone rocks

Millstone makers of the past were not geologists. They nonetheless displayed a keen, intuitive sense of the qualities of rocks. The principal properties sought out for grinding were hardness and abrasiveness. The rock had to be either porous or contain inclusions or crystals that could "bite" the grains while, at the same time, release as few grit particles as possible, as grit in the bread was detrimental to the teeth (and ultimately, the health) of the consumer. To ensure a higher quality of grinding, querns and millstones of some rock types, required dressing either by pecking or carving furrows. These peckings and carvings were arranged in a wide array of patterns.

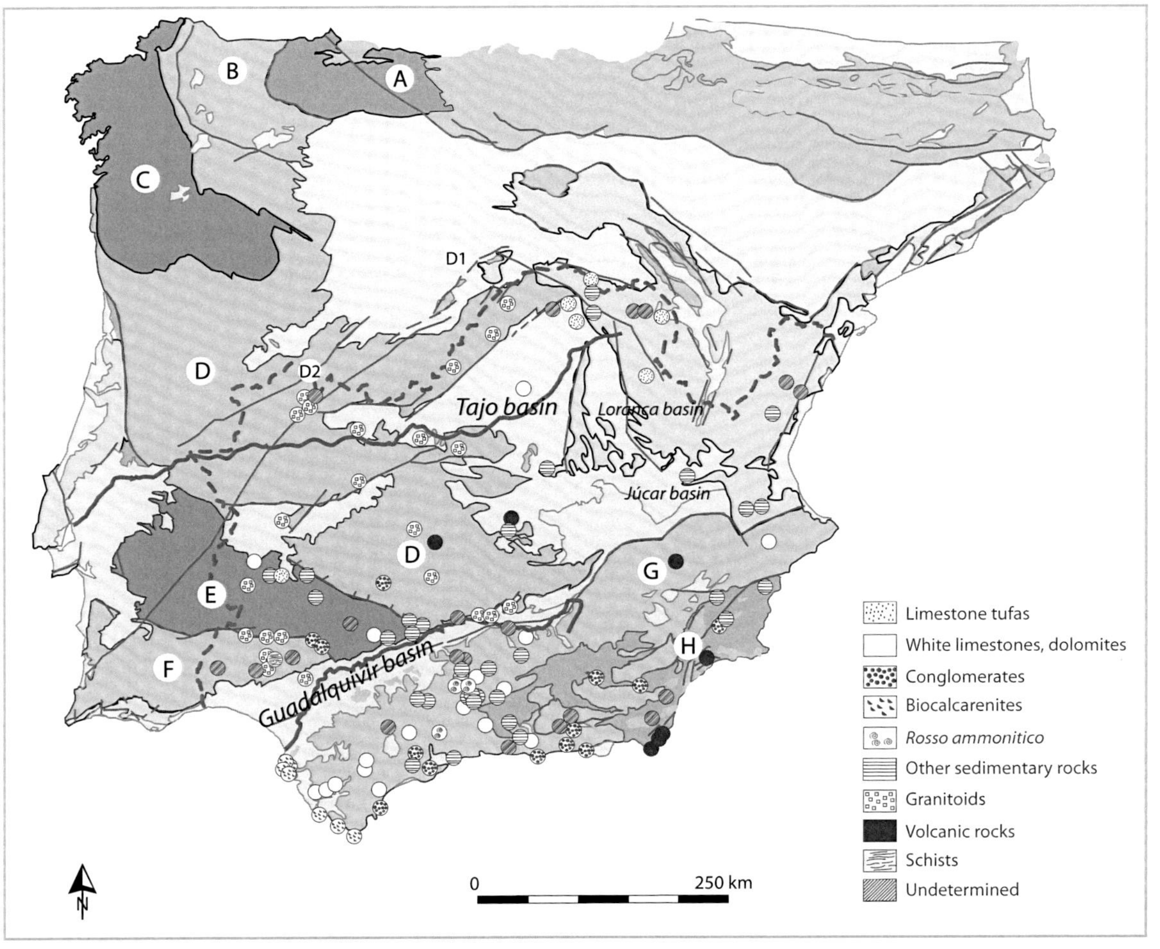

Fig. 4.4: Distribution map of the millstone quarries by rock type. The base is the geological map (see fig: 4.1). The broken line indicates the border of the study area.

Based on a combination of observations at millstone quarries themselves and of querns and millstones in museum depositories, as well as the scrutiny of geological texts (both old and new) and geological maps (*IGME*), nine basic millstone rock types stand out. The spread of the millstone quarries throughout the southern Iberian landscape following these nine categories is illustrated in fig 4.4. Most millstone rocks are sedimentary. The remaining are igneous, notably volcanic and granitoid, metamorphic (one case of schist) or undetermined.

The portraits of each of the rock types were brought together, for the most part, from dictionaries and the following geological websites: http://geology.com/dictionary/glossary-t.shtml; http://college.cengage.com/geology/resources/geologylink/glossary.html; http://www.webref.org/geology/l.htm. When relevant, I have completed these descriptions with my own observations.

4.2.1. Sedimentary millstone rocks

4.2.1.1. Limestone tufas or travertines

This type of rock is cream-coloured, highly porous and formed by calcium carbonate precipitation (springs, caves, or streams carrying large amounts of carbonate solution). It is a rock that was relatively easy to carve and once dry became very hard. In Protohistory it was exploited as surface material or in quarries in the southern mountains of Córdoba, as seen by the example of Almedinilla, Córdoba (CO-15). Several recent quarries are reported in the northern sector of the study area, notably at the town of Tobes (Guadalajara) (GUA-5). The name of this town derives from the Spanish name of the rock (*toba*).

4.2.1.2. White limestones and dolomites

These are fine, hard, white (or slightly grey) sedimentary rocks, composed mainly of calcium carbonate or dolomite. They sometimes present themselves in the form of massive layers that allow deep quarry exploitation such as at Moclín, Granada (GR-1). They are also in the form of layers approximately one metre thick, as is the case at Alhama de Granada (GR-6), Lachar by Jimena (J-2) and Colmenar de Oreja (M-2). They appear as well in the form of small, highly eroded surface outcrops (karsts), exploited individually or in small groups as seen at the Camino del Calvario quarry near Loja (GR-3) .

4.2.1.3. Pebble conglomerates

These very hard detrital rocks possess rounded clasts of varying lithology, generally measuring a few centimetres in length (up to 10 cm) bounded by a fine cement matrix. These were exploited both in the form of surface blocks and true extractive quarries. Layers of this type of rock, although not thick (approximately 1 m), extend over large surfaces and were exploited in extensive true extractive shallow quarries at the Puerto de la Cadena in Murcia (MU-1), Rambla Honda (AL-3) or Playa de Carchuna (GR-9). They were also exploited by means of detaching angular blocks, such as the trench quarry at the Cantera de la Rambla near Caniles, Granada (GR-11).

4.2.1.4. Biocalcarenites

Biocalcarenite is a yellowish, highly porous sandstone composed predominantly of marine shells. This type refers, in particular, to numerous outcrops along the Bay of Cádiz commonly known as *ostionera* (meaning oyster). The outcrops, ranging from fine-grained to very coarse-grained deposits are spread over a corridor about 100 km long between the cities of Tarifa and Chipiona. They were exploited for querns and millstones, as well as construction material, in several extractive quarries along the Atlantic coastline of the Province of Cádiz, notably at the Bay of Trafalgar (CA-1).

4.2.1.5. *Rosso ammonitico* limestones

These are fine, hard, pink limestones with large fossil inclusions (notably ammonites). Like the white limestones or dolomites, these outcrops present themselves both in the form of massive formations exploited by true extractive quarries such as that at Cantera de los Frailes, Cabra, Córdoba (CO-1), or in the form of small, highly eroded surface outcrops or karsts such as at El Torcal in Málaga (MA-1).

4.2.1.6. Other sedimentary rocks

This large group corresponds to millstone workings found in areas dominated by sedimentary rocks that although identified by written sources (by old texts, for example), have not been pinpointed either in the field or on a map and therefore do not benefit from a precise geological portrait.

4.2.2. Metamorphic millstone rock (schists)

Schist, the only metamorphic rock known to have been exploited for millstones in the study area, is coarse-grained, highly foliated and rich in mica. Characterised by flat, parallel slabs that split with ease, this rock also contains crystals, such as garnets or staurolites, that play an important role in the grinding process. In the south-east of the Iberian Peninsula, many Prehistoric and Protohistoric saddle querns are fashioned from this material. Evidence of quarry workings is minor. Only one recent exploitation, El Campillo Province of Huelva (HU- 1), has been identified. Elsewhere in Europe, schists were exploited massively for querns and millstones in Norway, notably at Hyllestad and Selbu (Grenne *et al.* 2008). In Roman Switzerland they were an alternative, in the form erratic blocks, to *grès coquillier* quarries (Anderson *et al.* 2003: 64-67).

4.2.3. Igneous rocks

4.2.3.1. Granitoids

The term granitoid covers a range of very hard, granular, crystalline rocks mainly of quartz, mica and feldspar. The colour of these millstones is mostly white but can veer to pink. Often called "*roca berroqueña*", granitoids present themselves in the form of rounded surface blocks (called *bolos*), small bedrock outcrops exploited individually or in small groups or massive extensions exploited in large and deep pit quarries such as those around the town of Gerena in the Province of Seville (SE-7).

4.2.3.2. Volcanic rocks

Volcanic rocks comprise a range of black, grey, or reddish-purple vesicular rocks. Rotary querns and human- or animal-driven mills were highly prized in Roman times and exploited in quarries in the two volcanic provinces in the study area: 1) the south-eastern Spanish volcanic district, covering the Provinces of Almería, Murcia and part of Albacete (Fig. 4.5); and 2) the vast Campo de Calatrava district in the Province of Ciudad Real (Fig. 4.6). Although they can be identified as volcanic by the trained eye, painting their precise portrait (rhyolites, dacites, basalts, lamproites, melelinites) requires petrographical and geochemical analyses.

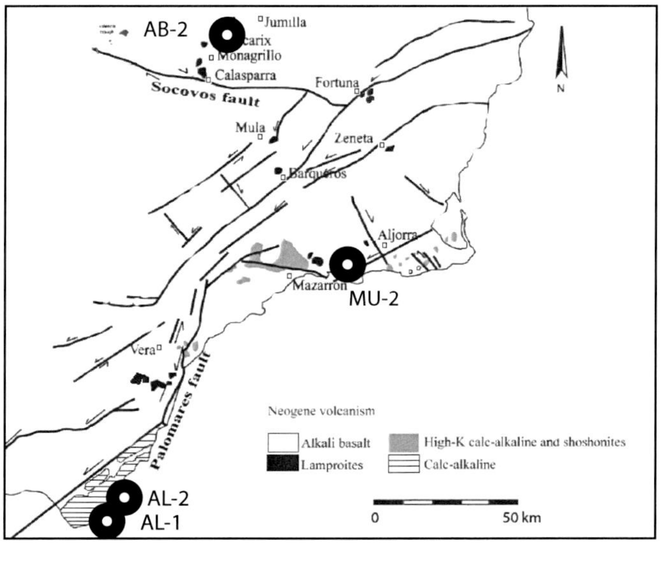

Fig. 4.5: Detail of the south-eastern Spanish volcanic district with the location of the reputed Roman quern and millstone quarries (adapted from Cambeses 2010: 8, fig. 1.3).

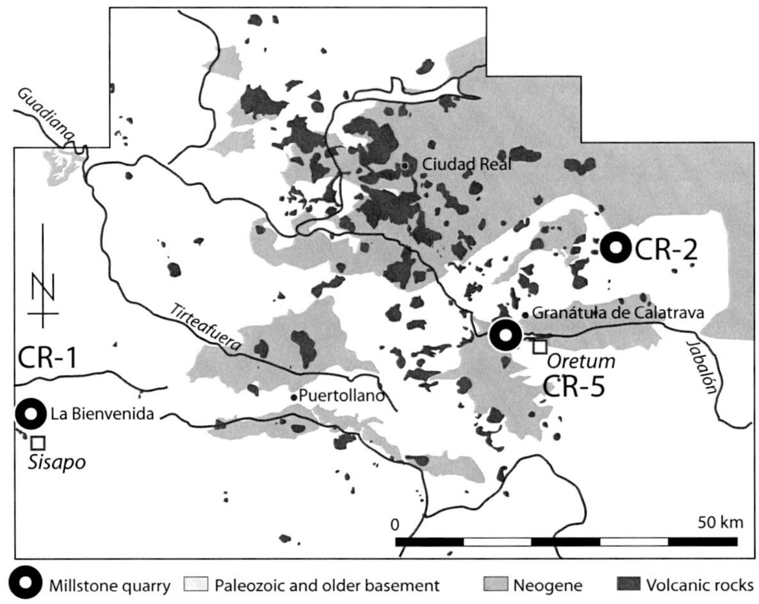

Fig. 4.6: Detail of the Campo de Calatrava volcanic district with the position of the three Roman quern and millstone quarries (adapted from Anderson et al. *2011a: 158, fig. 9).*

4.3. The spread of millstone quarries

The spread of millstone quarries throughout the landscape of southern Spain, in contrast to the commercial distribution of their products (see Chap. 12), reveals that they were not distributed equally. Quaternary and Tertiary deposits, usually too soft for millstones, were avoided. This could explain the feeble number of millstone workings along the Guadalquivir Basin in the south, and in the Tagus, Loranca and Zúcar basins in central-eastern Spain (Jane H. Scarrow, pers. comm.).

The inner mountainous zones, due to their inaccessibility, were also apparently not exploited. This inaccessibility could explain the vacant spaces in large regions of the Baetic Range, in particular in the higher ranges of the Sierra Bermeja of Málaga, the Sierra Nevada straddling the provinces of Granada and Almería and the Sierra de Cazorla in Jaén. There are also no sites deep into the Guadarrama Mountains to the north-west of Madrid.

To conclude, the Molares (CU-1) quarry near Portilla at 1350 m, in the Sierra de las Majadas (Cuenca), is the highest millstone exploitation identified in the study area. Due to the complicated logistics involved in transporting products from this site, it must have furnished a fine millstone. In contrast, a number of sites, such as Trafalgar (CA-1), Rota (CA-3) and Chipiona (CA-6) in the Province of Cádiz, as well as Carchuna (GR-9) in Granada, are at sea level. This position along the coastline obviously facilitated maritime transport.

5. MILLSTONE QUARRIES:

TERMINOLOGY, TOPOGRAPHY AND TECHNIQUES

5.1. Terminology

The general terminology (Fig. 5.1) used to describe the features of millstone quarries is, for the most part, borrowed from traditional building block quarries. Other terms are adapted to the specific features inherent to millstone workings, especially in the case of exploitations where cylindrical extractions were scored directly from bedrock resulting in tubular quarry faces that contrast with the traditional straight, vertical rock faces of block quarries.

The *overburden* is a combination of topsoil (humus, sands, silts, etc.) and "unhealthy", brittle rock overlaying the layer of "healthy" or "live" rock desired for millstone production. The *quarry face* is the exposed vertical surface of rock resulting from extraction directly into the mass of rock. The height of the face varies, depending on the number of *extractions*. Quarry faces resulting from block extraction, following natural fissures, are *planar*. Cylindrical extractions scored directly from bedrock by means of *trenching* yield *circular hollows*. Multiple, superimposed circular extractions, like "stacks of coins", result in high *vertical tubular* faces. Los Guillares, Padul (GR-7), with up to five levels of horizontal extraction, is a good example.

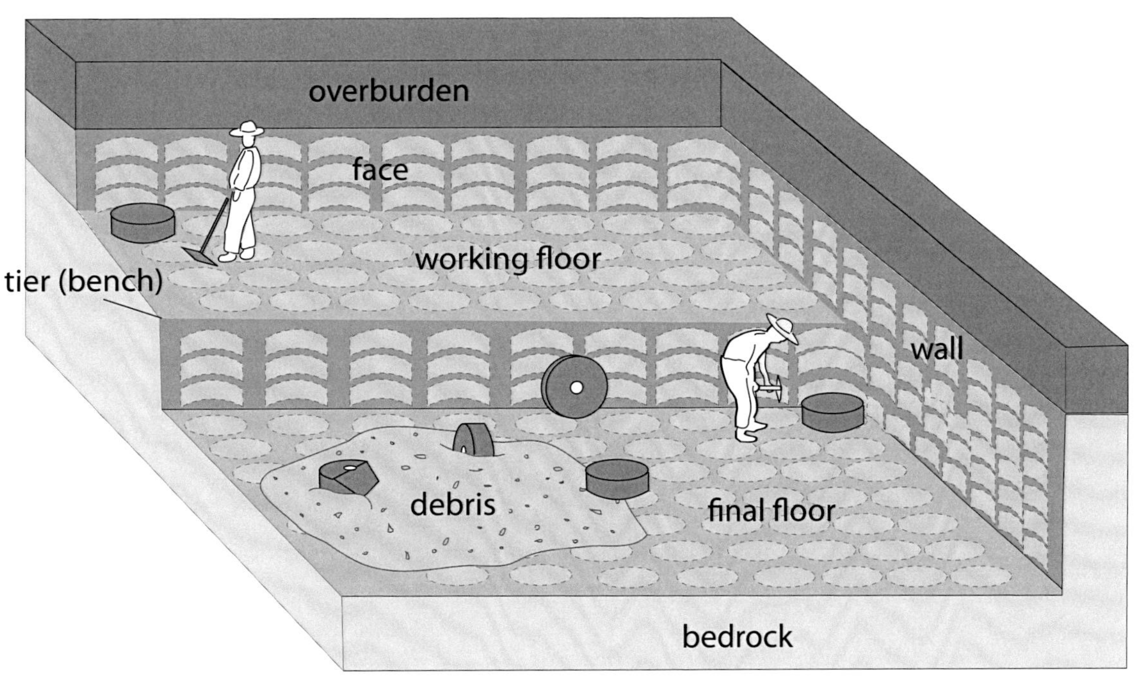

Fig. 5.1: Schema illustrating the terminology related to millstone quarries (drawing by T. Anderson).

Multiple vertical (or subvertical) extractions with a linear, horizontal forward progression into the heart of the bedrock, an extraction technique unknown in Spain at the moment but observed in the arkose subterranean quarries of the Mont Vouan in France (Belmont, unpublished), results in long *horizontal tubes* visible on the quarry *ceiling*.

The horizontal base of the quarry is the *floor*, a feature that is often covered with *tool marks* resulting from the *splitting* of the cylinders. Floors, after the abandon of the exploitation, are habitually *backfilled* with *working debris*, so it is often not possible to observe the tool marks that indicate the splitting techniques used to *detach* the cylinders. Certain quarries have several floor levels forming steps separated by *tiers* or *benches*. These differences of level permitted millstone makers to exploit more than one level of rock simultaneously. Tiers intended for exploitation are defined as *working floors*. The deepest floor of a quarry is the *final floor*.

Millstone extraction produces a great volume of *debris* or *spoil* that can be stored away from the working area in round or oval *mounds*, or *heaps*, or in elongated *cordons*. It is difficult (at times impossible) to distinguish between the debris generated from the *extraction phase* and debris resulting from millstone *fashioning*. Broken cylindrical *roughouts* and *blanks* are often discarded among the debris. These artefacts are essential to determine the *typology* and *morphology* of the site's product.

5.2. The topography of millstone quarries

Millstone quarries are associated with several types of natural topographical features (Walton 2004, chap. 4) (Fig. 5.2-5.3). Most often they are found on the slopes of mountains, hills or hillocks (Fig. 5.2a) or along the top of ridges or cliffs (Fig. 5.2b), where bedrock is exposed. They are also found where bedrock has been exposed by permanent or periodical passage of water, such as streambeds, riverbeds, ravines and torrents. In these cases, quarries could exploit both the centre of these features, for example, a thalweg or dry streambed (Fig. 5.2c), and the edge, like an embankment, where the rock is exposed (Fig. 5.2d). Millstone quarries are also found on flat features such as mountaintop mesas or plateaux or plains where rock layers are exposed by agents of erosion (*e.g.* flash floods) (Fig. 5.2e). Finally, certain quarries exploited bedrock exposed along the coastline (Fig. 5.2f). The trade of the products from these coastline quarries, today at times under high tides, obviously benefited from maritime transport.

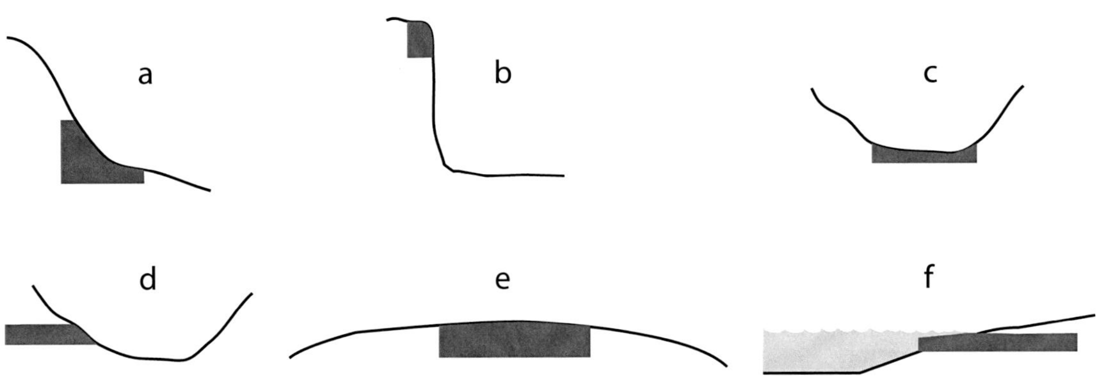

Fig. 5.2: Schematic sections of the topography of bedrock millstone quarries: a) slope quarry; b) cliff or scarp quarry; c) ravine bottom or riverbed quarry; d) valley or ravine edge quarry; e) plateau quarry; f) coastline quarry (drawing by T. Anderson).

Fig. 5.3: Examples of millstone quarries found at the six principal types of topographical formations: a) slope quarry of Loja, La Merced (GR-4b); b) scarp quern quarry of Zagra, Granada (GR-5); c) ravine bottom quern quarry of La Calzadilla, Almaden de la Plata (SE-1a); e) mountain-top plateau quarry of El Tajo, Teba (MA-2); f) the coastline quarry of Chipiona, Cádiz (CA-6).

5.3 "Prospecting" quarries

Millstone production required a rock with specific properties (hardness, porosity, inclusions) suited for grinding cereals. As obvious as this appears, the outcrop had to be visible and readily accessible. No written documents exist detailing how these outcrops were singled out and chosen. Millstone makers were obviously not geologists. They did, nonetheless, possess a keen knowledge of rock properties, in part acquired from other millstone makers, and built upon through experience.

It is possible to imagine these craftsmen, even in very early times, surveying the landscape, searching for new potential sources of rock and testing them by means of extracting one or two millstones. At this stage the site is a "prospecting quarry", a term formulated in recent research in Norway (Grenne *et al.* 2008: 51). If work prospered, the site became a veritable quarry. At times, work was abandoned owing to a defective rock or to other unidentified problems. Sites with only a few extractions, possibly "prospecting quarries", have been identified at La Merced 1 (GR-4a) (Fig. 5.4a), (GR-5a) and La Cuerda, Siguënza (GU-4).

This type of site could also pertain to the single abandoned large millstone at Zagra (GR-5b). A possible scenario at Zagra is that a millstone maker was aware of the presence of many earlier rotary quern extractions and supposed, since the rock was suitable for querns, that it would be appropriate for a large millstone (Fig. 5.4b). This attempt, nonetheless, was abandoned for no apparent reason as no break or unwanted fissure is visible.

Fig. 5.4: Examples of possible "prospecting" sites at a) Loja, La Merced (GR-4a) and b) Zagra (GR-5b). The large abandoned millstone extraction at Zagra is adjacent to an alignment of older quern extractions on the edge of the outcrop (to the right).

5.4. Production techniques (extraction and fashioning)

The first step of millstone production, after identifying a suitable rock, required either extracting rough cylinders directly from the rock mass or detaching angular blocks by means of levers. The two different extractive processes required different tools and skills and depended on the properties of the bedrock, such as the presence or not of natural cracks and fissures. Local extraction traditions also certainly played a role in the choice of techniques. As seen later in this chapter, some millstone quarries were not restricted to one single extraction technique. There are cases where the millstone makers showed versatility by adapting their techniques and tools to the different conditions of the rock.

The second major phase of production was fashioning the rough cylinders into a fine shape, cutting the eyes and different slots or sockets for driving fittings and preparing the grinding surfaces, a process that required other skills and, at times, other tools.

The observations of the working techniques presented in this study are based on a combination of sources such as specialised quarry literature, notably the research of J.-C. Bessac in southern France (Bessac 1982; 1996), as well as of the reading of the tool marks at the quarries themselves. This study is also based on observations gathered from discussions with other specialists during the excavation of several quern or millstone quarries in Switzerland and in France.

Fig. 5.5: Photograph dating to 1933 of moleros *(millstone makers) at the millstone quarry of Brañosera (Palencia) in northern Spain. The sketch below the photograph highlights the tools and the roughouts. Although the scene appears to be staged as seen by their poses and attire, it offers a rare glimpse into millstone making and the tools of their craft, highlighted in the sketch. Three of the workers wield picks, the main cutting tool. In the background two workers handle a lever to move or overturn a millstone. The millstones in the centre and the left have already been overturned so as to carve their second half into shape (photograph from Cuevas Ruiz 2006).*

5.4.1. The tools of the millstone maker

The millstone maker employed a wide variety of tools and *gestes* (actions) borrowed from the tradition of building stone workings. Most of the extracting and fashioning tools fall into two large categories defined by J.-C. Bessac (1996). The first, including the pick, is that of tools associated with the technique of "thrown" or direct percussion (*percussion lancée*). Examples of this type of tool are clearly visible in the 1933 photograph where the *moleros* of Brañosera (Palencia) carved their millstones directly with the pick (Fig. 5.5). The second tool category, associated with the technique of "indirect percussion" (*percussion posée*), comprises a wide variety of hammers and sledge hammers and chisels and punches, tools that work, by definition, in tandem with other tools. Both techniques can intervene in either of the two phases of millstone production (extraction and fashioning). They are complemented by other instruments such as levers or pry bars.

González Peralbo (2008) cites a notarial protocol in the local archives of Pozoblanco, Córdoba dating to 1619 that lists the possessions of the stonemason Juan de Bargas, a millstone maker, at the moment of contracting marriage with Francisca Ruiz. Besides a small quantity of money and no house or farm, the record lists the following tools: "*barras, almádena, picos, plomada, escuadras, reglas, nivel, cuñas*" (*i.e.* iron levers, sledge hammer, picks, plumb line, set-squares, rulers, level, wedges) (González Peralbo 2008).

5.4.1.1. Direct percussion tools (picks)

The pick is in itself the millstone maker's most important extraction tool. It is defined as a double-pointed iron head attached at a right angle to a long wooden handle. There are a number of variations, from elongated thin versions to stubby versions. Its points could also be single or, as in the case of the *escoude* in southern France, double (Bessac 1996: 207). Known since Antiquity, it appears engraved, for example, in a Roman quarry in the Rhineland of Germany (Fig. 5.6a) and on a capital of a medieval church in the Province of Zaragoza (Fig. 5.6b). The pick most often served to cut the trenches around cylinders. However, it was not exclusively a tool of extraction as it was also used in the fashioning process, as seen in the photograph of Brañosera (Palencia) where the *moleros* are carving the rough surface of the second unworked face (*cf.* Fig. 5.5).

Fig. 5.6: Examples of tools of direct percussion. a) Engraving from Antiquity of a quarryman with his pick, Kruft, Kr. Mayen (Germany) (photograph from Röder 1957, Plate 21, 1). b) Scene of a quarryman cutting or shaping a block on a capital of a column of the western entrance to the church of San Miguel de Biota (Zaragoza), 12th century (photograph by Ray Escámez Rivero, flicker, http://www.flickr.com/photos/adfinem/6849827592/).

A smaller version of the pick, the pick axe or mattock, with a short handle served to carve the holes at the base of the cylinder to lodge wedges. These smaller picks could also have served to rectify the surfaces of millstones during the fashioning phase.

5.4.1.2. Indirect percussion tools (hammer, chisel)

Hammers and chisels in different forms were the principal tools of "indirect percussion". In the initial extraction phase they served to carve the wedge holes at the base of the cylinder. In the fashioning phase they served for the finer work, such as carving the edges and the surfaces of the cylinder into a regular circular shape and piercing the eye of the upper stone, the spindle hole of the lower stone and ultimately the other different sockets and slots for the handle and rynd fittings. Each millstone maker had a set of tools adapted to his specific needs. For the more delicate work he also probably had a wooden mallet.

The socketed gad (*pointerolle* in French), another indirect percussion tool, is a sort of chisel with a wooden handle. This tool, struck with a hammer, is most often associated with mining, permitting the cutter to adopt postures and attain surfaces (such as extraction from ceilings along vertical planes) that were not adapted to the use of the larger traditional pick. The socketed gad allowed a high degree of precision and resulted in concentric linear marks at the base of the trench, as seen on the ceiling of the excavation of the subterranean quarry of La Corbière at Mont Vouan (Belmont unpublished). Although not identified in quarries in southern Spain, the socketed gad cannot be excluded from the panoply of tools of *moleros* owing to the long mining tradition in Spain.

5.4.1.3. Leverage tools

Levers and long pry bars also played an important role in millstone production, in particular in quarries where blocks were extracted or pried from rock faces profiting from natural fractures. Yet levers also remained essential in true extractive quarries where millstones were hewn directly from bedrock. If fashioning took place at the quarry itself, as was often the case, it was necessary to dislodge the roughout from the quarry face a distance of about 50 cm so as to comfortably access all of its sides.

During fashioning, levers served to shift and overturn the roughout to access its opposite face. This is apparently what the two men are doing with the wooden lever measuring about three metres in the background of the photograph of the Brañosera *moleros* (*cf.* Fig. 5.5). According to Maestro, levers in the Palencia mountains, such as that in the photograph, were of oak, a material that is much harder and resistant than beech (Maestro 2011: 41). Maestro quotes oral testimony that although moving or turning roughouts was usually undertaken by a team, a single *molero* could do the job, on the condition that he be adept with the lever and know where to place the *leva* (a rock acting as a fulcrum) (Maestro 2011: 41-42).

5.4.1.4. Other tools

Other tools intervened in millstone work. The compass served to trace the initial diameter of the cylinder. Brushes, rakes and hoes removed debris that hindered work progression. Shovels of different sizes served to fill wicker baskets or wheelbarrows to evacuate debris from the workings. Wooden troughs retained water to wet down the working areas (to reduce the quantity of dust), soak tool handles (to keep them snugly attached to their iron head) and soak wooden wedges to make them swell to split rock.

5.4.2. Extraction: cutting cylinders directly from bedrock

The most common technique of millstone extraction was cutting cylinders directly from bedrock (Fig. 5.7). This method resulted in a characteristic feature seen on many millstone quarry faces: the circular hollow. This technique was widespread throughout Europe since Antiquity and identical (except for the resulting shape) to that of quarries for construction blocks.

In the context of quern quarries, the technique of cutting a circular trench is illustrated in detail in the study of a Roman quern quarry excavated at Châbles in the Canton of Fribourg, Switzerland. Here the ancient tool marks on the rock surface are particularly well-conserved permitting the reconstruction of the operation sequence of rotary quern extraction, from the choice by the quern maker of the position of extraction, to pecking the centre to serve as a reference point to trace the contour of the cylinder with a compass (or simply an iron nail attached to a string), to cutting the trench with the pick, and finally, to splitting the cylinder from the bedrock (Anderson *et al.* 2003, chap. 5).

Recent excavations in the French Jura under the direction of Luc Jaccottey (2011), as well as excavations under Alain Belmont at the sites of Claix near Angoulême (Charentes) (Belmont 2011; Belmont *et al.* 2011) and Mont Vouan (Haute-Savoie) (Belmont 2010, unpublished), reveal subtleties of this extraction technique.

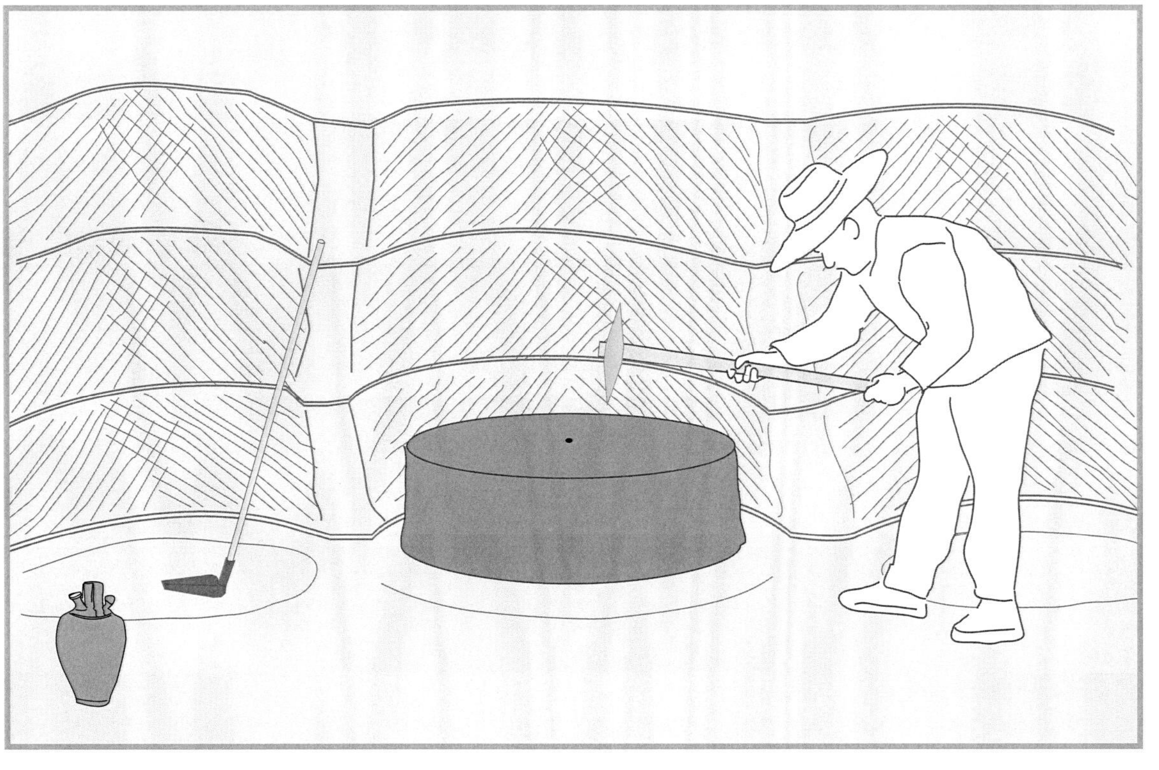

Fig. 5.7: Reconstruction of the technique of extracting a large (Middle Ages to Contemporary), horizontal cylinder directly from bedrock by means of circular trenching with a pick. This technique results in tubular quarry faces with opposite groups of diagonal tool marks as seen behind the worker (drawing by T. Anderson).

5.4.2.1. Horizontal extractions

The technique of extracting millstones along horizontal planes following the rock's natural bedding plane required the cutting of a circular trench, a task most often done with a pick. The trench was always cut following an inclined, diagonal working plane, a technique that facilitated the process by reducing the resistance of the rock (Fig. 5.8).

The trench was cut by multiple, contiguous impacts called "passes", placed along a predetermined axis. This technique, from the French *passes (*Bessac 1996), results in the linear tool marks that, in favourable conditions of observation, appear as multiple, diagonal parallel lines both along the edge of the cylinder and along the quarry face (Fig. 5.9), as illustrated in detail in figure 5.10.

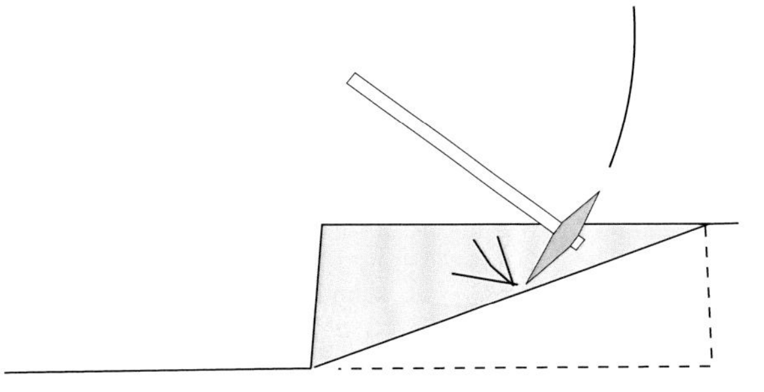

Fig. 5.8: Schema of the diagonal inclination of the pick work for a horizontal cylindrical extraction. With this inclination, the rock presents less resistance (drawing by T. Anderson).

The trench had to be sufficiently wide so as to allow access to the base of the cylinder both for the millstone maker to be able to place one of his feet (to maintain his equilibrium) and, when the trench was complete, to be able to cut the wedge holes. In the case of small rotary quern extractions (where the quarryman need not place a leg in the trench), the width of the trench

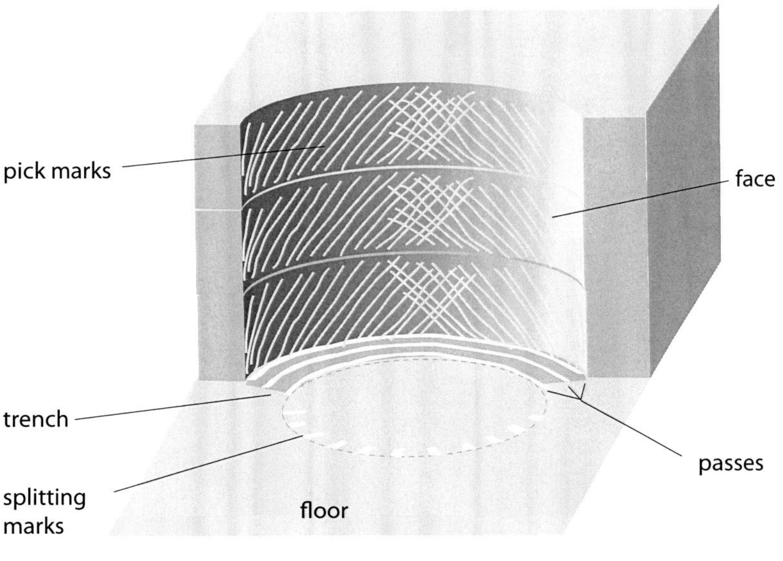

Fig. 5.9: Tubular quarry face resulting from the superimposition of horizontal extractions scored by means of cutting circular trenches with a pick. The quarry floor, in optimal conditions of observation, will reveal at the base of the trench parallel lines corresponding to the passes of the pick and radial splitting marks (wedge holes, chisel marks) (drawing by T. Anderson).

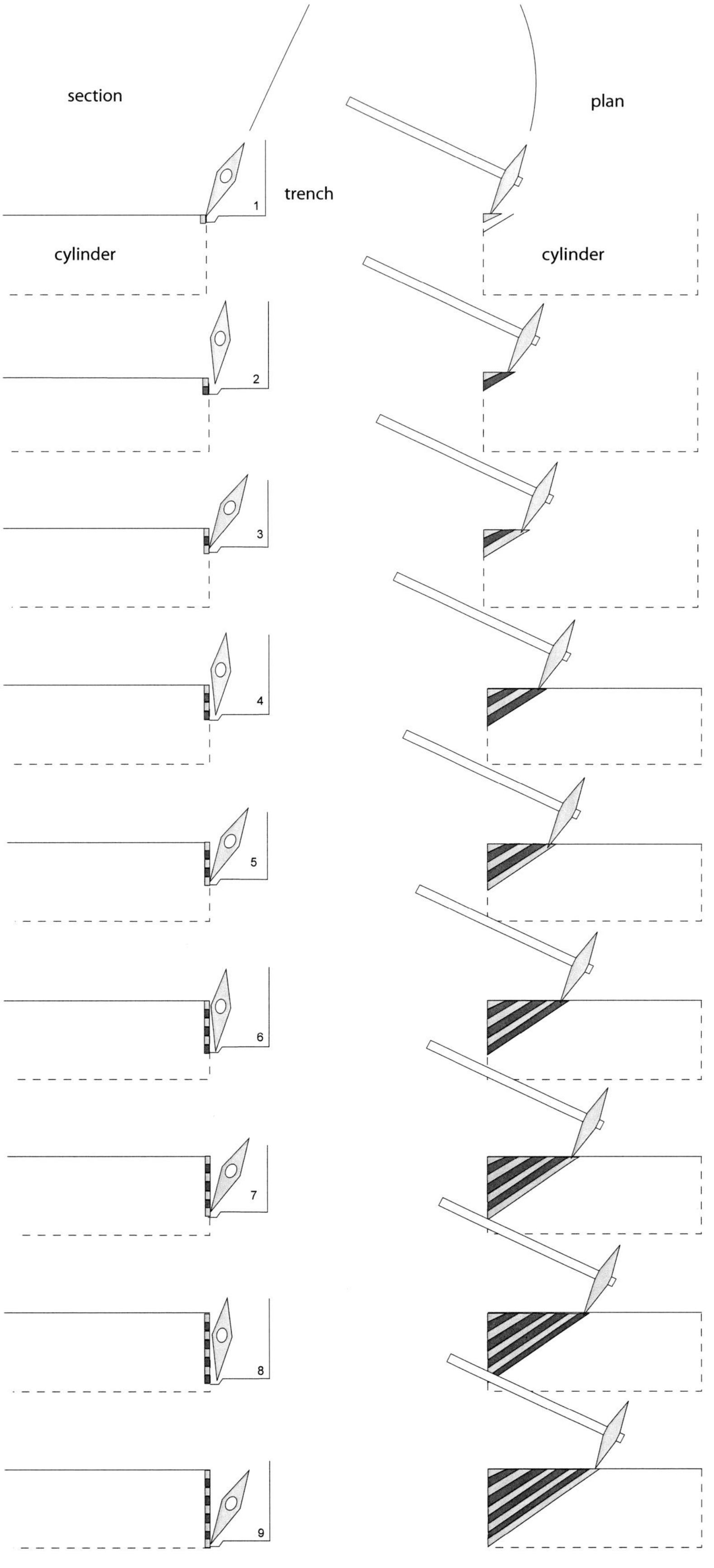

Fig. 5.10: Schematic representation of the cutting of the trench with a pick (direct percussion), a process that produces diagonal lines on both the quarry face and the edge of the millstone. The lighter lines result from direct impact of the pick, whereas the darker colour corresponds to where the rock has broken off (drawing by T. Anderson).

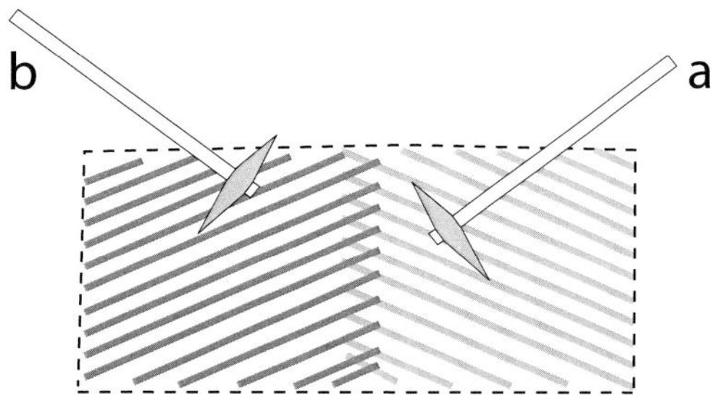

Fig. 5.11: Schema of the diagonal pick marks visible on millstone quarry faces. It is typical to observe two groups of opposite marks in "mirror effect" that result from the millstone maker's change of position while cutting the circular trench. a) original cutting position b) secondary cutting position. The marks of (b) overlap those of (a) (drawing by T. Anderson).

was small, from 10 to 15 cm. Large cylinders measuring up to 1.50 m in diameter and 60 cm in height, however, required a trench between 30 and 40 cm wide. The number of "passes" of the pick depended on the width of the trench. In the case of quern extractions there were usually three, whereas larger trenches had from four to five.

In the field, the criteria to determine this trenching technique is the presence of multiple diagonal lines on the quarry faces. It is typical to observe two groups of multiple parallel lines in "mirror effect" (Figs 5.11 and 5.12) resulting from the millstone maker's change of position (for example, when his stance is hindered by his proximity to the quarry face). The regularity and symmetry of the tool marks suggest that millstone cutters developed ambidexterity.

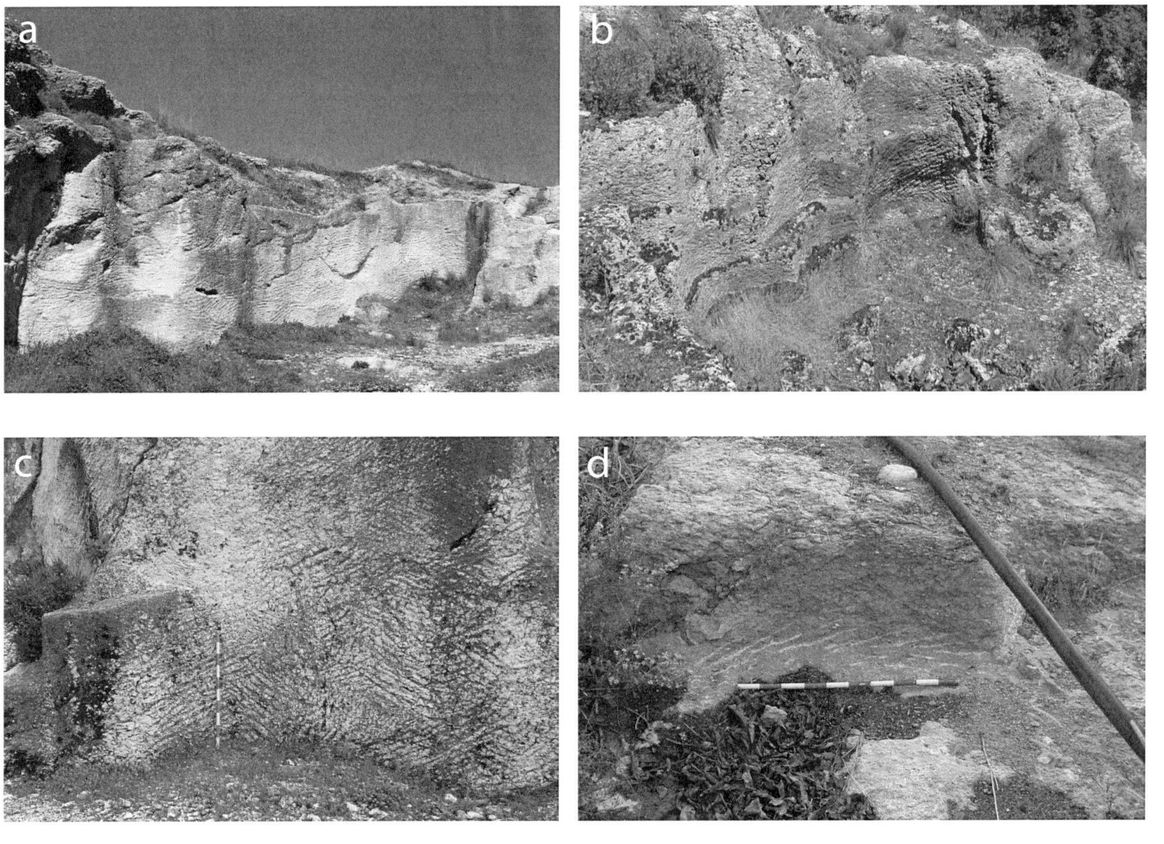

Fig. 5.12: Examples of multiple linear diagonal tool marks on tubular quarry faces indicating horizontal extractions from bedrock using a pick: a) Cabra (CO-1); b) Los Guillares, Padul (GR-7); c) Moclín (GR-1); d) Los Guájares (GR-8).

5.4.2.2. Vertical extractions

There is less evidence in southern Spain of the technique of extracting millstones following vertical planes (resulting in vertical circular hollows). This technique has been, nonetheless, identified at the following sites: Moclín, Las Pedrizas (GR-1b); Los Guillares (GR-7); Vélez de Benaudalla, GR-10; Peña Harpada (CA-10); El Campillo (HU-1); and Castillo de Locubín (J-1) (Fig. 5.13). In general, nonetheless, this technique represents only a small proportion of the total extractions.

Vertical extraction is often related to the bedding plane of the rock. At Campillo Viejo (HU-1) the vertical orientation of the extractions followed the natural vertical (or subvertical) bedding planes of the upright schist blocks (Fig. 5.14). A different case is seen at certain sectors of Castillo de Locubín (J-1) at the Mont Vouan in France (excavations directed by Alain Belmont), where cylinders were scored along planes perpendicular to the natural bedding. In these cases the sedimentary rocks were sufficiently homogeneous and hard and could support detachment along a plane opposite to their natural horizontal bedding.

Fig. 5.13: Examples of vertical circular extraction hollows at the sites of a) Padul, Los Guillares (GR-7) and b) Castillo de Locubín (J-1). Extracting along a vertical plane, perpendicular to the natural bedding of the rock, was not common in millstone quarries of southern Spain.

A variation is seen at the modest site of Las Pedrizas (GR-1b) near Moclín where the blocks, after breaking off of the cliff, ended up in a position with their original horizontal bedding plane in a vertical sense obliging the millstone makers to score the cylinders vertically.

This extraction technique presupposes different *gestes* (actions) by the millstone maker and possibly the use of other tools. Direct percussion with a small pick or indirect percussion with hammer and chisel, permitting the cutter to remain with his back upright, would seem best adapted to this type of work. At Mont Vouan, for example, concentric linear tool marks were clearly visible at the base of the circular trenches suggesting the use of a socketed gad (Belmont unpublished).

Apart from El Campillo (HU-1), there is no site in Spain where vertical extractions are the dominant extraction technique. In each of the Spanish sites that do show several vertical extractions, the hollows always correspond to single carvings, never multiple extractions that result in long horizontal "tubes" such as those at Mont Vouan that, when combined with a large number of other vertical extractions, develop into underground exploitations. This would appear to indicate that in southern Spain the technique was avoided when possible, probably because the position was not natural or comfortable for the *molero* and the majority of the rocks were not sufficiently hard or homogeneous.

Fig. 5.14: Examples of vertical extractions on large blocks of schist at El Campillo Viejo (HU-1). This is the only site where the vertical extraction technique dominates that of horizontal extraction (photographs by Alonso García Veiga).

5.4.3. Splitting cylinders from bedrock

After the millstone maker cut the trench around the future millstone, the cylinder had to be split from the bedrock by means of a variety of techniques involving either direct or indirect percussion. This was a delicate operation that was not always successful, as can be seen at certain sites by parts of the broken cylinders still attached to the bedrock (Fig. 5.15).

The principal splitting technique identified in the south of Spain, as elsewhere in Europe, consisted of carving a series of evenly spaced cavities at the base of the cylinder destined to lodge a battery of wedges. These wedges, under pressure, exerted an upward and downward force on the rock that triggered the horizontal fracture (Fig. 5.16a). This type of splitting, following a long tradition recorded since Antiquity, endured until very recent times.

Although the method is common, it is rare to be able to observe the marks of this technique in southern Spain because quarry floors are usually covered by debris or grown over by vegetation. A rare example is that of the site of El Campillo (HUE-1) where holes are visible along the base of a vertical schist cylinder that was never detached (Fig. 5.17). Tool marks of this technique have also been discerned at *Sisapo* (CR-1), Rambla Honda (AL-3) and Plasencia (CC-1) (Fig. 5.18) on different types of rocks. Yet the marks at these sites are not sufficiently conserved to determine the original shape of the cavity (square, trapezoidal) .

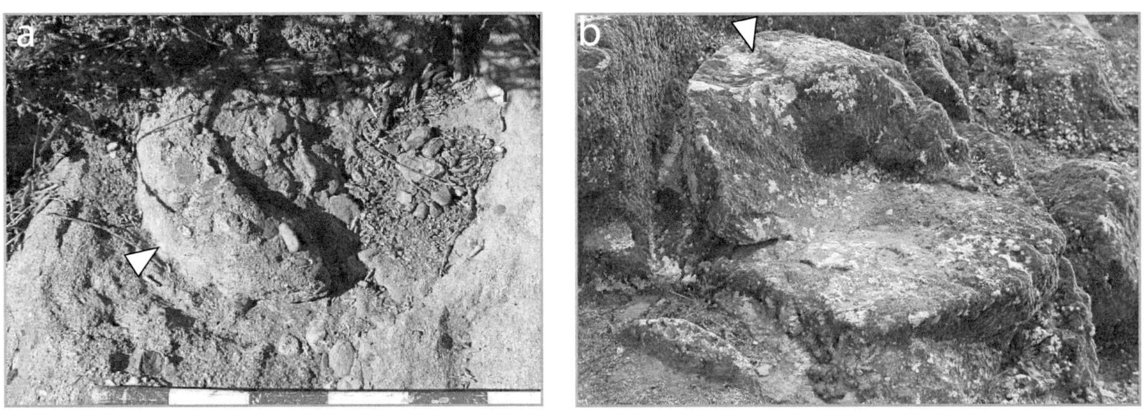

Fig. 5.15: Examples of defective splitting from bedrock at a) Almadén de la Plata (SE-2) and b) Sisapo *(CR-1). The horizontal fracture at the base of the cylinder was not completed, probably due to the presence of a diagonal micro-fissure, and part of the cylinder remained connected to the bedrock.*

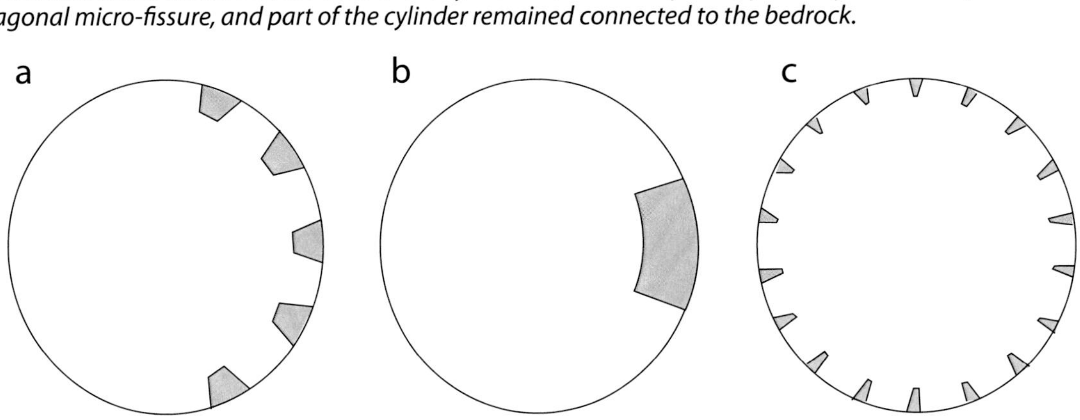

Fig. 5.16: Schema of different types of marks on quarry floors indicating the different splitting techniques: a) evenly spaced multiple wedge holes; b) large single wedge hole; c) multiple radial pick impacts or chisel marks (drawing by T. Anderson).

Fig. 5.17: Detail of the holes at the site of El Campillo Viejo (HUE-1) carved to lodge wedges to split the cylinder from the rock mass (schist) (photograph by Alonso García Veiga).

Fig. 5.18: Extraction hollows with tightly spaced multiple small wedge holes at a) Rambla Honda (AL-3) and b) larger spaced holes at Placencia (CC-1). The sites exploited, respectively, conglomerate and granite.

A second technique, different from that of multiple wedge holes, is that of a large single wedge hole (*cf.* Fig. 5.16b and Fig. 5.19). This method, identified in the study area only at the Roman volcanic site of *Sisapo*, is known at Claix in south-western France in the context of quern workings from the Early Middle Ages (Belmont 2011: 11, fig. 21).

A question that remains vague regarding this splitting technique is whether the wedges were of iron, lodged between metal shims and struck by a hammer (Fig. 5.20), or whether they were of wood that when swollen with water split the cylinder (Fig. 5.21). There is no physical evidence to suggest one or the other. The advantage of iron wedges is that the result was more immediate. Wooden wedges, with their slow, gradual pressure, required a longer period of time. This second method may, in fact, have been more apt for splitting certain delicate rocks.

Another splitting technique is that of a single, long continuous channel (in French *saignée*) surrounding a third, half or, at times, the totality of the circumference of the cylinder (Fig. 5.22). This method, adaptable for both iron or wooden wedges and observed in France at Claix (Belmont 2011: 11, fig. 22) and at Mont Vouan (Belmont 2010, unpublished) has yet to be identified in Spain.

Fig. 5.19: Views of a large single wedge hole at the base of a quern at the volcanic quarry of Sisapo *(CR-1). This technique contrasts with that of multiple small wedge holes.*

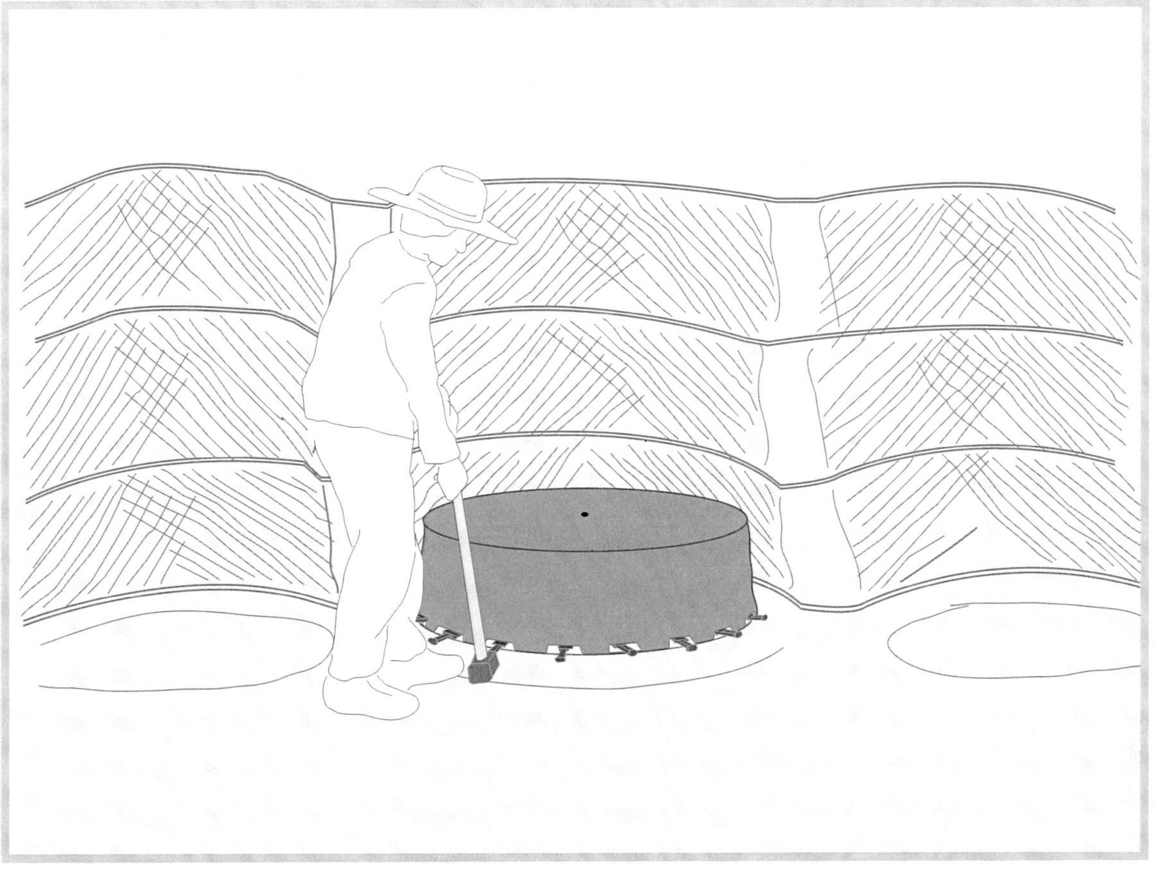

Fig. 5.20: Reconstruction of the technique of splitting a large (Modern to Contemporary) cylinder from bedrock by means of a series of multiple metal wedges (drawing by T. Anderson).

Fig. 5.21: Reconstruction of the technique of splitting a large (Modern to Contemporary) cylinder from bedrock by means of swelling multiple dry wooden wedges with water (drawing by T. Anderson).

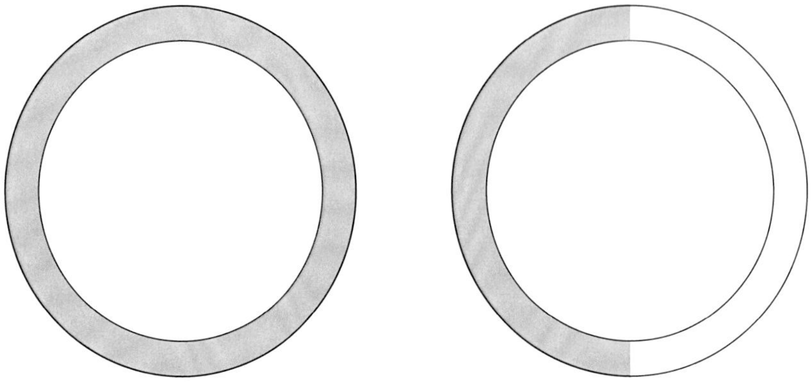

Fig. 5.22: Schema of the long continuous channel technique for wedges at the base of the cylinder along either part or the whole of the circumference (drawing by T. Anderson).

An alternative to the use of wedges, was splitting the cylinder directly with a pick. This technique consisted of provoking a fracture by placing a number of strikes of the pick along the base of the cylinder (Fig. 5.23b). This technique yielded a series of short, evenly spaced, radial marks along half of the cylinder's circumference on the quarry floor. It was the method used to split the querns at the Roman site of Châbles (Canton Fribourg) in Switzerland (Anderson *et al.* 2003) (Fig. 5.24).

The only technique observed in southern Spain resulting in extraction marks analogous to those of Châbles is at the Medieval quern workings of La Calzadilla, Almadén de la Plata (Seville) (SE-2) and La Atalayuela, Zagra (Granada) (GR-5). The multiple, evenly spaced, marks surround the totality of the cylinder's circumference (*cf.* Figs 5.16c and 5.25).

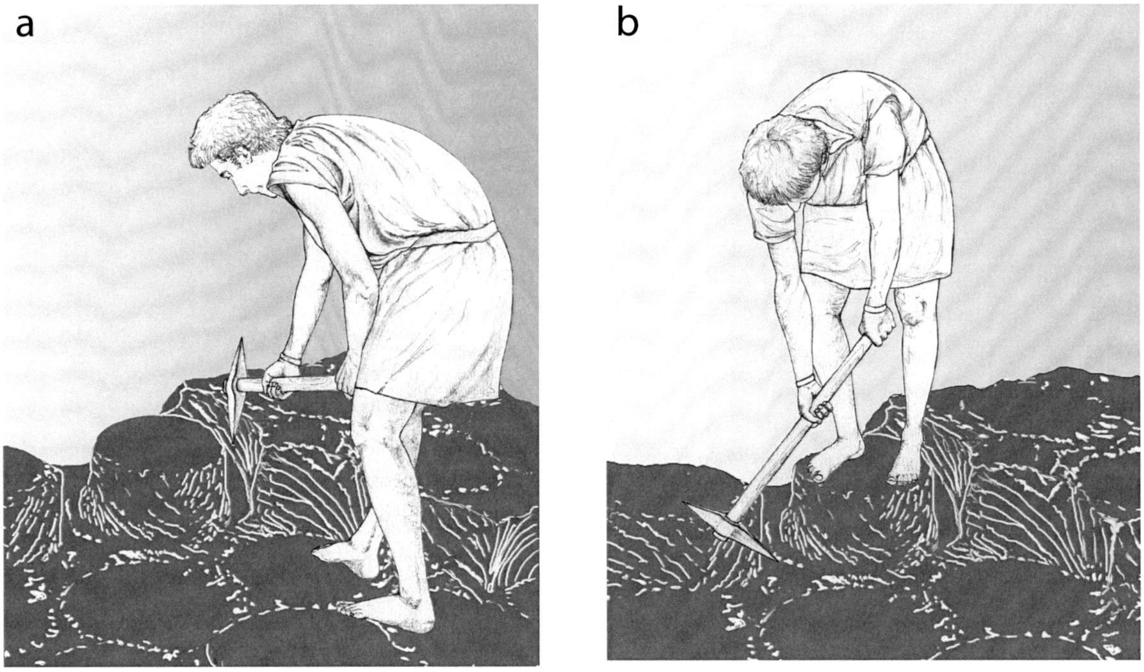

Fig. 5.23: Reconstruction of a) cutting the trench with a pick followed by b) splitting the cylinder directly with the pick, avoiding the use of wedges (from Anderson et al. 2003: 49, fig. 44, drawings by Aurora Pulido).

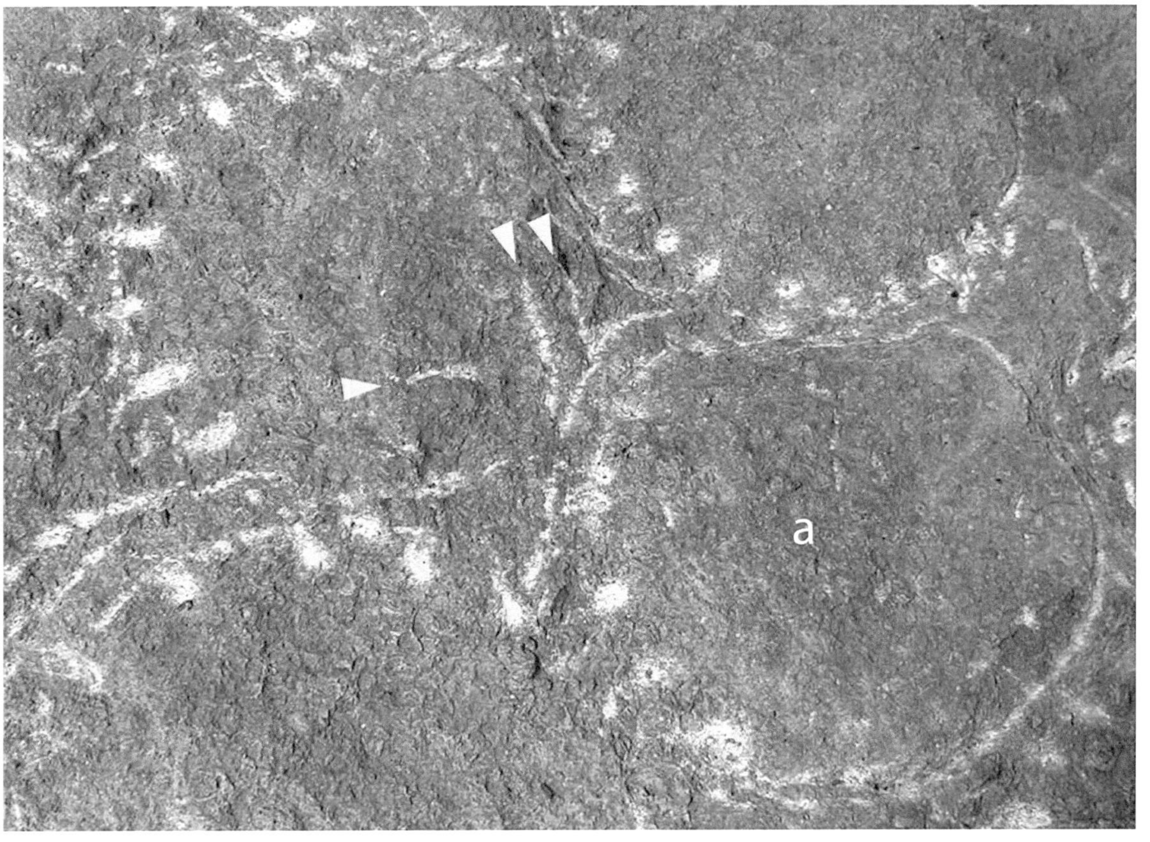

Fig. 5.24: Examples of tool marks on the quarry floor at the quern quarry of Châbles, Fribourg (Switzerland) corresponding to the technique of splitting cylinders directly with a pick (avoiding cutting wedge holes). The pick marks, resembling the units of time of a clock dial, occupy about one half of the perimeter of each extraction. The white arrows indicate three cutting lines (outside the perimeter of cylinder "a") resulting from the removal of a bulge from the quarry floor that hindered access of the pick to the base of cylinder "a" (from Anderson et al. *2003: 51, fig. 52). I thank Jean-Claude Bessac for pointing out this significant detail that proves the use of the pick as an alternate means to split cylinders from bedrock.*

Unfortunately, they are too poorly preserved to determine securely if they resulted from direct percussion with the pick or indirect percussion by chisel. In any case, similar tool marks interpreted as chisel work are known in quern exploitations from the Middle Ages in locations as diverse as those of the Balearic Island of Minorca (Sánchez 2006:190, fig. 10) and Norway (Grenne *et al.* 2008: 57, fig. 17).

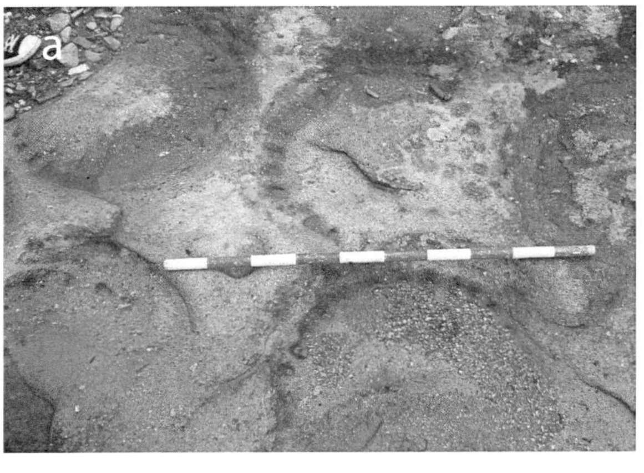

Fig. 5.25: Quern extraction hollows with multiple splitting marks at a) Almadén de la Plata (SE-2) and b) Zagra (GR-5). The small marks, interpreted possibly as chisel work, are tightly spaced and surround the totality of the cylinder.

5.4.4. Extraction: detaching angular blocks from bedrock

The most simple means of block extraction from bedrock, demanding less infrastructure and equipment, is the technique of prying out angular blocks or slabs following the natural fractures of the rock. This technique was carried out with long levers of wood and, in more recent times, of iron (pry bars). To insert the levers, the fractures had, at times, to be enlarged by cutting holes with picks, a technique put to use by the modern millstone makers of the Palencia mountains and described by Maestro (2011: 38). This extractive work took place both following horizontal (Fig. 5.26) and vertical cleavage planes (Fig. 5.27). It served as a means for extracting both large blocks for monolithic millstones and smaller blocks for composite millstones and querns (both saddle querns and rotary querns).

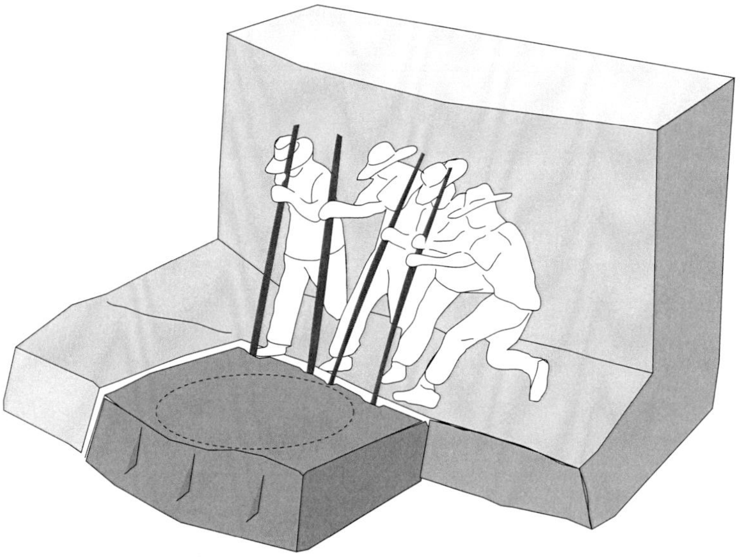

Fig. 5.26: Reconstruction of the technique of detaching an angular block or slab from the quarry floor for millstone production (drawing by T. Anderson based on a photograph from http://burgess-shale.rom.on.ca/en/transcripts/slideshow_1998.html).

Fig. 5.27: Reconstruction of the technique of detaching angular blocks from the quarry face for millstone production (drawing by T. Anderson).

Quarries using these techniques are not as simple to identify in the field as true extractive sites because their extractions leave little or no characteristic trace on the quarry face. Furthermore, depending on the type of rock, this method may have required moving a large quantity of overburden to attain the desired rock layer. Removing blocks or slabs from vertical quarry faces, as opposed to working blocks or slabs on the quarry floor, was a dangerous procedure. It is possible to imagine that the millstone makers placed mats or some other type of material along the base of the quarry face to cushion the fall of the block.

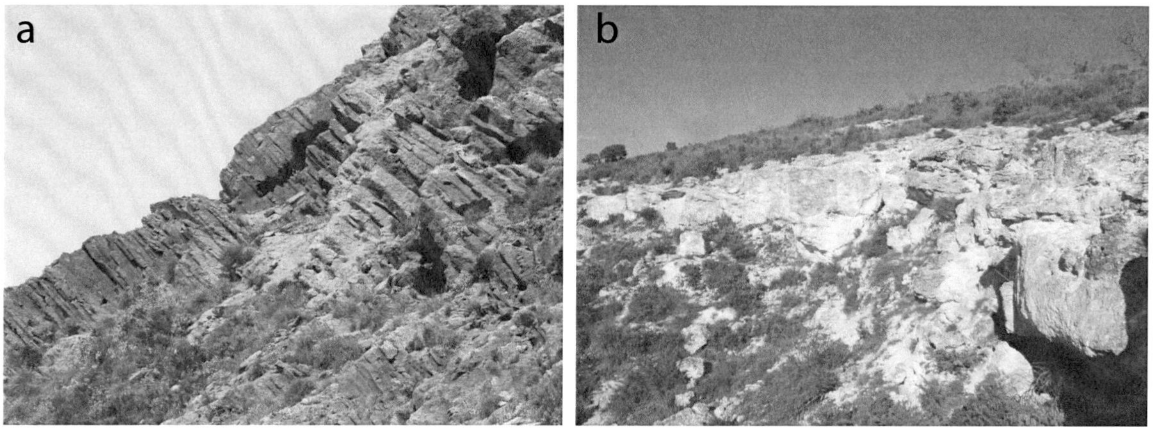

Fig. 5.28: a) Strongly inclined columnar jointing at the volcanic quern quarry site of Cerro de Limones (A-1); blocks slightly larger than the size of the querns were probably detached from these "columns" by means of levers or pry bars; b) the quarry face of the Alhama de Granada (GR-6) millstone quarry, where angular blocks of a specific layer were detached.

Fig. 5.29: Different views of a partially detached angular block abandoned in an early phase of fashioning at the trench quarry of El Lachar (Jaen) (J-6). The upper right photograph shows that the rocks have slipped into the crack provoked by the pry bar.

Early examples of these block extraction quarries are recorded in the volcanic district of the Eifel in Germany. These saddle quern workings date to the Neolithic and Bronze Age (Harms & Mangartz 2002: 75). The technique was presumably adopted for saddle querns in southern Spain at the sandstone quarry of Zujaira (GR-12) and the volcanic site of El Barronal (AL-10). In Roman times the block working examples that stand out are the volcanic sites of Cerro de Limones (AL-1) (Fig. 5.28a) and Hoya del Paraiso (AL-2).

In Contemporary times block extraction is certified at the conglomerate trench quarry of Caniles (GR-11) and the limestone cliff of Alhama de Granada (GR-6) (Fig. 5.28b). The best Spanish examples, however, are the fine white limestone trench quarry of Lachar (Jaén) (J- 2) and a sector along the yellow sandstone cliff of Castillo de Locubín (J-1). In the case of Lachar the technique is illustrated by a detached block at floor level in the process of being fashioned into a millstone (Fig. 5.29). At Castillo de Locubín the evidence of vertical block detachment is certified by an extraction face with angular hollows bearing a striking difference of colour (bright yellow) that contrasts with an adjacent smooth and grey naturally weathered sector (Fig. 5.30).

Fig. 5.30: Views of the face of one sector of the quarry of Castillo de Locubín (J-1). a) The angular fractured surface and fresh light coloration of this sector where blocks were detached, contrasts with the b) untouched and weathered, natural cliff face. A fashioning workshop is located beside this quarry face (see Fig. 5.31).

Fig. 5.31: Examples of querns at a) Cerro de Limones (AL-1) and a millstone at b) Teba (MA-2) that were discarded after they broke during the fashioning process.

5.4.5. Quern and millstone fashioning

The fashioning of querns and millstones into a final or near-final shape was the last major phase of the production sequence before final transport and commercialisation. The phase consisted of shaping the different surfaces of the roughout into a perfectly symmetrical form, carving the different driving and tentering fittings and adjusting the grinding surfaces. The advantage of fashioning directly at the quarry was that the object, if defective, could be immediately discarded (Fig. 5.31) avoiding an intermediary transport. It is likely, however, that fashioning of certain productions took place beyond the quarry in specialised workshops. One such workshop has recently been identified by in the Roman city of Autun, France (Luc Jaccottey, pers. comm.).

The initial task of fashioning at the quarry, applicable to all periods and millstone types, was to arrange an amenable work space or "workshop". The advantage of rotary querns, as opposed to larger models, was that they could simply be picked up by one or two men and placed in the workshop. Larger models, as mentioned previously, were too heavy to be lifted and had to be moved little by little with levers.

The millstone maker then had to decide if the rough cylinder was apt for an upper or lower stone. The proportion of height and diameter of the roughout was also a factor that entered into this choice. Certain blanks bearing rough or slightly broken edges were suitable for lower stones but, since their faces lacked symmetry, could not serve as upper stones. For a Roman *sombrero* lower stone, for example, the blank had to be about twice as thick of that of an upper stone.

Fashioning Prehistoric and Protohistoric saddle querns consisted of knapping the boulder into a rough shape and then preparing the grinding surface by means of *bouchardage* or pecking. For a detailed description of saddle quern production, I refer to the recent experimental work in the French Jura (Jaccottey 2011: 294-298). In the case of rotary querns, different techniques were applied depending on the rock type and whether the roughout was procured as a surface boulder, a detached angular block or a bedrock cylinder. For quern fashioning techniques I have relied heavily on examples from the Roman period (there is little to no documentation regarding the fashioning of medieval querns). For the fashioning of more recent millstones, my arguments are based on observations of abandoned products at quarry sites.

5.4.5.1. Fashioning rotary querns

Fashioning rotary querns from small surface boulders

The first action, after choosing a suitable boulder, was to give it a rough shape. This could be done by splitting the boulder into two halves. If the fracture was "clean", one of the halves served as the upper stone and the other as the lower stone (Fig. 5.32). At an early stage in the process the surfaces were regularised by knapping, chipping or pecking. At times, the original surface of the stone coincided with the desired shape of the quern and therefore remained unpecked. After giving the stone a general circular shape, its upper and lower surfaces were carved before piercing the eye. Unpecked natural surfaces revealing the original patina of the boulder, along with the general hemispherical shape, are characteristic of this type of work.

Although this technique was certainly present in the study area, it has not been unequivocally identified. It has, however, been recorded in La Rioja of northern Spain in a Roman context with the exploitation of riverbed blocks near a Roman villa (Pascual & García 2011: 287-289).

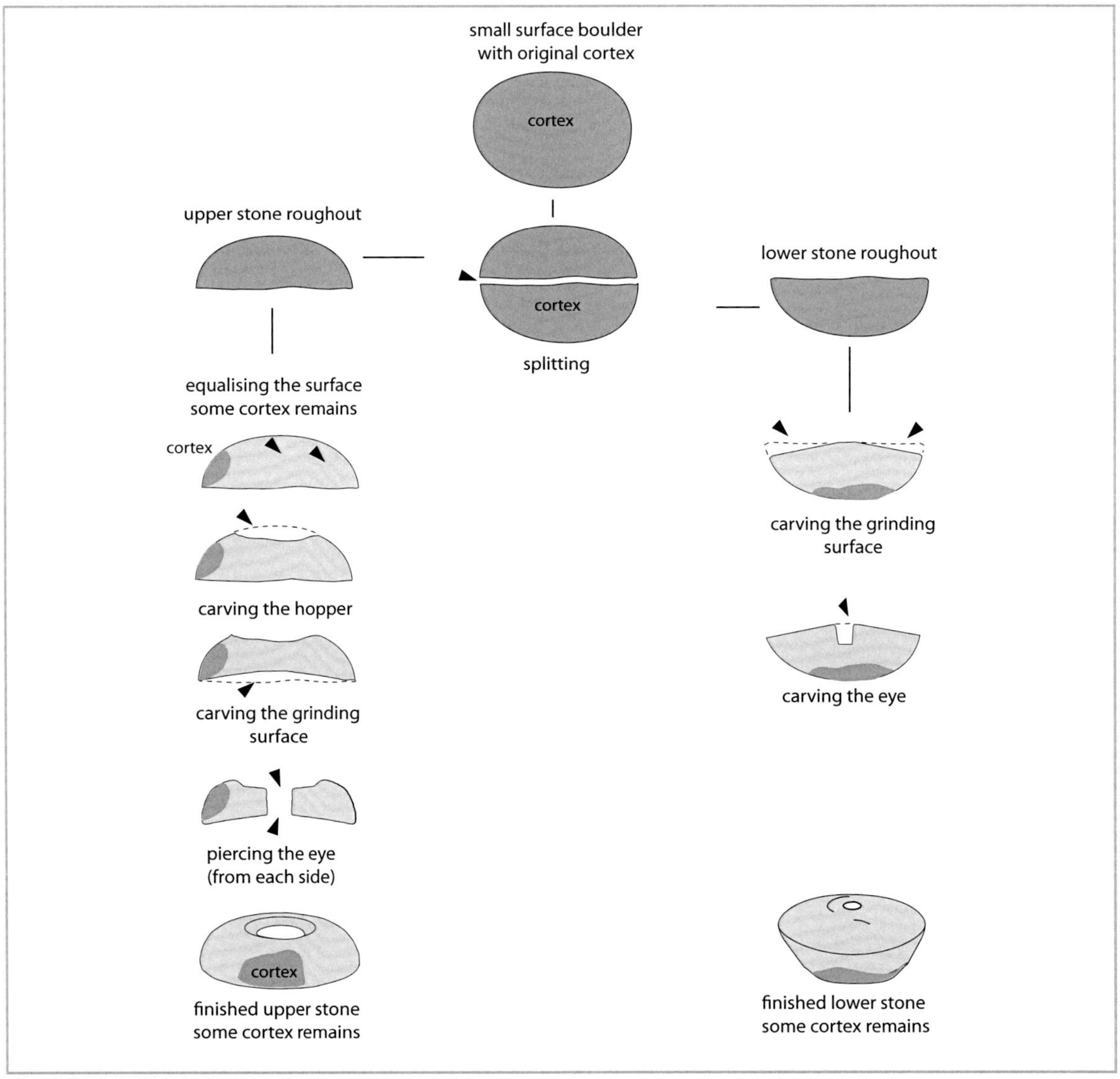

Fig. 5.32: Schema of the fashioning of rotary querns from rounded surface boulders. Patches of natural cortex indicated by a different surface patina betray the origin of the block as a rolled surface boulder (drawing by T. Anderson).

Fashioning rotary querns from angular blocks

The process of fashioning angular blocks into rotary querns is often associated with exploitations of very hard rocks. The detailed *"chaîne opératoire"* of this type of "knapping" or "chipping" has been defined recently by members of the Groupe Meule, based on the study of unfinished rotary querns at Portus (Collonge-en-Charollais, Saône-et-Loire) in France (Jaccottey *et al.* 2011). This detailed schema (Fig. 5.33) is applicable to workings in the south of the Iberian Peninsula.

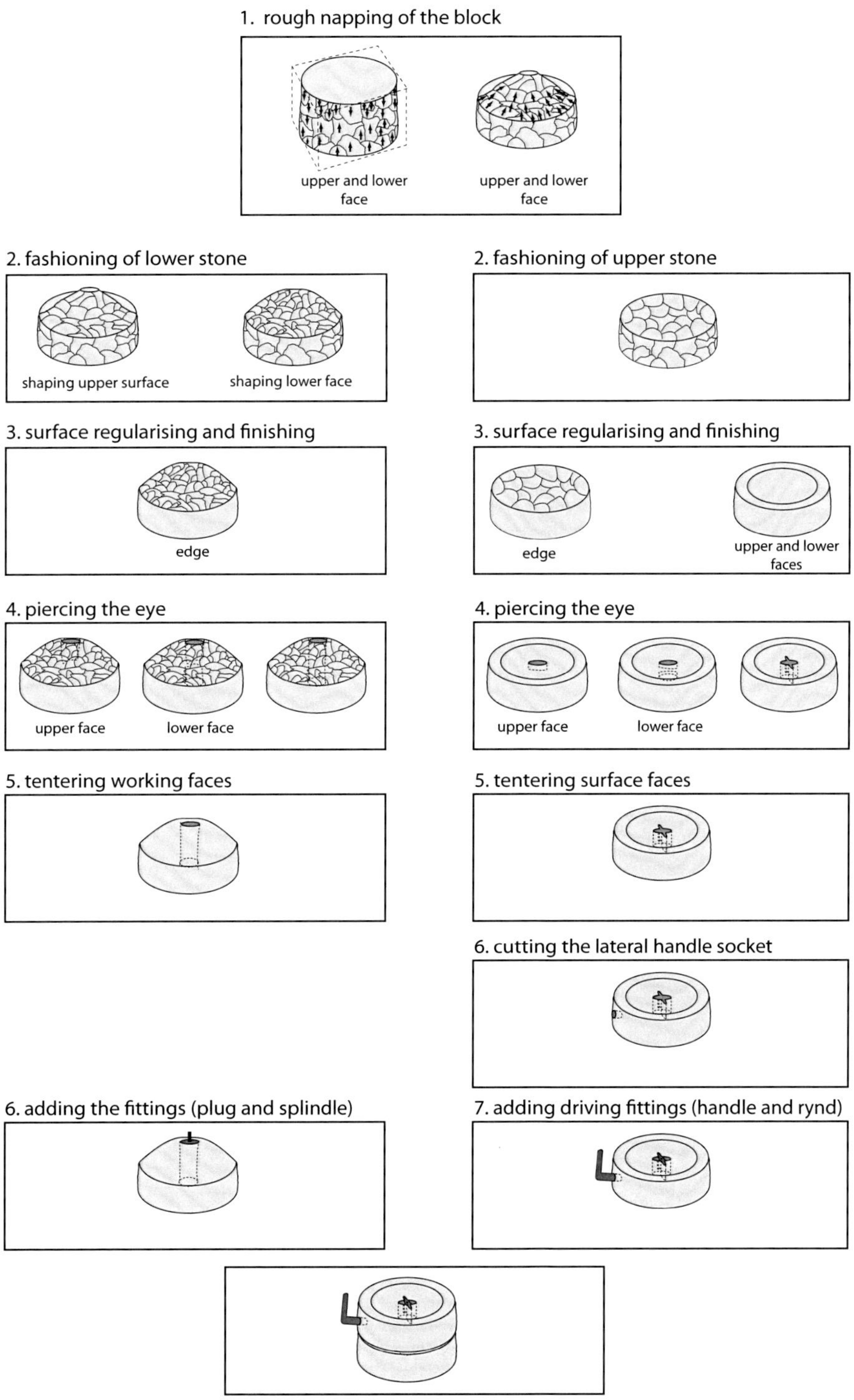

Fig. 5.33: Example of a "chaîne opératoire" *of rotary quern fashioning from angular blocks. The sequence is based on observations at the quarry site of Portus (Collonge-en-Charollais, Saône-et-Loire (from Jaccottey* et al. *2011: 194, fig. 39).*

A detailed study similar to that of the assemblage of Portus is still waiting to be undertaken in southern Spain. I propose a simplified sequence similar to that of Portus based specifically on observations of block fashioning at the Roman volcanic quarry of Cerro de Limones (AL-1).

The process began after first prying out angular blocks from volcanic columnar jointing. These were then knapped into a rough cylinder by means of direct percussion (with a hammer?), a phase that left rough scars on the surfaces (Fig. 5.34). Upper stones were scored from the thinner cylindrical blanks, whereas, lower stones, due to the central protuberance (the future collar around the eye of the *"sombrero"* model), required thicker cylinders (Fig. 5.35).

The finer fashioning work was carried out with hammer and chisel (Fig. 5.36). From what I have observed, eyes were systematically cut at the quarry site. This "dangerous" step, undertaken from each face and joined at the middle, explains the high number of unfinished broken querns strewn about the site.

The finer cuttings to lodge the rynd and the other tentering features did not systematically take place at the quarry. Of the hundreds of discarded querns, only one shows evidence of the cutting of the radial slot to lodge the horizontal crosspiece, an element that served both as the rynd and the base for the vertical handle (Fig. 5.37).

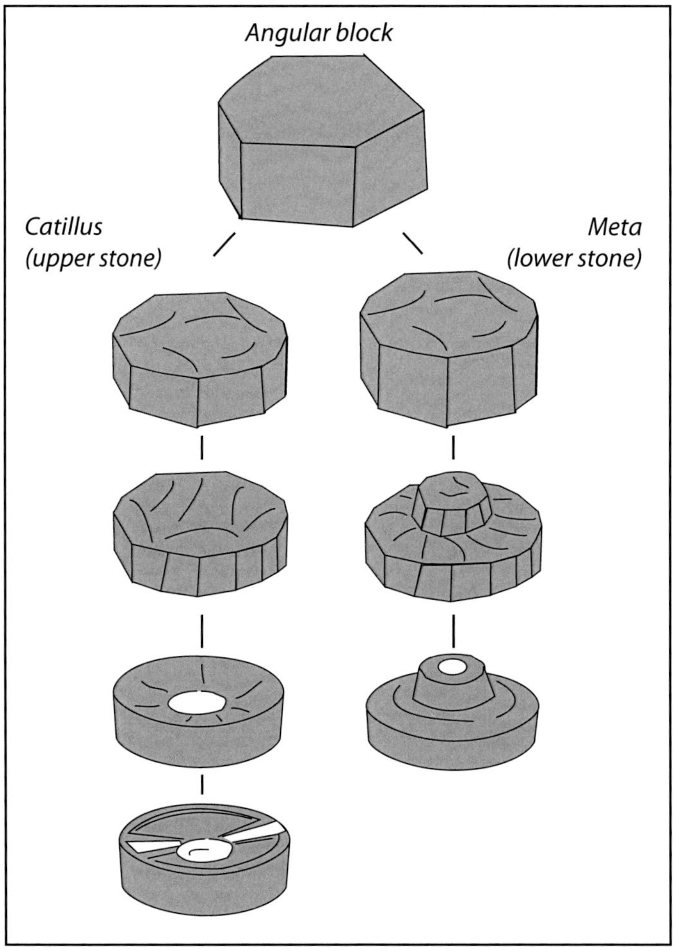

Fig. 5.34: Schema of the main phases of fashioning rotary querns from the angular block quarry of Cerro de Limones quarry (AL-1) where radial slot catilli *(left) and the "*sombrero*"* metae *(right) were hewn from blocks detached from volcanic columnar jointing (drawing by T. Anderson).*

Fig. 5.35: View of volcanic roughly knapped cylinders from a) Cerro de Limones (AL-1) and b) Hoya del Paraíso (AL-2). These thinner cylinders were to be fashioned into upper stones.

Fig. 5.36: Examples of thicker roughly knapped lower stone (sombrero type) roughouts at the sites of a) Cerro de Limones (AL-1) and b) Hoya del Paraíso (AL-2).

Fig. 5.37: Examples of abandoned catilli *at the quarry of Cerro de Limones (AL-1): a) broken* catillus *in its final stages of fashioning; b) unfinished eye carved from both faces; c) the fine circular line indicating the eye was carved with a pointed chisel; d) the outer rim and crosspiece slot in the process of carving.*

Fashioning rotary querns from true extractive quarries

The different steps of fashioning querns from extractive quarries also cannot be defined from evidence in southern Spain. The few sites of this type, for example, Trafalgar (CA-1) and Rota (CA-3) for the Roman period, and Almadén de la Plata (SE-1) and Zagra (GR-5) for the Middle Ages, are devoid of abandoned roughouts. To describe this technique I turn once again to the quarry of Châbles, Switzerland where the operational sequence was reconstructed based on the study of tool marks on a number of abandoned querns, as well as on experimentation undertaken by professional stonemasons (Anderson *et al.* 2003: 47, 54-56).

The principal element to note, related to the fashioning of softer rocks with more grainy textures (sandstones, conglomerates, limestones), is that they did not require a first phase of direct percussion and therefore do not bear the characteristic knapping scars (such as on flint artefacts). Instead the rough cylinders were worked by indirect percussion with hammers (iron head, wooden mallet) and chisels (punches, pointed and bladed chisels). The traces of these tools were either small peck marks corresponding to single chisel-point blows or lines corresponding to a series of contiguous chisel-point blows (Fig. 5.38).

An important detail in this step is the direction of the chisel work (Fig. 5.39). To avoid an unwanted break of the cylinder's edge the chisel always pointed toward the heart of the rock. This required fashioning one half of the stone at a time (*i.e.* one surface and one half of the edge) leaving a sort of "paunch" (from the French "*bourrelet*") along the contour. This technique is visible at Châbles and at a number of different production sites in different periods throughout Europe, (*cf.* Fig. 5.5).

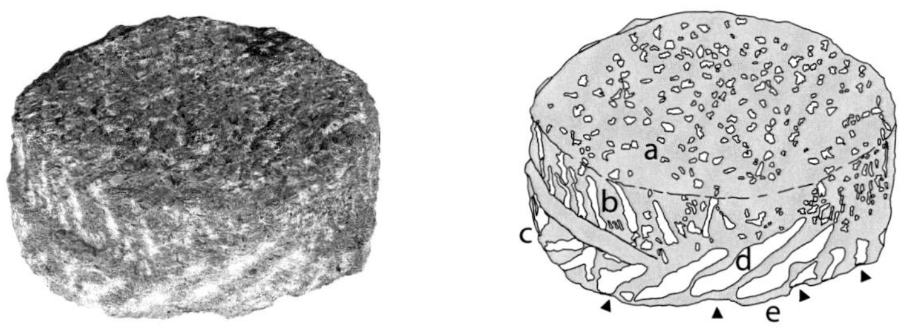

*Fig. 5.38: Cylindrical roughout from the true extractive quarry of Châbles, Fribourg, Switzerland. The pecks on the upper surface (a) are single pointed chisel blows, whereas the thin vertical lines on the edge (b) are the result of multiple contiguous chisel workings. The "paunch" (*bourrelet *in French) to the bottom left (c) indicates working one half at a time. The larger diagonal lines (d) are the original extraction pick marks and the splitting cavities, whereas the teeth (e) at the bottom are the original splitting marks (from Anderson* et al. *2003: 55, fig. 61, drawing by T. Anderson).*

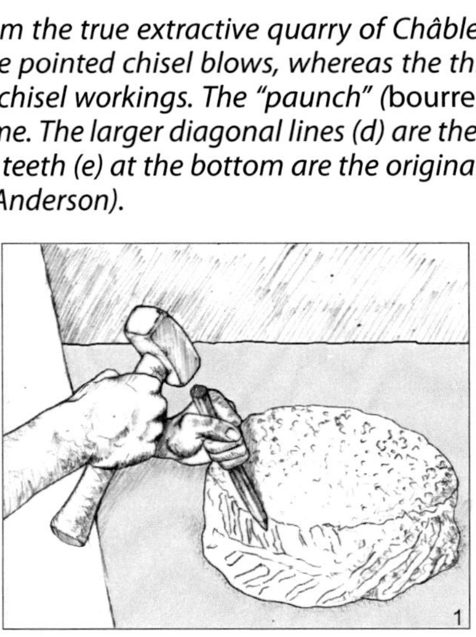

*Fig. 5.39: Example of fashioning a sandstone (*grès coquillier*) rotary quern with hammer and chisel at Châbles, Fribourg, Switzerland. One half was fashioned before turning the cylinder around to fashion the second half, resulting in a "paunch" along the base (from Anderson* et al. *2003: 54, fig. 60, drawing by Aurora Pulido).*

5.4.5.2. Fashioning Modern and Contemporary monolithic millstones

The process of fashioning large monolithic millstones in Modern and Contemporary times consisted of eliminating the unnecessary bulges, rectifying the surfaces and edges and piercing the eye. Since millstones from these periods were cylindrical with flat grinding surfaces, blanks could serve both for upper or lower stones. This differed from that of earlier quern work requiring a preliminary sorting to separate the roughouts better adapted for upper stones and lower stones. Fashioning in more recent times also differed from early work in that the first step, due to the size and weight of the millstone, was to prepare a work space at the quarry itself. This required displacing the stone, at least slightly, because its original splitting position was too near the quarry face to allow all of its edges to be fashioned comfortably.

While moving the millstone with levers, like those seen in the photograph of the Brañosera *moleros* (*cf.* Fig. 5.5), the workers took the opportunity to slip one or several large rocks under the stone to create a gap to facilitate inserting the lever during subsequent handling (Maestro 2011: 42). This propping up is seen, for example, at Moclin (GR-1) (Fig. 5.40) and Teba (MA-2) (Fig. 5.41). An example of one of these fashioning workshops is at Lora de Estepa (SE-4) where a group of larger rocks that served to prop up a millstone is surrounded by a cloud of fine fashioning debris (Fig. 5.42).

Fig. 5.40: Views of an abandoned millstone from the site of Moclin (GR-1). After splitting, and to enable subsequent work, the cylinder was a) moved away from the quarry face creating an access space of 50 cm and b) propped up on rocks to facilitate further handling.

Fig. 5.41: View of a millstone at Teba (MA-2) propped up by a rock.

Fig. 5.42: Lora de Estepa (SE-4). Millstone fashioning workshop as seen by a group of rocks that served to prop up a millstone surrounded by fine fashioning debris. The dashed circle indicates the approximate position of the millstone.

Carving or knapping first took place on one half of the stone, that is, the flat surface and half of the edge, before the stone was reversed to work its other side. The photograph of the millstone makers from Brañosera shows that millstone fashioning was undertaken with the pick, the same tool used for their initial extraction. This pick work was certainly completed by fine hammer and chisel working in the later stages of fashioning.

The technique of carving in the direction of the "heart" of the stone, observed since at least Roman times by "paunches" on roughouts (Anderson *et al.* 2003: 55-56) and illustrated in the photograph of the *moleros* of Brañosera (*cf.* Fig. 5.5), reduced the possibility of an unwanted fracture. This fashioning technique is commonly observed at different sites on abandoned millstones. A good example is the half-fashioned cylinder at the quarry of Castillo de Locubín (J-1), still propped up in a vertical working position (Fig. 5.43).

Most of the finer elimination of rock was carried out by hammer and a variety of pointed and bladed chisels (Fig. 5.44). When work was more delicate, the millstone maker could resort to a wooden mallet. To assure a perfectly circular shape, he resorted to a compass and to assure that the flat surfaces were perpendicular to the edge, he used a set square.

Another aspect that appears common to the more recent millstone quarries in the south of Spain, at least to those where abandoned roughouts are numerous (Moclín GR-1, Vélez de Benaudalla GR-10, El Torcal MA-1, Teba MA-2, Los Frailes CO-1, Cerro de Limones AL-1, Castillo de Locubín J-1, etc.), is that the eye was pierced at the quarry site itself (Fig. 5.45). This was probably done for two reasons: to "sound" or test the stone for possible defects and to facilitate future transport.

Fig. 5.43: Frontal and lateral views of stones abandoned after fashioning only one half: a) Castillo de Locubín (J-1); and b) Peña Harpada (CA-10). The Castillo de Locubín example is still in its original working position.

Fig. 5.44: Reconstruction of the technique of fashioning a large Modern or Contemporary millstone. The fashioning tools include hammers, chisels, a compass, a set square and a wooden mallet (drawing by T. Anderson).

Fig. 5.45: Examples of abandoned millstones with pierced eyes: a) Teba (MA-2); b) Vélez de Benaudalla (GR-10); c) Guardias Viejas (AL-4); and d) El Torcal (MA-1). The millstone at El Torcal probably broke during the piercing.

5.4.5.3. Fashioning composite millstones

The majority of the millstones examined in this survey are monolithic, that is, fashioned from single cylindrical blocks. Composite millstones, on the other hand, are made of segments or sections of angular stones assembled with plaster (or cement) and bound along their girth by iron bands. Millstone workings producing these millstones, however, are hardly known in southern Spain.

The 16th-century Codex of Juanelo Turriano records that composite millstones were destined to mills either found in difficult terrain, where the heavy monolithic stones could not be attained overland by carts, or to mills in areas where no quarry produced monolithic stones (Vol. III, transcription of 1996: 351). Turriano proposes two means of assembling these sections (Fig. 5.46). The first, using five stones, joined the sections (B-E) at right angles around a centre-piece (A). The second, comprising four stones, assembled the sections around a central triangle (Juanelo Turriano 1996: Vol. III: 351, fig. 204). It is the first type of assemblage, according to the Turriano, that results in a more solid millstone.

An early reference to a composite millstone in southern Spain is in a contract related to the lease of a watermill dating to 1114 (during the Islamic domination). The document states that the new stone must measure a diameter of four *empans* (about 1.06 m) and be made of eight sections (Lagardère 1991: 109). Other brief references to composite millstones in the study area date to 1491 and 1492 and are related to the transport of "two-piece stones" (*i.e.* half sections) to mills in the city of Córdoba (Córdoba de la Llave 2003: 305). The quarry of Albaida (CO-7) is reported to have produced composite millstones, according to another written source (Cordoba de la Llave 2003: 306 footnote 26).

In more recent times composite millstones in the mountains of the Alpujarra (Almería) are referred to as "*cuarterones*" (Rodríguez Monteoliva 1989: 705). Contrary to what their name (quarters) implies, they were made up of between five to seven sections. Composite millstones were presumably assembled at Las Canteras (GR-14), near Ugíjar in the Alpujarra mountains (Francisco Moreno, pers. comm.) and in the Province of Cádiz during the difficult economic period after the Spanish Civil War at the sites of Aguadulce (CA-4) and Roa Martín (CA-5) (Juan Antonio Martínez, pers. comm.).

The only site in the study area that I have observed composite millstone workings, as well as monolithic models, is Fuente de los Morales (GR-6) near Alhama de Granada (confirmed in a second field survey of the site in 2015). The site comprises about a dozen abandoned millstone half and quarter sections in its northern sector at the foot of the outcrop (Fig. 5.47).

The single straight interior edge of the millstone halves and the two straight interior sides forming a right angle of millstone quarters facilitated their subsequent binding, following the first schema of Turriano. Besides the question of transport, fashioning millstone segments implies a more efficient and economic use of the rock. Smaller blocks, unfit for monolithic millstones, were not wasted.

An advantage of this type of production, as certified in France, was that the fee for the transport of sections destined to be composite millstones at road tolls was much more reasonable than that of finished monolithic millstones (Alain Belmont pers. comm.). This and the other factors

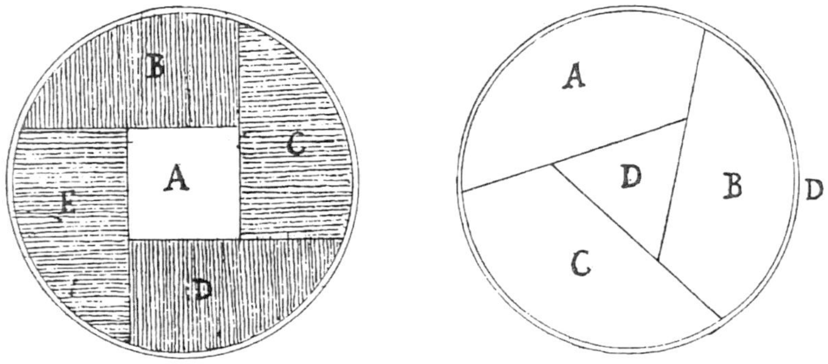

Fig. 5.46: Schema of the two types of composite millstones proposed in the pseudo Juanelo Turriano Codex. The assemblage to the left with five sections is more solid, according to Turriano, than that to the right (from pseudo Juanelo Turriano 1996, Vol. III: 351, fig. 204 1996. http://issuu.com/juaneloturriano/docs/21librosingeniosymaquinastomo_iii).

related to transport imply that the millstones were not bound at the quarry. This step, requiring the intervention of a blacksmith to fix the iron bands, may have taken place at the mill itself or at a specialised workshop in town, as is the case in La Ferté-sous-Jouarre in France.

Most of the composite millstones seen today in southern Spain, often decorating public spaces, are *muelas francesas*. These are the French burrstone imports (Fig. 5.48), vestiges of the last generation of traditional watermills, before their disappearance with the introduction of the industrial methods (steel rollers).

Fig. 5.47: Composite millstone production at Fuente de los Morales, Alhama de Granada (GR-6). Abandoned semi-circular and quarter segments. The unfinished curvilinear outer edge (see arrow) of the bottom left triangle indicates that the interior sides (forming a right angle) were carved before the outer edge, proof of composite millstone production.

Fig. 5.48: Example of a composite, siliceous French burrstone. This millstone decorates a park in the small town of Moclín, Granada. Ironically this imported stone was in use in a flour mill in the town that has the most celebrated millstone quarry in the Province of Granada (GR-1).

5.5. Quantifying the duration of making a millstone

The time invested in making a millstone, from its extraction to its fashioning, depended on the size of the product, the hardness of the rock and the skill of the millstone maker. From an experiment undertaken in Switzerland on a rotary quern 40 cm in diameter, the total time of manufacture by a professional stonemason, including extraction and fashioning, was between two and three days (Anderson *et al.* 2003: 47, 59). This experiment was carried out on a shell-rich sandstone (*grès coquillier*) that was relatively easy to carve. I assume that if the experiment had been repeated, the working time could have been reduced.

Extraction of rough blocks with a lever for Roman rotary querns at the Cerro de Limones (AL-1) or Hoya del Paraíso (AL-2) was probably undertaken quickly, in a matter of less than an hour. Knapping or chipping the block into a rough shape, was also relatively quickly done, probably less than a day. However, the fine carving of the perfectly circular shape and the cutting of the eyes, receptacles, handle and rynd fittings and adjustments of the grinding surfaces, probably added another day. Scoring a cylinder from the bedrock with a pick, such as *Sisapo* (CR-1), on the other hand, took longer than prying out a block with a lever. This longer investment in time of the extraction phase, would nonetheless, reduce the time of fashioning. The total time to make a rotary quern in a block or true extractive would therefore be roughly equivalent.

Based on interviews with descendents of millstones makers, Maestro Hernández records that in the Mountains of Palencia the process of manufacture by a small team of specialists of a single millstone (not a couple) over a metre in diameter took between four and seven days (Maestro: 2011: 43). This suggests that the time to make a large millstone was roughly double that of the smaller handquerns.

I have only identified one old written source that alludes to the question of quantification of millstone production. The quarry of El Berrueco (CA-8), according to Madoz, with five different extraction sectors manned by 23 men, produced 64 millstones for water and windmills and 480 millstones for *tahonas* (animal-driven mills) on a yearly basis (Madoz 1846, Vol. 4: 290).

Here, as in the case of Palencia, quantification is based on producing single stones, not pairs, since it was common in the 19th century, due to the problem of wear, to replace one stone at a time. The numbers presented by Madoz suggest a total of 544 millstones a year, or about 1.5 millstones a day (based on working every day of the year). Based on the assumption that the teams were composed of four to five men (deduced from Madoz), this suggests that each team, working continuously every day of the year, could produce slightly more than 100 millstones, which estimated on a weekly basis would come to about one a week per team, a number similar to that of the Palencia *moleros* recorded by Maestro. Since they certainly did not work every day of the year, and since the rock was probably softer and easier to carve than the conglomerate of Palencia, weekly production was certainly higher, possibly about two per team.

All said, the exact amount of time invested in the making of querns and millstones is difficult to quantify. These first results based on data as disparate as an experiment in Switzerland, oral information from the north of Spain and a 19th-century text need to be weighed against the results obtained in future research.

6. MILLSTONE QUARRY CLASSIFICATION

From observations in the field of about 40 millstone quarries (MQ) in the south of Spain, as well as consulting diverse building block quarry literature from elsewhere in Europe (Bessac 2003; Walton 2004; Martínez Torres 2008), I propose that millstone quarries be classified into two major categories, contingent on whether the rock exploited was loose, surface material (MQ-1), or, more commonly, bedrock (MQ-2) (Fig. 6.1). These two major categories can, in turn, be subdivided into two parts. The first subdivision, MQ-1a and MQ-1b, is conditioned by the size of the surface block, whereas MQ2a and MQ-2b, depend on whether cylinders were scored directly from bedrock (yielding circular hollows) or if they were carved from angular blocks detached with levers from bedrock.

This proposal of classification has recently been expanded beyond the borders of Spain in the colloquia of Arbeca, Lleida, Catalonia (Anderson 2014) and Reims, France (Anderson & Jaccottey, in preparation). These different classifications, the first of their type for quern and millstone workings, need to be measured against the results of future quern and millstone production research.

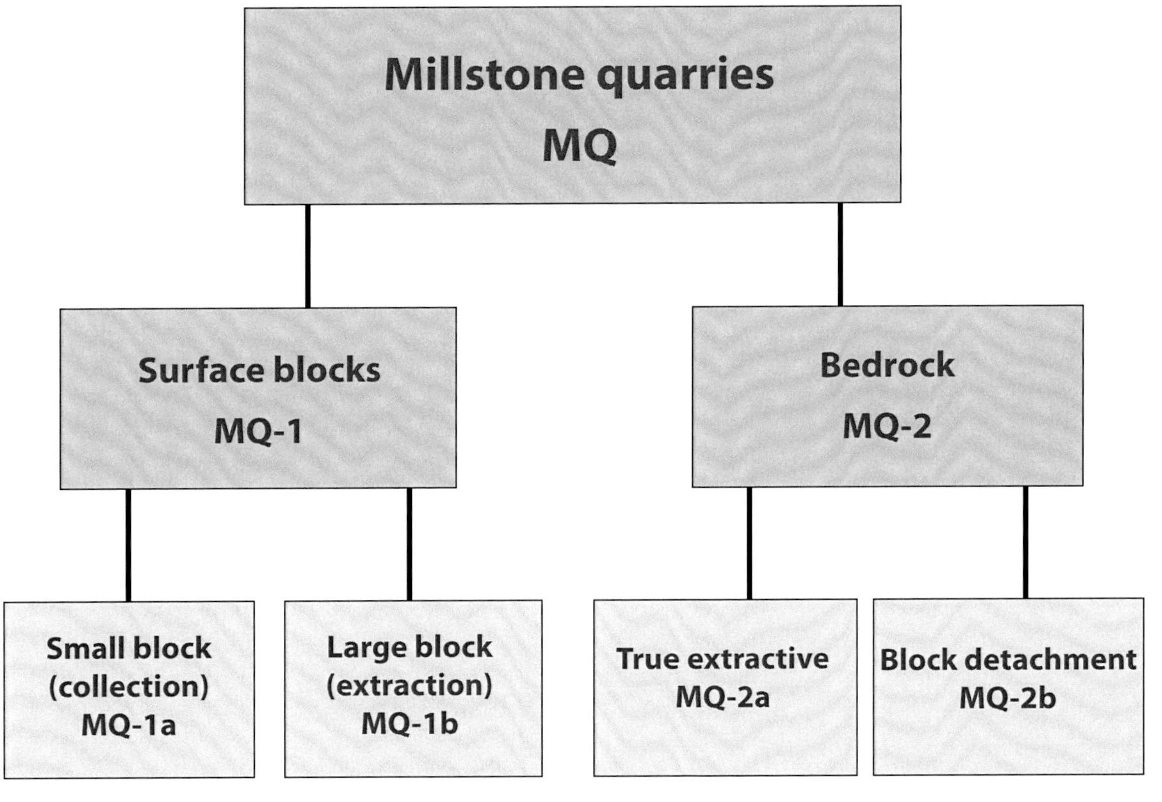

Fig. 6.1: Organisational chart of the classification of quern and millstone quarries identified in southern Spain.

6.1. Surface block millstone quarries (MQ-1)

6.1.1. Small surface block collection (MQ-1a)

A single, small boulder or slab collected on the surface and hewn into a quern or millstone does not constitute a quarry. In contrast, certain geographical features with concentrations of transported material such as moraines, riverbeds, ravines, screes and taluses where blocks were repeatedly collected or "harvested" (a term borrowed from Tom Heldal, NGU) for millstones can be considered quarries (Fig. 6.2).

Due to the natural shape of the small boulders, elongated in the case of saddle querns and rounded for rotary querns, no veritable extraction was required and the process of fashioning was reduced to a minimum. This type of site obviously did not demand the type of tools and skills of the more complex extractive sites where millstones were hewn directly from bedrock. As will be noted at the end of this chapter, a criterion that can often be applied (although not always) to identify rotary querns carved from small surface blocks in a hemispherical section, whereas the characteristic shape of products from true extractive quarries is the cylinder.

Since surface block quarries leave little or no material evidence, no quarry of this type has been positively identified in southern Spain. This type of site must, nonetheless, have been common, especially throughout Prehistory and Protohistory. Surface block workings account for the many garnet-mica schists and conglomerate saddle querns (Delgado Raack 2008: 138) recovered in the settlements of the Millares and Argar Cultures throughout south-eastern Iberia since schists and conglomerates are common surface materials. The same model of exploitation applies to the granite fields of south-western Spain abounding with rounded granite boulders sculpted by nature (*bolos*). Furthermore, surface collection from screes and taluses in the volcanic district of Cabo de Gata certainly accounts for part of the pre-Roman saddle querns of volcanic material.

Fig. 6.2: Schematic reconstruction of an MQ-1a type of millstone quarry where millstone makers "harvested" small surface boulders that were fashioned directly into millstones. This type required no extractive phase (drawing by T. Anderson).

Small surface block harvesting quarries are also known in Roman times. In the Province of La Rioja in northern Spain, for example, there is evidence that rotary querns were hewn from small rolled boulders collected in nearby alluvial deposits. This is based on the observation of unworked surfaces or cortex bearing the original patina of wear resulting from alluvial transport (Pascual & García 2010: 287-288). These locally manufactured querns were certainly a more economical alternative to imports from a larger scale in bedrock quarries.

In the study area, part of the production of the Roman volcanic quern quarries of Cerro de Limones (AL-1) and Hoya del Paraíso (AL-2) was most likely the result of harvesting surface material from screes and taluses (Fig. 6.3).

Fig. 6.3: Examples of the scree and talus at the Cerro de Limones quarry (AL-1). Part of the quern production could have been fashioned from the loose surface material collected in the talus.

6.1.2. Extracting from large surface blocks (MQ-1b)

This second quarry type differs from the first in that the surface blocks, much larger than those harvested in MQ-1a sites, required a preliminary extractive process. The blocks were large enough to permit the scoring of one or several millstones and required both the use of extraction tools (mostly picks) and the mastery of the skills to cut through rock. This quarry type is divided into several subtypes. In the case of subtypes MQ 1b/1-3, the large blocks were transported or formed *in situ* by natural means, whereas the large blocks exploited of the MQ 1b/4 subtype, originally construction material, were transported by manually.

MQ-1b/1

Millstone quarries of subtype MQ-1b/1 exploited large surface blocks accumulated naturally in taluses at the base of cliffs or sharp slopes (Fig. 6.4). Examples of this type of production are sectors of the quarries of Moclín (GR-1b), Castillo de Locubín (J-1) and Vélez de Benaudalla (GR-10) (Fig. 6.5). The volume of production at these sites depended on the quantity of suitable blocks. In general, however, these sites did not deliver large quantities of millstones.

Fig. 6.4: Schematic reconstruction of an MQ-1b/1 quarry where millstones were extracted from large surface blocks, such as in taluses (drawing by T. Anderson).

Fig. 6.5: Examples of an MQ-1b/1 quarry exploiting taluses: a) Castillo de Locubín (J-1) and b) Vélez de Benaudalla (GR-10).

MQ-1b/2

This subtype corresponds to the exploitation of large single blocks transported, presumably, by landslides or other natural means (Fig. 6.6). This would also include glacial erratics, a type that is not documented in southern Spain but is common elsewhere in Central and Northern Europe (notably in Switzerland and Norway). The two cases, Arbuniel, Jaén (J-4) and Molino de la Piedra, Córdoba (CO-5) (Fig. 6.7), are located in valley bottoms. Because of the restricted volume of the single block, these workings could not develop into anything but small, local productions.

Fig. 6.6: Schematic reconstruction of a MQ-1b/2 quarry where millstones were extracted from large single blocks displaced by landslides (drawing by T. Anderson).

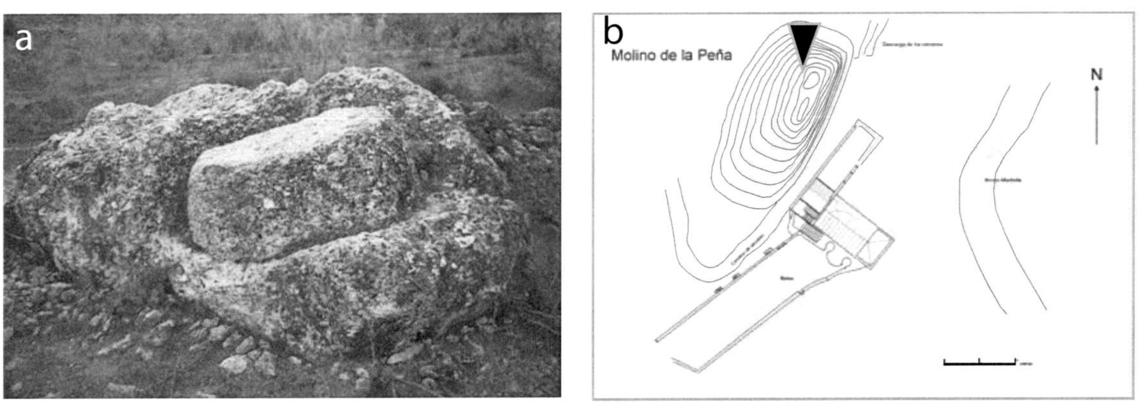

Fig. 6.7: Examples of MQ-1b/2 quarries where millstones were hewn from large single surface blocks: a) Arbuniel (J-4) (from López & Cabrera 2004) and b) Molino de la Piedra (from Córdoba de la Llave & Varela 2011: 109).

MQ-1b/3

This millstone quarry subtype exploited naturally detached, non-transported, surface blocks resulting from *in situ* erosion. This is the case of certain karstic landscapes where numerous surface blocks and slabs were sculpted over time (Fig. 6.8). Examples of these workings have been identified at El Torcal (MA-1), Teba (MA-2), Sierra de Utrera (MA-8) and Fuensanta, Loja (GR-2) (Fig. 6.9). This type also applies to exploitations of granite fields cluttered with naturally rounded surface boulders (*piedras caballeras* or *bolos*). Sites such as Miraflores de la Sierra (M-5), exploiting granite, are potentially extensive and can develop into very large productions.

The schist workings of El Campillo Viejo (HUE-1), also falling into this type, is slightly different. These large surface blocks were not rounded by erosion but were broken up into large slabs, presumably by tectonic movements.

Fig. 6.8: Schematic reconstruction of a MQ-1b/3 quarry where millstones were scored from large, naturally detached, non-transported slabs formed by agents of erosion. These exploitations took place, for example, in karstic limestone and granitic boulder landscapes (drawing by T. Anderson).

Fig. 6.9: Example of a MQ-1b/3 millstone quarry: The karstic landscape of the Natural Park of El Torcal (MA-1) comprises numerous naturally sculpted and detached surface blocks and slabs that were well-suited to be hewn into millstones. a) general view; b) detail of a roughout in vertical position.

MQ-1b/4

This quarry subtype corresponds to the exploitation of anthropic features such as abandoned constructions (Fig. 6.10). Few cases of this subtype have been confirmed in southern Spain. The most notable is at the Roman city of *Baelo Claudia* (CA-2) where an unfinished Pompeian upper stone (Fig. 6.11) was hewn from a biocalcarenite ashlar or column drum originally transported from a nearby extractive quarry. Column drum segments from the Roman quarry of Cerro Bellido (SE-4) are also reputed to have been recycled into millstones in the Middle Ages. Two excellent examples of this subtype are beyond the Spanish border: at Chardy (Orsennes), France, where a millstone maker attempted to carve a cylinder from the table top of a dolmen (Gérard Coulon, Database Meulières.eu, no. 159) and at the Temple of Zeus at Olympia in Greece where abandoned column drums served to produce a number of rotary querns (Alain Belmont, Database Meulières.eu, no. 232).

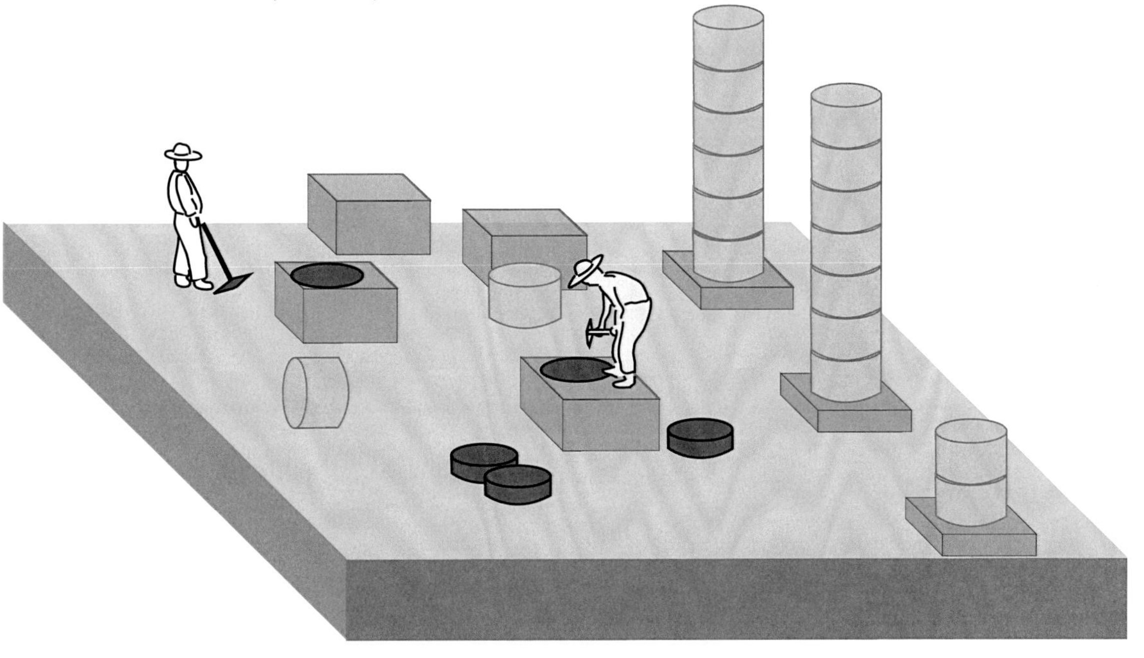

Fig. 6.10: Reconstruction of an MQ-1b/4 exploitation where millstones were hewn from recycled construction material such as blocks or column drums (drawing by T. Anderson).

Fig. 6.11: Example of an unfinished Pompeian millstone (catillus) *recycled from a biocalcarenite ashlar or column drum. The original source of the rock is probably the neighbouring quarry at Punta Camarinal (CA-2a) or Punta Paloma (CA-2b).*

6.2. Bedrock millstone quarries (MQ-2)

The second major group of quarries, those exploiting bedrock, is divided into two major categories. The first of these extractive workings (MQ-2a) corresponds to "true extractive" sites where cylinders were scored directly from bedrock, whereas the second, block extractive workings (MQ-2b), correspond to bedrock exploitations where angular blocks were first detached (following natural fissures). Each of these types can be subdivided according to morphological criteria.

6.2.1. True extractive quarries (MQ-2a)

In "true extractive" millstone quarries (term borrowed from Runnels 1981: 72), labelled MQ-2a, cylinders were scored directly from bedrock by means of trenching (Fig. 6.12). Cutting directly into the heart of the rock required specific skills and tools, in particular command of the use of the pick. These workings resulted in circular extractive hollows and, in favourable conditions of conservation, resulted in numerous tool marks on both the quarry faces and floors.

Multiple, superimpose extractions, like a "stack of coins", took place for the most part on horizontal (or slightly inclined) extractive planes, following the rock's natural bedding resulting in characteristic vertical "tubular" faces. Sites of this type, among others, are Los Guillares, Padul (GR-7), Montesa, Valencia (V-2) (Fig. 6.13), Los Frailes, Cabra (CO-1) and Moclín (GR-1) (Fig. 6.14).

Fig. 6.12: Schematic reconstruction of a true extractive millstone quarry (MQ-2a) where horizontal scoring produced quarry faces with superimposed, tubular-shaped hollows (drawing by T. Anderson).

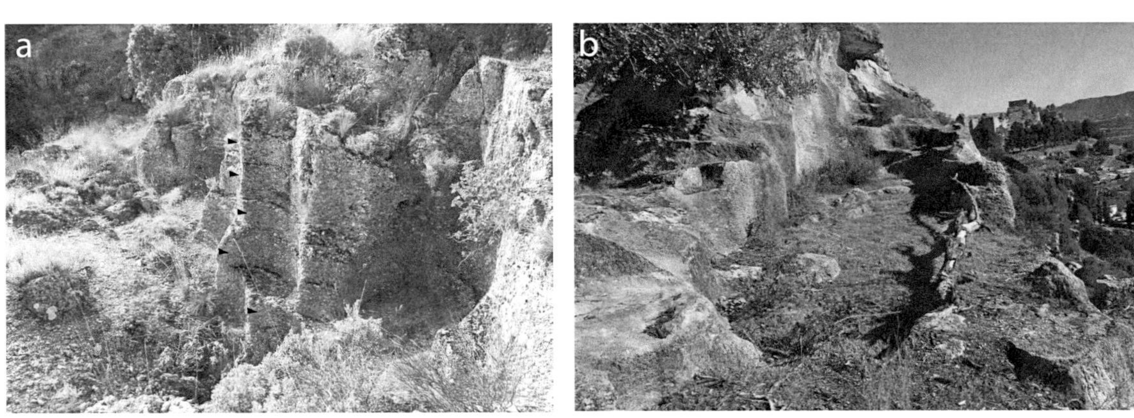

Fig. 6.13: Examples of true extractive millstone quarries with high vertical tubular faces (MQ-2a): a) Los Guillares, Padul, Granada (GR-7) with up to five levels of extraction and b) Montesa, Valencia (V-2).

Fig. 6.14: Examples of true extractive millstone quarries with high vertical tubular faces (type MQ-2a): a) Los Frailes, Cabra, Córdoba (CO-1); and b) Moclín, Granada (GR-1).

Workings where extraction took place along vertical (or sharply inclined) planes, perpendicular to the rock's natural bedding or cleavage are an alternative to this type (MQ-2a-2) (Fig. 6.15). This technique required a particularly compact and homogeneous rock that could withstand this perpendicular fracture. Only four sites in the study area made use of this technique: Barranco de los Molinos, Ibi (A-2), Vélez de Benaudalla (GR-10), Los Guillares, Padul (GR-7) and Castillo de Locubín, Jaén (J-1) (Fig. 6.16). It must be noted that, contrary to certain sites such as at Mont Vouan in France, for example, extractions along horizontal planes prevail at each of these Spanish quarries.

Fig. 6.15: Schematic reconstruction of a true extractive quarry scoring millstones following vertical planes (MQ-2a-2) resulting in quarry faces with vertical extraction hollows (drawing by T. Anderson).

Fig. 6.16: Examples of vertical extractions from true extractive quarries: a) Los Guillares, Padul, Granada (GR-7) and b) Castillo de Locubín (J-1).

6.2.2. Block detachment quarries (MQ-2b)

Block detachment quarries differ from true extractive quarries in that the extractive process consisted of detaching angular blocks or slabs from the rock mass with levers and crowbars, profiting from natural fissures and clefts, as opposed to scoring directly from the rock by cutting trenches. The quarry faces of these sites were therefore not "tubular" but "planar" or irregular, with few, if any, visible tool marks (Fig. 6.17). Although the basic extractive tool was the lever, picks and chisels were certainly used at times to enlarge the fissures to facilitate the insertion of the lever. Although extraction took less time than cutting the bedrock, the subsequent fashioning phase required a larger investment of time since a larger quantity of mass had to be removed to obtain a cylindrical form. Identification of these sites in the field is not simple because they can be mistaken for outcrops where natural mechanical processes, such as gelifraction, detached the rock.

An example is Fuente de los Morales near Alhama de Granada (GR-6) (Fig. 6.18a) where slabs were pried out from a specific layer of hard, white limestone. This technique was also used at certain sectors of the Roman quarry of Cerro de Limones (AL-1), where small blocks were pried from the summit of volcanic columnar jointing. Part of the production of Miraflores de la Sierra (M-5), in the granite fields of the Sierra de Guadarrama, profited from previously detached angular blocks (Fig. 6.18b).

Fig. 6.17: Schematic reconstruction of an MQ-2b-1 millstone quarry where angular blocks or slabs were detached along horizontal planes before being carved into millstones (drawing by T. Anderson).

Fig. 6.18: Examples of quarries where angular blocks were detached from the bedrock for millstone production: a) Fuente de los Morales, Alhama de Granada (GR-6); and b) Miraflores de la Sierra (M-5) (photograph left by Fernando Colmenarviejo).

6.3. Mixed exploitations: Surface block and bedrock quarries

Not all millstone quarries can be comfortably classified into the previous categories. Quern makers at the Cerro de Limones (AL-1) exploited both loose surface blocks (MQ-1a) and columnar jointing (MQ-2b). At El Campillo (HU-1) they had the choice of large surface slabs (MQ-1b) or bedrock (MQ-2a) (Fig. 6.19 a-b), whereas at Castillo de Locubín (J-1), they chose between working large surface blocks in the talus (MQ-1b), cutting cylinders directly from the bedrock (MQ-2a) or detaching angular blocks with levers (MQ-2b) (Fig. 6.19 c-f). There were also sites labelled "opportunistic quarries" (D. Peacock, pers. comm.) producing millstones, among other products, when the form and size of the extractions were suitable.

Fig. 6.19: Example of quarries with mixed extraction techniques: a-b) El Campillo Viejo (HU-1): a) loose surface block exploitation; and b) vertical bedrock outcrop extraction; c-f) Castillo de Locubín (J-1): c) unfinished cylinder carved from a surface block in the talus; d) vertical extraction cut directly from bedrock; e) "planar" quarry face (lighter colour) where blocks were detached along natural fissures; f) The natural stratification of the rock of the abandoned millstone (arrows) indicates a vertical extraction. The absence of vertical extractions in this sector of the quarry, suggests this cylinder was hewn from an angular block detached from the adjacent quarry face (photographs a and b by Alonso García Veiga; c-f by T. Anderson).

6.4. The morphology of extractive millstone quarries

Independent of the extractive techniques (true extractive or block detachment), bedrock millstone quarries (MQ-2a and MQ-2b) take on different forms depending on the configuration of the terrain and if they are open air or subterranean. Other factors affecting their morphology are their size and depth, direction of progression, inclination of the rock layer, accessibility and thickness of the desired stratum, and if the extractions are contiguous or not. Some of the larger quarries are difficult to classify from a morphological point of view because they integrate different extractive techniques and different types of topographical features. The following designations, with terms at times borrowed from construction quarries (notably Bessac 2003), are meant to facilitate the general description and definition of millstone quarries.

6.4.1. Bench quarries

Bench quarries (Fig. 6.20) are the result of a contiguous extractive process following a forward and lateral progression into the heart of the slope of a hill or mountain. These sites, among the most common in southern Spain, are basically U-shaped with rectangular-shaped faces and triangular-shaped walls. They are both deep and shallow. Los Frailes, Cabra (CO-1) and specific sectors of Peña Harpada (CA-10), Zagra (GR-5) (Fig. 6.21) and Loja (GR-5) are good examples (Fig. 6.22). The term "bench" alludes to the general shape of the quarry and should not be confused with the term that defines the steps (or tiers) between different quarry floors.

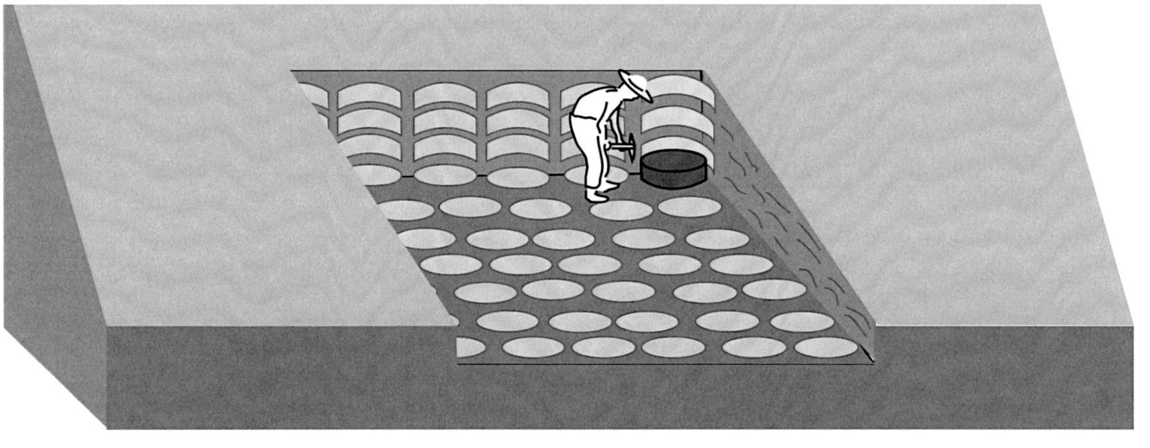

Fig. 6.20: Schematic representation of a bench millstone quarry (drawing by T. Anderson).

Fig. 6.21: Example of a bench millstone quarry at Zagra (GR-5).

Fig. 6.22: Example of a bench millstone quarry at Loja, Cerro Fuensanta (GR-3).

6.4.2. Pocket quarries

Pocket quarries (Fig. 6.23) are simply smaller versions of bench quarries. Instead of extracting in one single location, resulting in a large bench quarry, these exploitations comprise a number of workings spread over the slope of the hill or mountain and evoke the leasing of concessions to different teams. At times their faces present a slight overhang, as if extraction was going to be pursued underground. The best example is El Hacho near Lora de Estepa (SE-3) (Fig. 6.24).

Fig. 6.23: Schematic representation of a pocket millstone quarry (drawing by T. Anderson).

Fig. 6.24: Example of a pocket quarry. El Hacho, Lora de Estepa (SE-4).

6.4.3. Edge quarries

The factor that characterises edge quarries is the contiguous lateral progression of extractions along the edge of specific topographical features, notably cliffs, ridges and ravines. This type of exploitation is opposite to that of bench quarries that are characterised by a progression directed into the heart of the slope. These workings adopt both true extractive (Fig. 6.25) and block detachment techniques (Fig. 6.26). Examples are La Merced (Gr-4c), Fuente de los Morales (GR-6) and Cudillas (CO-3) (Fig. 6.27), as well as sectors of Los Guillares, Padul (GR-7).

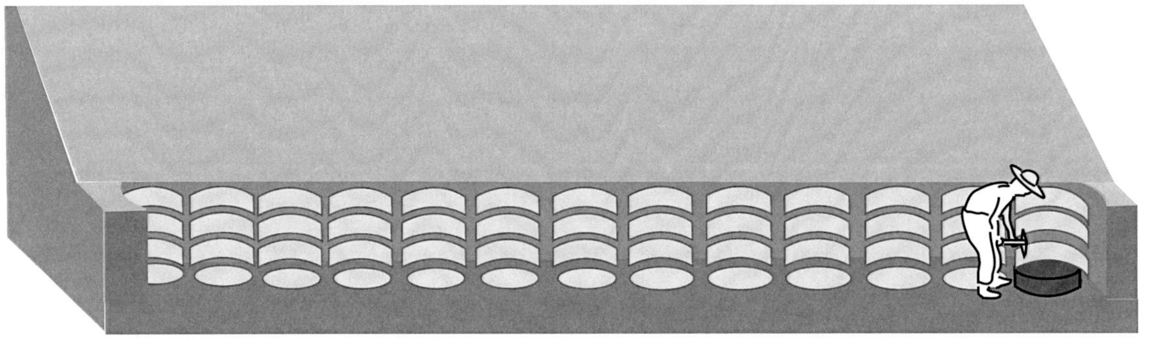

Fig. 6.25: Schematic representation of an edge (true extractive) millstone quarry (drawing by T. Anderson).

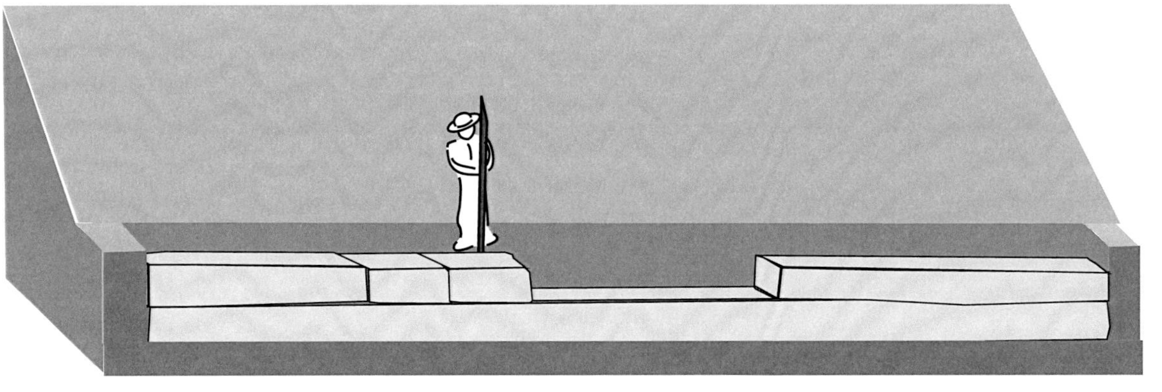

Fig. 6.26: Schematic representation of an edge (block detachment) quarry (drawing by T. Anderson).

Fig. 6.27: Example of an edge quarry (Cudillas (CO-3). A specific layer of rock was sought at this site.

6.4.4. Pit quarries

Pit quarries, characterised by vertical and lateral progression, are mostly dug in flat or slightly inclined terrain so as to attain a specific layer of rock (Fig. 6.28). These pits in millstone working contexts are generally relatively shallow, just deep enough to attain the desired stratum. There are few examples in the study area. A sector at Cerro de Chispas (AL-8) on a terrace, today resembling a crater, exploited a layer of puddingstone. The large, shallow pits on the hilltop of Cerro de Limones (AL-1), now cluttered with working debris, presumably exploited the summit of volcanic columnar jointing. The northern sector of Moclín (GR-1) is a deep pit (Fig. 6.29) that opens up to a slope where debris was evacuated.

Fig. 6.28: Schematic representation of a pit quarry (drawing by T. Anderson).

Fig. 6.29: Examples of pit quarries: a) Cerro de Limones (AL-1) and b) the northern sector of Moclín (GR-1).

6.4.5. Trench quarries

Trench quarries are characterised by a narrow, linear, forward progression (Fig. 6.30). These sites, comprising at times more than one trench, are found in flat or slightly inclined terrain. In the study area there are two examples, Cantera de la Rambla (GR-11) near Caniles, and Lachar (J-2) by Jimena (Fig. 6.31) exploiting, respectively, layers of pebble conglomerate and white limestone. These sites are also characterised by more than one trench bordered by elongated cordons of working debris.

Fig. 6.30: Schematic representation of a trench millstone quarry (drawing by T. Anderson).

Fig. 6.31: Example of a trench quarry, El Lachar, Jimena (J-2).

6.4.6. Extensive contiguous shallow quarries

Extensive contiguous shallow quarries are exploitations where the bedrock stretches over a large area along or near the surface. Millstone makers at these sites simply progressed along the surface without having to tackle the complications inherent to vertical extraction on multiple levels (in particular the problem of removing debris or the rise of the water table) (Fig. 6.32). These sites, of varying size, do not always present clearly defined boundaries, especially when they are partially covered by topsoil or when they are found on the coast and covered (at least in part) by water. Examples in the study area include the coastal sites of Trafalgar (CA-1) and Playa de Carchuna (GR-9), the hill sites of Las Calzadillas 1 (SE-1a), Guajar Faragüit (GR-8), Puerto de la Cadena (MU-1) and the upper sector of the Rambla Honda (AL-3a) (Fig. 6.33).

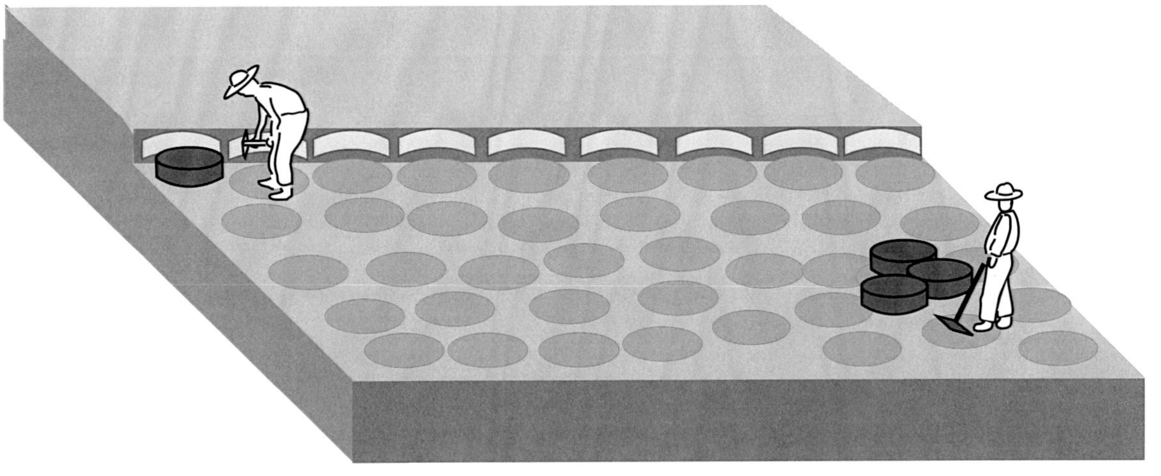

Fig. 6.32: Schematic representation of an extensive contiguous shallow millstone quarry (drawing by T. Anderson).

Fig. 6.33: Example of extensive shallow quarries. a) Puerto de la Cadena, Murcia (MU-1) and b) Cantera de la Rambla Honda (AL-2).

6.4.7. Extensive dispersed shallow quarries

Extensive dispersed shallow quarries correspond to sites where millstone makers exploited a number of different small, dispersed bedrock outcrops. Work at each outcrop was never deep and usually comprised a single or a handful of extractions. Distances between the different sectors varied, with intervals of up to *c.* 50 m to 100 m. The borders of these workings, like extensive contiguous quarries, are ill-defined. They can at times extend, theoretically, over very large areas. These workings differ from pocket quarries in that the extractions are limited to one, or at most two, levels.

The most characteristic extensive dispersed shallow sites are found in karstic limestone landscapes or granite fields. Examples of karstic limestone workings are El Torcal (MA-1), El Tajo, Teba (MA-2), Estepa (SE-8) and Camino del Calvario, Loja (GR-3) (Fig. 6.34). Granite exploitations include Miraflores de la Sierra (M-5) and Guijo de Galisteo (CC-5) (Fig. 6.35).

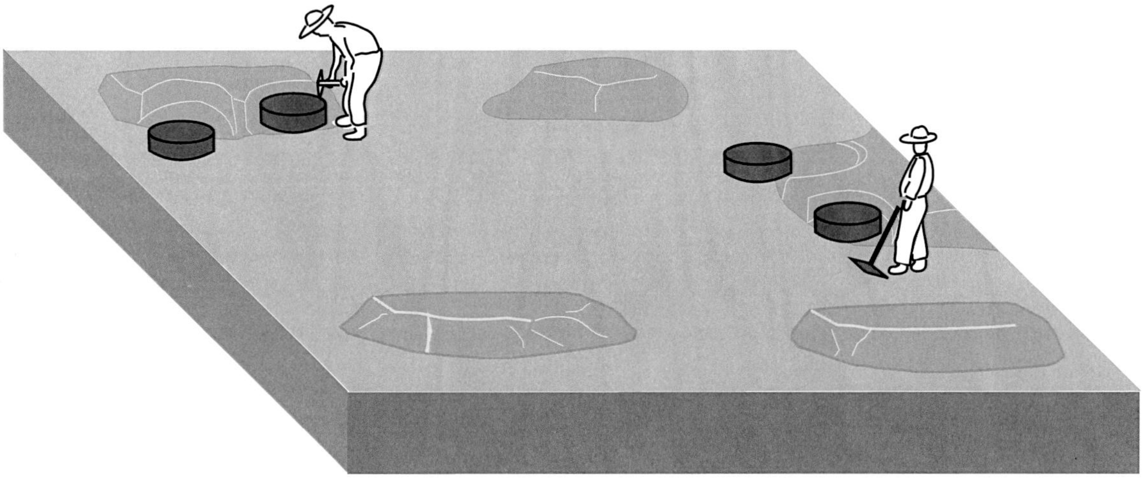

Fig. 6.34: Schematic representation of an extensive dispersed quarry (drawing by T. Anderson).

Fig. 6.35: Examples of extensive dispersed outcrop quarries: a) Camino del Calvario, Loja (GR-3), worked karstic limestone outcrops, and b) Guijo de Galisteo (CC-5), worked granite outcrops.

6.4.8. Subterranean quarries

Millstone makers, to pursue their work, were at times obliged to follow a specific layer of rock underground since the overlaying levels of overburden ("unhealthy" rock and soil) were too voluminous to remove (Fig. 6.36). These subterranean workings resulted in artificial caverns of varying size and depth. La Merced (GR-4b) in Loja is, to date, the only quarry of this type identified the study area (Fig. 6.37). The absence of underground workings is surprising considering the long tradition of underground mining in Spain.

The most celebrated subterranean quarry near the study area is across the Strait of Gibraltar at the Cave of Hercules (Tangiers, Morocco) where the late millstone specialist, Cecil Curwen, witnessed quern extraction in 1956 by candle light (Williams & Peacock 2011a: vii). In Portugal there is also an important underground site at Alvito (Lidia Catarino, pers. comm.). In France a number of subterranean *meulières* have been identified, notably at Mont Vouan (Haute-Savoie), where three have been the object of recent archaeological work under the direction of Alain Belmont. In Germany, there is the Hörer Berg in Bayern (Czysz 2014). The most spectacular underground workings are in the Eifel volcanic district of Germany, where millstone scoring resulted in kilometres of underground galleries (Harms & Mangartz 2002).

Fig. 6.36: Schematic representation of a subterranean millstone quarry (drawing by T. Anderson).

Fig. 6.37: The small exploitation at La Merced (GR-4b) is the sole subterranean millstone quarry identified to date in the south of Spain. In the picture to the right the compact layer (a) exploited for millstones contrasts with the overhanging, brittle rock (b). The arrow indicates the limit between the strata.

6.5. Quantifying the types of millstone quarries

6.5.1. Surface block and bedrock quarries

Of the total millstone quarries (138) presented in the catalogue, about half (70), for the most part sites I have either surveyed in the field or sites that have been described to me in detail, can be classified into the four main categories that I have described above (Fig. 6.38). Although most fall into a single category, a few are mixed exploitations and are therefore represented in more than one category.

On the whole, even if the proportions are only indicative, I can conclude that bedrock quarries, that is, true extractive (MQ-2a, 46 cases) and block detachment workings (MQ-2b, 14 cases), together far outnumber the other categories. The main quarry category, almost double that of all the other categories combined, is that of true extraction (MQ-2a). This category, known since Roman times, reflects the abundance of extensive homogeneous rock outcrops suitable for millstones. Quarries exploiting surface blocks (MQ-1a and MQ-1b), totalling ten, are in clear minority. Of these, only the two volcanic quern workings of Almería (AL-1 and AL-2) show evidence of the working of small surface blocks (MQ-1a).

The dominance of bedrock quarries (MQ-2a and MQ-2b) over surface block exploitations (MQ-1a and MQ-1b) merits further qualification. About half of the sites exploiting large detached blocks, such as Las Pedrizas (GR-1b), Castillo de Locubin (J-1) and Vélez de Benaudalla (GR-10), are incidental since they are secondary talus workings in the framework of larger true bedrock sites. The limited quantity of MQ-1a sites, by contrast, as previously noted, certainly does not reflect their real number, especially in Prehistory and Protohistory, since they do not leave perceptible evidence in the field.

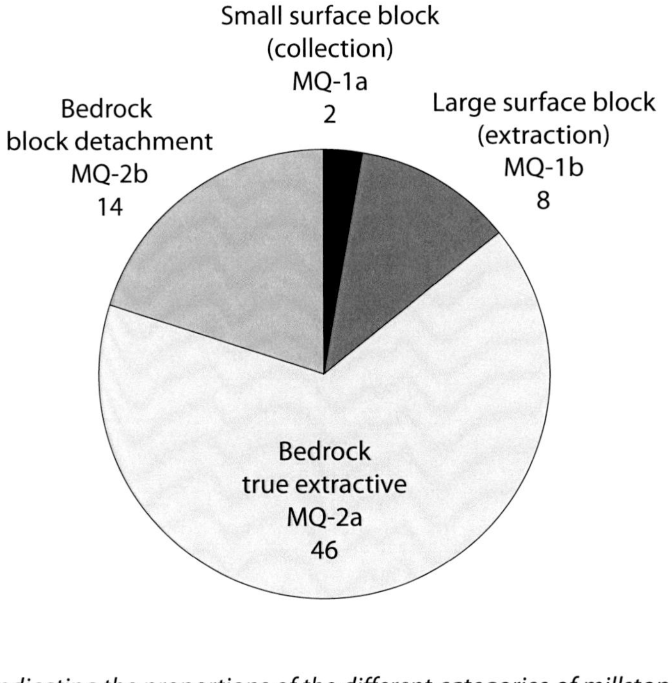

Fig. 6.38: Pie chart indicating the proportions of the different categories of millstone quarries. It is apparent that bedrock exploitations dominate by far those of surface block exploitations.

6.5.2. Quantifying the morphology of millstone quarries

Only 56 of the 138 sites can be assigned into one or more of the eight morphological classes that have been defined. On the whole, their morphology is homogeneous. Only a few, such as Los Guillares (GR-7) with both Bench and Edge workings, share more than one morphology.

From the pie chart (Fig. 6.39) it is evident that Extensive Contiguous Shallow and Bench workings, corresponding to different types of topographical features, together make up about half of the sites. Extensive Contiguous workings reflect a freely accessible source of rock. Bench workings, on the other hand, working within a restricted space, seem to indicate the opposite.

Less frequent are Extensive Dispersed, Edge and Pit quarries. The first are limited to particular geological features (karsts, granite fields), whereas the second are dependent on the topography and could reflect the initial stage of a Bench quarry.

Pit quarries in Roman and earlier times were relatively shallow as the desired rock layer was easy to attain. In more recent times, deeper pits required a complex infrastructure, in particular to evacuate debris. They were thus limited to large productions, such as the deep granite pits at Gerena (SE-7) in Seville. It is noteworthy that no millstone quarry district in the study area resembles the extensive fields of craters around La Ferté sous Jouarre in France, exploiting siliceous *meulière* (burrstones). These workings had to face evacuating both rain and groundwater, a problem less common in the drier regions of Spain.

Pocket sites, small versions of bench quarries, are rare. As noted they could reflect small concessions to different working crews. Trench quarries, also rare, lend themselves to exploitation of a specific layer of rock. This type, especially when in the form of multiple parallel trenches, could also reflect different concessions.

There is only one subterranean quarry in the study area, which is surprising, owing to the long tradition of underground mining in Spain. It appears that the prized rocks in the south of Spain did not require this onerous and complex means of exploitation. Future research will probably bring to light other underground workings.

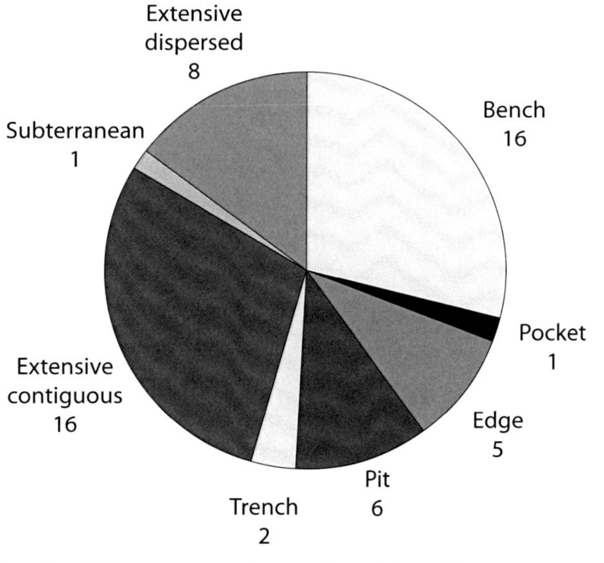

Fig. 6.39: Pie chart indicating the different proportions of the eight millstone quarry morphological classes.

7. MILLSTONE QUARRY TOPONYMY

The study of place names is a proven means of acquiring information about ancient millstone workings. In France, the names *Molière* and *Moulière* and a number of derivatives rooted in the Latin *"mola"* (meaning mill) account for the identification of a large number of millstone quarries. In the former Province of Dauphiné alone, there are 68 cases, some as old as the 13th and 14th centuries (Belmont 2006a, Vol. 1: 56-59). Moreover, a recent study in northern Spain has corroborated the utility of place names to identify millstone quarries, as *Molares*, also deriving from the Latin *mola*, has proven to coincide with a number of millstone workings in La Rioja and Castilla León (Pascual & Ruiz 2011: 285-286).

Although I have undertaken place name analyses in the south of Spain, the results have not been as conclusive as in France or northern Spain. *Molares* and its derivatives can, in fact, have other connotations. Furthermore, early *mola* based names probably disappeared during the long Islamic domination in the south of Spain. Other terms such as *cantera*, meaning simply "quarry", *piedra* (meaning rock) and *berruecos* (most often designating granite outcrops) also, at times, coincide with millstone workings. Although these names can have very old origins, they were probably introduced after the Islamic domination.

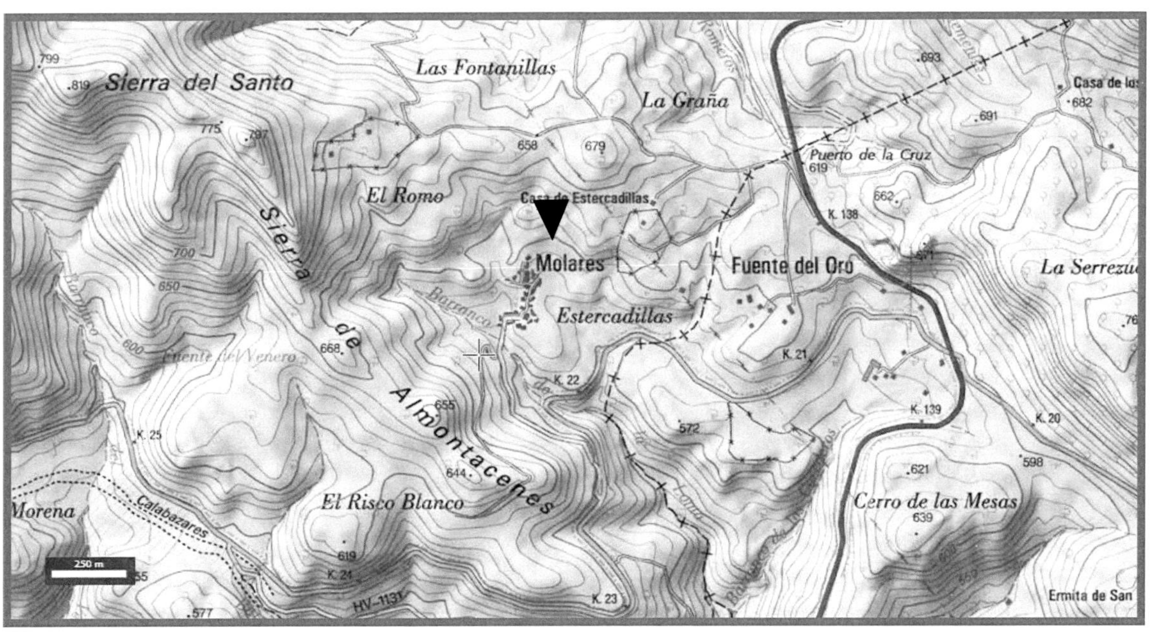

Fig. 7.1: The location of the place name Los Molares *corresponds to a hamlet in the Municipality of Almonaster la Real, Huelva. This is an example of a* Los Molares *place name that is directly related to a millstone quarry (extract from SIGPAC).*

7.1. *Molares* and its derivatives

Three sites in the south of Spain are linked to the *Molares* place name and can be unequivocally linked to millstone workings. These are found in the Municipalities of Almonaster la Real in Huelva (HU-8), Portilla in Cuenca (CU-1) and Sienes in Guadalajara, at the hamlet of Tobes (GUA-05) (Fig. 7.1).

Other derivatives, such as *Moles, Mola, Molar*, are also at times linked to millstone workings. This is the case of the sites of Canals (V-1), Montesa (V-2), Fuentealbilla (AB-1), Barranc de l'Escuera (A-1), Llerena (BA-2) and Hornachuelos (CO-13). The variant *Piedras Moleras* (literally "millstone rocks"), on the outskirts of Villanueva de Córdoba (CO-11), is highly evocative of these workings (Fig. 7.2). Although granite block working is known at the site, and Roman and Medieval granite querns and millstones are deposited in the local museum, the millstone workings have yet to be identified.

As Sanz Elorza has pointed out in his recent study of mill-related toponyms in the province of Burgos, *Molares*, *Molar* and *Muelas* (literally millstones) more often relate to natural topographical features, notably a butte or flat topped hill (Sanz Elorza 2012: 14). Therefore, it is not certain if the name *Moles* at Montesa (V-2), for example, refers to the butte or to its millstone production. Likewise, no millstone workings have been confirmed at three other locations with the *Molares* toponym, namely a town near Utrera in the Province of Seville, a valley near Rute, Córdoba and a mountain near Mazarrón, Murcia.

The case of *Mulera* by Alozaina, Málaga, is unique. Literally, the term appears related to "mules", notably the "mule keeper". However, in this case, it is likely that it is a variation of *Molares* due to the confirmed existence of millstone workings (Alozaina, MA-6).

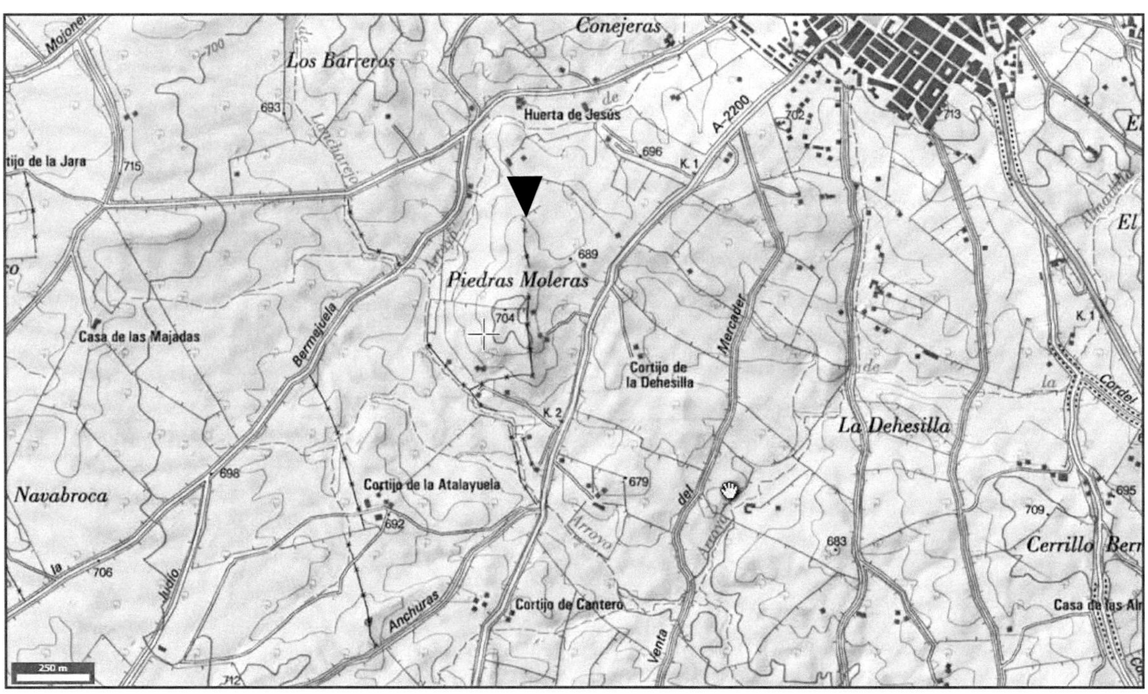

Fig. 7.2: The place name Piedras Moleras *outside of Villanueva de Córdoba is highly evocative of millstone workings and coincides with granite quarries. Millstone work, however, is yet to be confirmed (extract from SIGPAC).*

A number of other millstone exploitations, certified either by fieldwork or by texts, bear the name *Morales* (inverting the "R" and the "L") which means "mulberry", a bush or tree that is common to the south of Spain. This name is, for example, linked with workings on a hill above the town of Montarrón (GUA-8) and to a site near the city of Andújar (J-5). It also and forms part of the names of *Fuente de los Morales* by Alhama de Granada (GR-6), and *Vega de los Morales* in the Municipality of Priego de Córdoba (CO-4).

In the cases of Alhama and Priego, where *Morales* is linked respectively to the names fountain (*fuente*) and plain (*vega*), a botanical origin seems reasonable, in spite of the confirmed millstone workings. The inversion of the "R" and "L", however, is common in spoken Andalusian. It is therefore possible that certain *Morales* (mulberry) place names are actually *Molares* (millstone quarries). This is probably the case of Montarrón (GUA-8) where there is a clear inconsistency of transcription between the *SIGPAC* and cadastral *SEC* maps: the first records *Morales* and the second *Molares* (Fig. 7.3). This discrepancy can therefore also be envisioned for other cases of *Morales*.

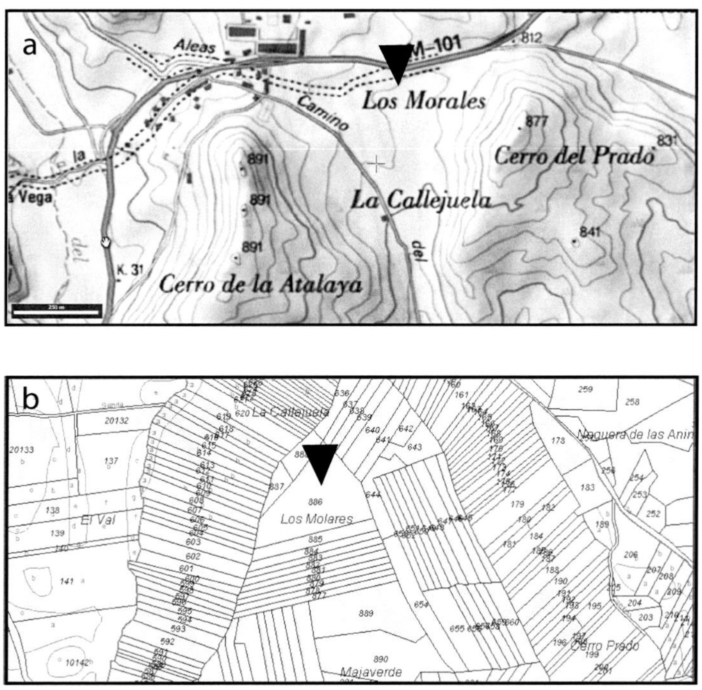

Fig. 7.3: The problem of interpreting the place name of Montarrón (GUA-8). There is a discrepancy between the first spelling "Los Morales" in the a) SIGPAC map, meaning "mulberry", and "Los Molares" in the b) cadastral map (SEC), meaning "millstone quarry".

A second *mola* related place name, *Las Amoladeras*, is associated with the millstone quarry of Alhama de Granada (GR-6). This name, not infrequent throughout the Spanish landscape, routinely denotes whetstone workings. I cannot explain its presence in Alhama de Granada because the rock, a fine white limestone, differs greatly from the abrasive fine sandstones that usually served to make sharpening stones.

In any case, as mentioned previously, it is plausible to imagine that *mola* place names disappeared during the many centuries of Islamic domination in the south of Spain. This would explain why these names are not as useful to identify millstone quarries as in France and northern Spain.

7.2. *Cantera* toponyms

The name *Canteras* derives from the Latin "corner" and is related to the word "quarry" coming from "squaring off". This obviously evokes rectangular building stone extraction (such as ashlars). The toponym is common in southern Spain and is associated with construction workings, whether it be solid rock, sand or gravel. In the online database of toponyms of Andalusia (http://www.ideandalucia.es/nomenclator/), for example, the word *cantera* or *canteras* scored 131 hits.

There are, nonetheless, a dozen cases in southern Spain where the name can be directly linked to millstone making: *Las Canteras*, Colmenar de Oreja (M-2); *Las Canteras*, Granátula de Calatrava (CR-5); *Cantera Honda*, Paterna (CO-8); *Las Canteras*, Moclín (GR-1a); *Rambla de las Canteras*, Caniles (GR-11); *Las Canteras*, Ugíjar (GR-14); *Cantera de las Pilas* (CA-9); *Las Canteras*, Castillo de Locubín (J-1); *Cantera de los Frailes*, Cabra (CO-1); *Cerro de la Cantera*, Huelma (J-3); *Las Canteras*, Guadalquitón (CA-13) and *Las Canteras*, Alhaurín el Grande (MA-3).

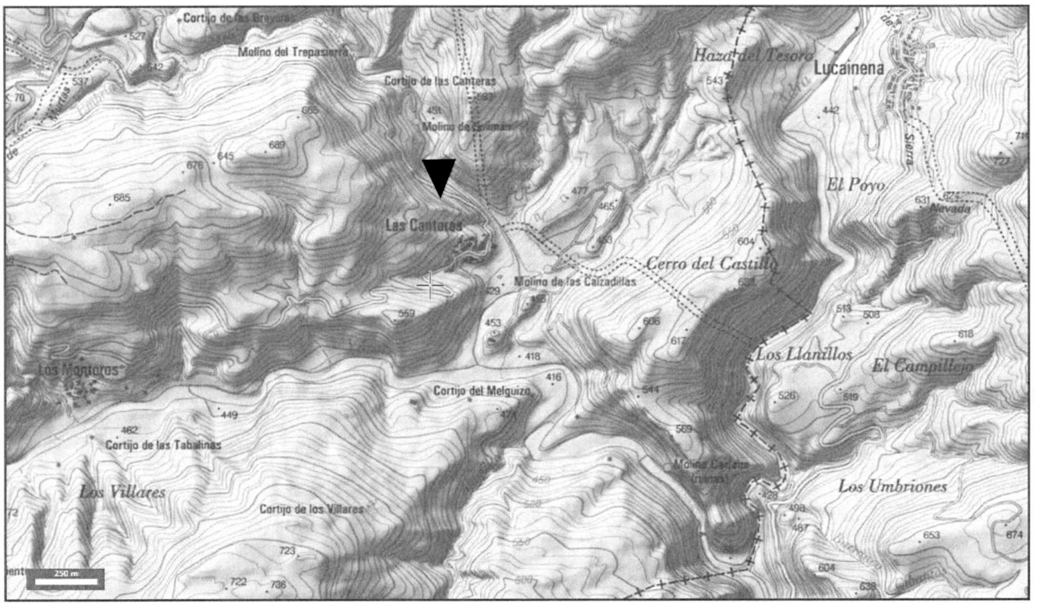

Fig. 7.4: The place name Las Canteras *near Ugíjar (GR-14) (Granada). This quarry produced both building blocks and millstones (extract from SIGPAC).*

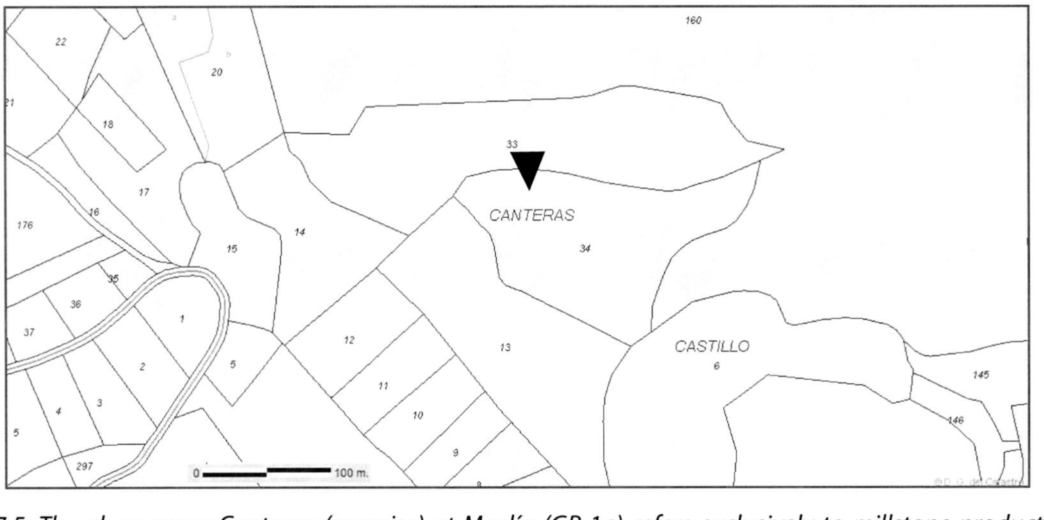

Fig. 7.5: The place name Canteras *(quarries) at Moclín (GR-1a) refers exclusively to millstone production. There is no evidence of extraction for building material (SEC).*

At some of these sites, such as Ugíjar (GR-14) (Fig. 7.4), there is evidence of both building blocks and millstone workings. Other sites, however, such as Moclín (GR-1a), Castillo de Locubín (J-1) and Cabra (CO-1) show no evidence of building material scoring, proving that *Cantera* is also a valid name to identify millstone production (Fig. 7.5).

Other place names similar to *cantera* could be millstone related or simply coincidental. *El Tajo* at Teba (MA-2), meaning both "quarry" and "steep cliff", could either signal the millstone quarry or the dominant topographical feature of the site. *Los Agujerones* (the "holes") near Moclín (GR-1a), could also be an allusion to the extraction hollows. *Los Cortaores* near Cabra (CO-2), meaning "the cutters", or *Barranco Baena or Pedrero* (AL-5) (*Pedrero* meaning "quarryman"), could also be linked to millstone making.

7.3. *Piedra, Berrocal* and other toponyms

Toponyms that are derivatives of *Piedra*, *Berrocal* and *Lanchar* (connoting slab formations), though not directly indicative of quarry work, are at times unquestionably associated with millstone workings.

Pedrera (from *piedra,* meaning rock) is synonymous with "quarry". In its plural form it is found outside of Alconera (BA-1) in Badajoz, and in the form of *Pedreros* (meaning quarrymen) at Alcolea (AL-5), Almería. Other *piedra* (rock) related names coinciding with millstone workings are *Barranco de las Piedras* at Vélez Benaudalla (GR-10), Granada, *El Pedroso* at Andújar (J-6), Jaén, *Las Pedrizas* (GR-1b), near Moclín, Granada and *Pedrizas de Piédrola* (CR-4), near Alcazar de San Juan, Ciudad Real.

The name *Berrocal* and its many derivatives is very common throughout certain regions of Spain. It is a very ancient name, presumably pre-Roman, and has been identified in a recent article at 212 different localities throughout the Central System mountains between Madrid and the Portuguese border (Llorente 2011). It habitually connotes granite outcrops, both in the form of loose rounded granite boulders, crags and extensive bedrock. In any case, the name coincides with several millstone workings in the granite fields of the Autonomy of Madrid, around Berruecos (M-1) and Colmenar Viejo (M-3), and in Extremadura at Jerez de los Caballeros (BA-5).

The name of the *Berrueco* exploitation near Medina Sidonia (CA-8) Cádiz, one of the largest in southern Spain, is assuredly a misnomer. The rock is not granite but a hard white limestone. The closest granite fields, in fact, are more than 100 kilometres away to the north. Since this name is not common in southern Spain, it is possible that it was imported when the Al Andalus countryside was repopulated by Castillians after the fall of Islamic rule.

Another rock related name is *Tobes*, the name given to a hamlet near Siennes (GUA-5) in the Province of Guadalajara. It evokes the *toba* (limestone tufa) that was reputedly exploited for millstones around the hamlet.

Similarly, the names *Lachar* and *Lanchar* related to the millstone workings of Jimena (J-2) in Jaén, Los Cortaores (CO-2) in Córdoba and Chapinería (M-4) in Madrid could relate to the exploitation of long slabs (*lanchas*) subsequently carved into millstones.

7.4. Toponyms potentially indicative of quarry infrastructure

A few millstone workings are associated with toponyms linked to elements of quarry infrastructure, such as roads and smithies. The direct association of these names with the millstone workings is probably incidental in these cases.

7.4.1. Roads

The *Camino de las Pedreras* (road of the quarries), for example, designates a road that connects the working(s) of the Albardado Stream (CO-10) with the town of Belmez (Córdoba) and certainly served to transport millstones. Similarly, the *Camino de la Cantera* designates a road leading north-west from the Colmenar de Oreja (M-2) to the town of Chinchón. This name probably derives from the transport of construction blocks from Colmenar to the monumental constructions undertaken by the royalty in the cities of Madrid and Aranjuez. The millstone production at Colmenar de Oreja (M-2) would have nonetheless benefited from this thoroughfare to transport its products to the *tahonas* of Madrid.

7.4.2. Smithies

The names *Las Herrerías* (the smithies) of Bolaños de Calatrava, (CR-2) and *Las Fraguas* (the smithies) at Montarrón (GUA-8) suggest the presence of old smithy workshops, a craft that goes hand-in-hand with rock work due to the necessity of repair and maintenance of the stone cutter's iron tools. For these names to have survived over time, the workings must have been important. It is therefore likely that *Las Fraguas* (GUA-8) was at one point a large workshop that did not depend exclusively, if at all, on the millstone workings.

Las Herrerías (CR-2) by Bolaños probably bears no relation with the old Roman millstone workings. This place name, probably traced back to Late Medieval or Modern times after the Islamic domination, could in fact be related either to a later phase of stone work at the site or, from its position along a traditional thoroughfare passing through Bolaños de Calatrava, could evoke a waystation providing the services of blacksmiths for animal drawn carriages and carts or farriers for equines (Casillas Antúnez 2010: 278).

7.5. Toponyms deriving from Arabic

The Islamic domination left a strong stamp on toponyms in Spain as seen, for example, by many geographical names such as Andalusia (*Al-Andalus*). In the tradition of milling in Spain, Arabic has also contributed the names *aceña* for watermill, *almazara* for oil mill, *tahona* for animal-driven mill (or bakery). Nonetheless, no Arabic toponym can be positively associated with millstone workings. This is certainly due to the lack of research in this field.

The closest mill related Arabic toponym coinciding with millstone workings is *Sierra Altahona* (or *Altaona*) (Porches, MU-3, Murcia). It is more likely, however, that the place name, from *"al-Tahuna"* meaning mill (Grandal 2004: 252-253), is linked to an old mill in the vicinity and not to the millstone workings.

8. MILLSTONE QUARRY INFRASTRUCTURE

Research in southern Spain regarding millstone quarry infrastructure is nonexistent. Furthermore, old written sources, although at times informative about commercial transactions, never delve into how millstone makers maintained and repaired their tools or how millstone crews were lodged. They never enter into the question of how the tonnes of stone debris were evacuated from the extraction zones, or how the heavy millstones were hauled through difficult terrain either to the mill or a staging point where they were loaded onto carts or boats for their final transport. Furthermore, as I have noted in other chapters, no site in the study area has benefited from an archaeological excavation that could cast light on features of millstone quarry infrastructure. For these aspects I have had to resort to the results of excavations elsewhere in Europe, notably in Switzerland and in France.

8.1. Tool repair and maintenance

The iron tools of millstone makers (in particular the points of picks and chisels) required constant maintenance and repair, a task that had to be undertaken by individuals possessing the skills of a blacksmith. Archaeological excavations uncovering millstone workings are sparse, and those with smithies are even more rare. Although there is no concrete evidence linking smithies and millstone productions in southern Spain, it is logical to assume that smithies were at times located directly or near certain millstone workings. There are also indications that certain millstone makers resorted to independent blacksmiths, whereas others possessed the skills and infrastructure to maintain their tools themselves.

The most noteworthy cases of smithies related directly to millstone quarries are at the Roman complex of Châbles-Les Saux in western Switzerland (Anderson *et al.* 2003, chap. 6), at Gardom's Edge, Baslow, Derbyshire in England (Radley 1964) and at Vioménil (Vosges) in France. The smithy workshop at this last site, dated to the 17th century, was under a lean-to attached to the quarry face by mortises cut into the rock (see Atlas of European Millstone quarries http://meuliere.ish-lyon.cnrs.fr/php/results2.php). The recent excavation at the subterranean workings at the Grand'Gueule (Haute-Savoie) in France (under the direction of Alain Belmont) has also brought to light evidence of a smithy in the form of hundreds of small smithy slags and hammer scales.

Two quarries in southern Spain are potentially associated with smithy place names. The first is *Las Fraguas* (the smithies), beside the *Molares* place name at Montarrón, Guadalajara (GUA-8). The second is *Las Herrerías* (also meaning the smithies), the name given to a present-day quarry at Bolaños de Calatrava (CR-2) with evidence of older Roman millstone workings.

At Montarrón (GUA-8), a direct relationship between the Medieval or Contemporary millstone extractions and the smithy work, is perfectly conceivable since *Molares* (or *Morales)*, suggesting millstone work, and *Fragua*, indicative of iron work, appear side by side in the cadastre. But the smithy here, although probably in part serving the millstone quarry, must have been independent since the millstone workings would not have provided sufficient work to keep it active on a permanent basis.

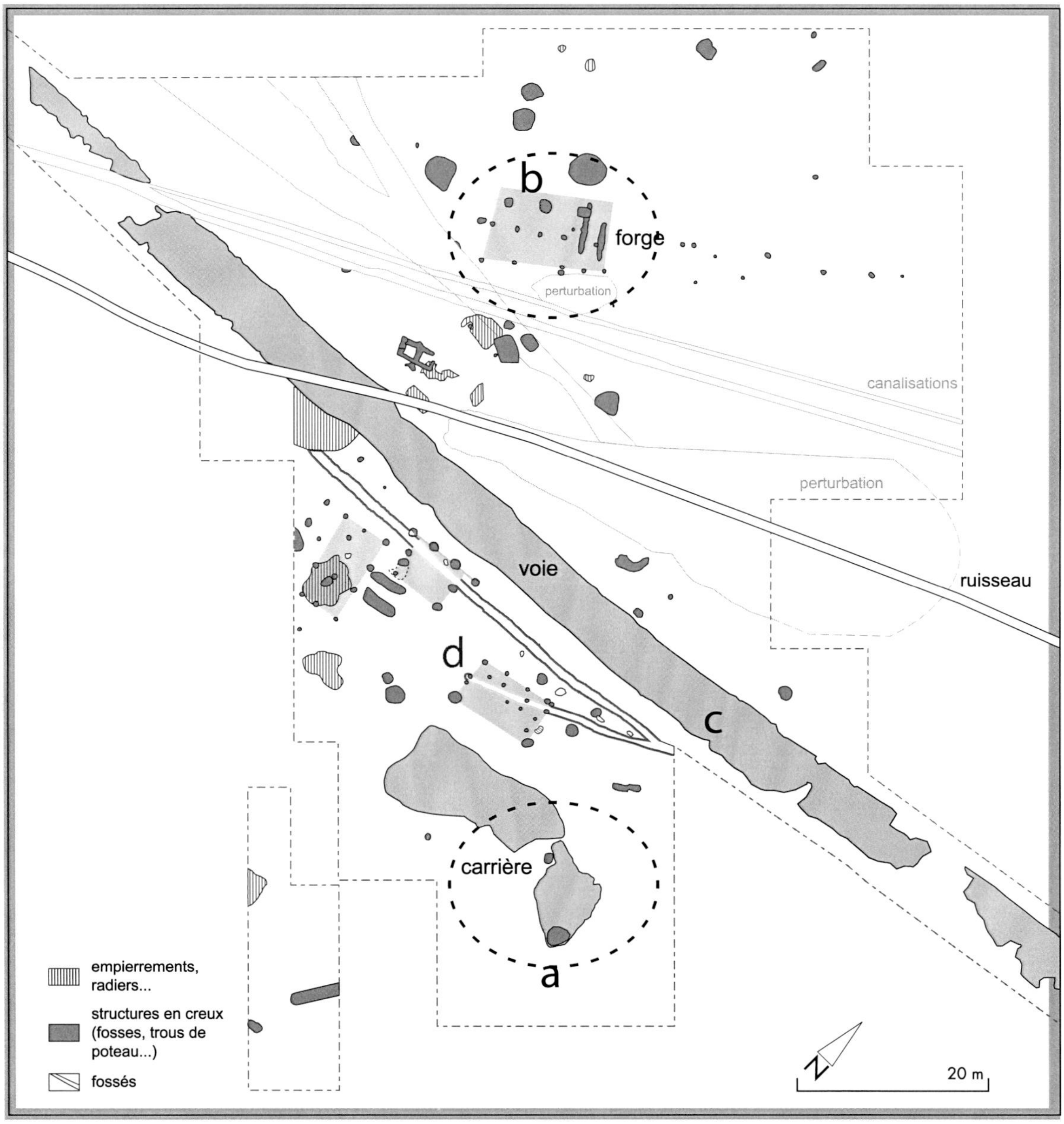

Fig. 8.1: Plan of the Gallo-Roman site of Châbles-Les Saux dating to the 1st-2nd century AD (Canton of Fribourg, Switzerland). The study of this excavation revealed a tight link between the millstone quarry a) and the smithy b) (adapted from Anderson et al. *2003: 34, fig. 23). The other noteworthy features are the road c) and the house d) interpreted as the residence of the artisan.*

In the case of the presumed smithy (or smithies) of *Las Herrerías* (CR-2), a direct relationship between the two crafts is unlikely because the name *Herrerías* probably has its origins in Late Medieval or Modern times, after the Islamic fall, whereas the millstone workings are Roman. It is more plausible that the hypothetical *Herrerías* smithy was either linked to iron workings that served a later, post-Roman, phase of stone extraction (probably for construction), or to a waystation where farriers could shoe horses and blacksmiths repair carriages and carts.

The excavation of Châbles-Les Saux in the Canton of Fribourg, Switzerland, brought to light a Roman smithy about 80 metres from a small Roman quern quarry (Fig. 8.1). Between the quarry and the road passed a major Roman road six metres wide. The detailed study of this site reveals a direct chronological relation between the road and the quern and iron workings (Anderson *et al.* 2003). However, as seen by the quantitative study of the iron artefacts, as well as the nature of the products (some unrelated to stone work), the maintenance of the quarry tools did not generate enough work to sustain the smithy. The smithy, in fact, besides serving the quarry, attended to a local and regional clientele that gained access to the site by means of the road.

The more recent excavation in France at a subterranean millstone quarry at the Grand'Gueule (Mont Vouan), under the direction of Alain Belmont, uncovered a large quantity of smithy slag and hammer scales associated with a black, carbon-rich, layer of sediment at the rear of the cavern (Belmont & Anderson 2010: 104-108). The quarry, exploiting a specific layer of rock on a steep slope halfway up the mountain, was accessible only by a narrow winding trail. Radiocarbon dating places the smithy work in the 15th-16th centuries. The remote location of the site, about a half-hour hike uphill from the nearest town, suggests that this smithy operated exclusively for the millstone workings (Belmont & Anderson 2010: 104-108). This assessment seems compatible with a constant, every day need of repair and maintenance of the tools of the massive workings.

In northern Spain, at Barruleo de Santullán (Palencia), a Municipal Ordinance dating to 1571 provides an interesting perspective as to the connection between smithies and millstone quarries. According to chapter 10 of the Ordinance, blacksmiths were obliged, under threat of a heavy fine, to drop all activity at the request of a millstone maker to repair his pick so that he could resume his work as soon as possible (Maestro 2011: 17-18). This record reveals not only the importance bestowed upon and the privileges granted to millstone makers, but also that the smithy was, at least in this case, independent of the quarry (probably located in town) and subordinate to the millstone workers.

Millstone research in the Palencia mountains by Maestro Hernández reveals that in the first half of the 20th century, millstone cutters remained stationed at the mountain quarries for up to four days at a time and only descended to town when they needed to repair their tools (Maestro 2011: 38). The author goes on to state that some of them possessed their own smithies (equipped first with traditional bellows and later electric blowers) so as to be self-sufficient (Maestro 2011: 38, 43).

It is therefore possible to conclude that larger millstone productions in southern Spain probably benefited from their own smithies at the site itself. Quarrymen at smaller productions, by contrast, appear to have either relied on independent blacksmiths or acquired iron working skills themselves for their tool maintenance.

8.2. Debris management

The quantity of working debris and where it was deposited has a direct correlation with the morphology of a quarry. True extractive sites produced more primary extractive debris at the place of extraction and less secondary fashioning debris. Sites where angular blocks were detached with levers, by contrast, produced little primary debris, but more secondary debris where the angular block was carved into a cylinder. When the fashioning of blocks took place away from the quarry face, which seems to be the case, then the debris was not an immediate hindrance to further extractive work. In true extractive quarries, by contrast, the primary debris had to be removed immediately to be able to continue extractive work.

Pit, bench, pocket and subterranean quarry debris therefore hindered work and thus required constant removal. Edge, extensive contiguous shallow and extensive dispersed quarries required little debris management because it never attained a volume large enough to hinder work progression. In the case of coastal sites such as Trafalgar (CA-1), Chipiona (CA-6), Rota (CA-3) and Carchuna (GR-9), no spoil heaps are visible and the debris was washed away. The same has been observed at the coastal quarries of the Island of Menorca (J. Sánchez pers. comm.).

In certain cases, debris was put to use to backfill retaining walls creating ramps that served to enter and exit the quarry. At the Roman site of Châbles in Switzerland, the larger aborted querns served to repair the foundation of the adjacent road, whereas the finer gravel debris served to pave the road's surface (Anderson *et al.* 2003: 181-182).

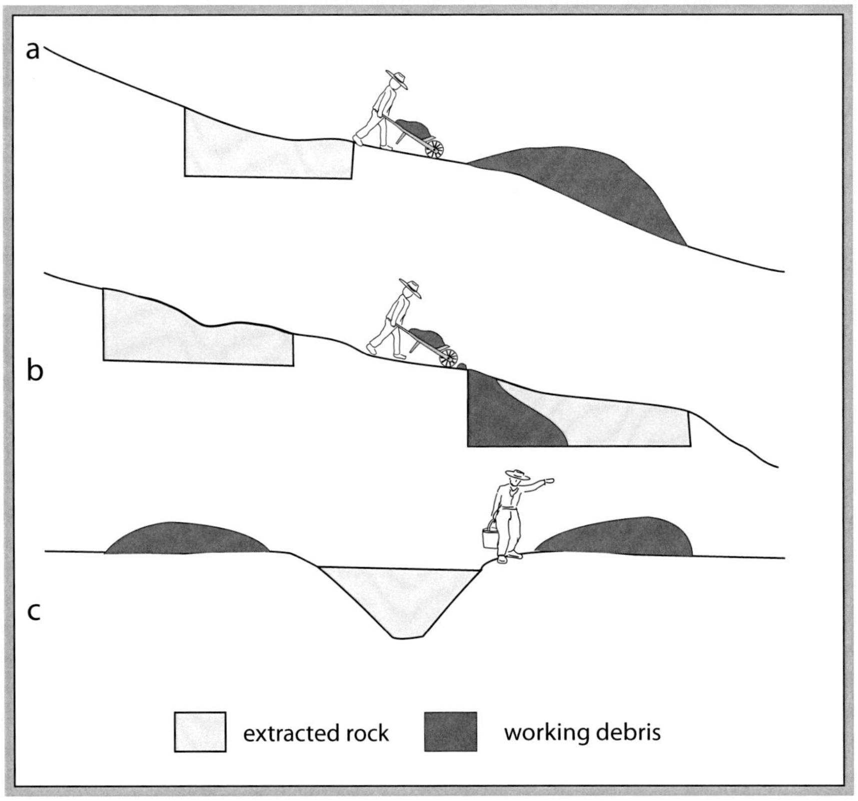

Fig. 8.2: Schematic representation of sections of quarries and the systems of deposit of working debris. a) downslope heap, b) debris backfilled in finished extraction area and c) debris placed in long cordons along the edge of a trench quarry (drawing by T. Anderson).

In other contexts, by contrast, millstone makers invested a larger amount of energy and time into managing debris. Certain unusable superficial layers of overburden required removal before initiating proper extractive activity, an activity that generated at times a large amount of spoil. The nature of this debris differed from that of extraction or fashioning in that it could include humus and/or organic topsoil giving it a darker colour. It is therefore plausible that in a future (hypothetical) excavation of one of these sites, archaeologists will observe an "inverted" stratigraphical sequence, that is, layers of dark humus and topsoil under layers of stone carving debris.

Extraction debris could also include large rock fragments mixed with the finer material, resulting from the elimination of unsuitable rock sectors to make way for subsequent work. In the debris there are also, at times, discarded cylinders. These are "treasures" for archaeologists, in particular at sites with many phases of extraction, because they allow a correlation of specific stratigraphical layers with specific extraction zones.

8.2.1. Downslope debris heaps

Millstone makers, so as to avoid moving the spoil a second time, had to assure that debris was placed in sectors that would not be subsequently worked. For this reason, spoil heaps were situated beyond the working areas, most often downhill so as to profit from gravity during transport (Fig. 8.2).

At Ibi (A-2), Alicante, the debris from quern extraction along the steep slope required little or no management. It was simply blown away by the wind or washed downhill by rain. At Moclin (GR-1a), debris attained a huge volume and was carted away and deposited downhill where it can still be seen today (Fig. 8.3). The presence of half-fashioned roughouts at Moclín indicates that the spoil originated both from extraction and fashioning.

In the case of the edge quarry of Alhama de Granada (GR-6), the debris, made up of a number of interlocking mounds, formed a wide continuous cordon more than 100 metres long (Fig. 8.4). This spoil originated from the fashioning workshops set away from the quarry face.

The oval and elongated heaps between 10 and 20 metres long at the Sierra del Hacho near Estepa (SE-3) are each located downslope and were probably formed by debris transported in baskets or in wheelbarrows (Fig. 8.5).

Fig. 8.3: General view and detail of the downslope working debris generated by both extraction and fashioning beside the deep extractive area of the Moclín quarry (GR-11).

Fig. 8.4: General view of the extended face, block detachment, quarry of Alhama de Granada (GR-6). The fashioning of millstones from angular blocks detached from the quarry face produced a large amount of debris that was dumped downslope, resulting in one long cordon of stone debris.

Fig. 8.5: Oval shaped spoil heap downslope from a pocket quarry at the Sierra del Hacho, Estepa (SE-3).

8.2.2. Backfilling finished sectors

A second option for storing working debris was to backfill finished sectors of the quarry (*cf.* Fig. 8.2b). This technique, characteristic of large extractive quarries, was observed recently at the excavations undertaken by Alain Belmont at Claix in Charente and Mont Vouan in the Haute-Savoie. In the southern Spain area, however, this type of debris management has only been observed in the larger, more structured, sites of Moclín (GR-1a) and Cantera de los Frailes near Cabra (CO-1).

It must be noted that debris backfilled in finished quarry sectors is an optimal type of feature to explore in modern archaeological excavations because it offers the possibility to analyse, through vertical stratigraphy, the chronological sequence of a backfill. At Claix in Charente, for example, pottery and discarded quern finds in the backfill, helped to establish the relative chronology between two separate working sectors (Belmont *et al.* 2011: 207). Furthermore, as seen systematically in the excavations in France, backfilled debris protects the quarry floors and quarry faces from weathering, providing the archaeologist with a rare opportunity to observe and interpret well-conserved tool marks.

8.2.3. Lateral cordons (trench quarry debris)

Debris from trench quarries (*cf.* Fig. 8.2c) was stored in heaps along the edge of the trench that eventually formed long parallel cordons (Fig. 8.6), as seen at the sites of Cantera de la Rambla, Caniles (GR-11) and El Lachar near Jimena (J-2) (Fig. 8.7). To avoid having to move the debris a second time, subsequent parallel trenches were opened a reasonable distance from the original trench, beyond the debris cordons.

Fig. 8.6: Long cordons of fashioning debris parallel to the trench quarry of the Caniles (GR-11).

Fig. 8.7: Cordons of working debris along the edges of the trench quarry of El Lachar, Jimena (J-2).

8.2.4. Retaining walls

In one of the deeper sectors of a trench of the site of Lachar, Jimena (J-2) are two levels of drystone retaining walls backfilled with working spoil (Fig. 8.8). This type of feature, known for example in the basalt quarries of the Eifel (Harms & Mangartz 2002: 78, 85) and at Claix (Charente) in south-western France (Belmont 2011: 9), is unique, at least for the moment, in southern Spain. As is the case in France, these backfilled walls may also have served as paths for the workmen to enter and exit the trench, as well as a "slipways" to facilitate hauling the heavy cylinders out of the trench.

Fig. 8.8: Views of the retaining walls at the trench quarry of Lachar near Jimena (J-2). The lower drystone walls not only appear to have retained working debris, but served as paths to enter and exit the trench. They also served as "slipways" to drag millstones from the base of the trench.

8.3. Residences and shelters

8.3.1. Long-term residences

The residences or shelters of millstone working crews depended on whether the exploitation was permanent or seasonal. It is assumed that workers at sites such as Moclín (GR-1a), adjacent to town, could walk back and forth from their lodgings every day. But there were also sites too far from towns that required lodging crews for extended periods of time.

The only feature brought to light on an archaeological dig that is interpreted as the residence of a millstone maker is a rectangular post-construction beside the quern quarry at the Roman site of Châbles-Les Saux in Switzerland (Anderson *et al.* 2003: chap. 8) (Fig. 8.9). This modest building (45 m2), with roots in the regional Celtic architecture, is similar to the rudimentary cabins known on contemporary regional agricultural domains in Roman Switzerland (Anderson *et al.* 2003: 218). At this site, the question remains as to whether the craftsmen resided in this cabin on a permanent or seasonal basis.

One of the rare glimpses into the question of lodgings for millstone makers in southern Spain relates to the quarry of Berrueco in Cádiz (CA-8). A first written reference, dating to the 18th century, is by the priest of Medina Sidonia who records that the Berrueco millstone workers were residents of Medina Sidonia (Martínez y Delgado 1875: 129). It is therefore likely that at least some of these workers resided at the site, at least on a temporary basis, so as to avoid a daily 10 kilometre hike.

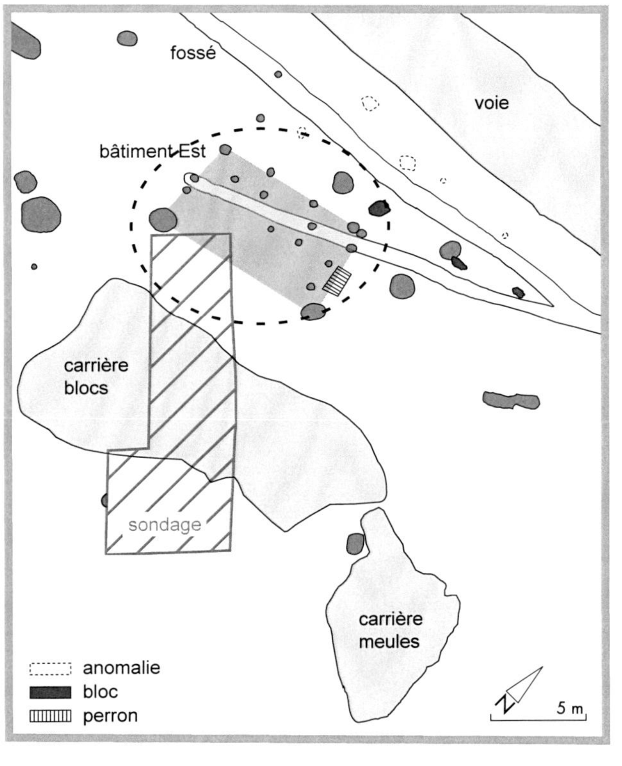

*Fig. 8.9: Plan of part of the Gallo-Roman site of Châbles-Les Saux dating to the 1st-2nd century AD (Canton of Fribourg, Switzerland). The rectangular house (*bâtiment est*), built with wood, was between the road and the quarry (*carrière*). This modest structure is interpreted as the residence of the quarryman (adapted from Anderson* et al. *2003: 196, fig. 241).*

A second author, Cruz y Bahamonde, in the first decades of the 19th century, provides more insight into the question of the Berrueco dwellings when he states that "... on top of the Berrueco mountain there are several buildings that house about 50 millstone makers" (Cruz y Bahamonde 1813: 91, note 1). With this number of workers lodged at the site it is obvious that production was at its height. Although there is no mention in the text, it is conceivable that the site also had a canteen to feed the work crews.

A last reference is that of Madoz who, thirty years later, noted that twenty-four men resided at Berrueco and worked in five different workshops (Madoz 1846, Vol. 4: 290). The extensive recent work at the site (blocks for construction of the port of Rota) has probably obliterated all traces of these and earlier lodgings.

In any case, it appears evident that millstone making crews at large workings removed from towns had to be housed at least on a temporary or seasonal basis.

8.3.2. Short-term shelters

Millstone makers also built simple huts or hovels as shelters for the short stints at a site. Crude drystone shelters are found around the sites of El Lachar (J-2), El Torcal (MA-1) and Ubrique (CA-12). At Lachar several of these features, called *caracoles* (meaning snail shells), are erected against the edge of the quarry trench and surrounded by abandoned millstones, indicating they were contemporary (Fig. 8.10). These hovels are very small and appear to have served as little more than a shelter to pass the night or provide a reprieve from the hot sun. Two similar drystone shelters were found during the excavation of the millstone quarry at Vic-le-Comte, Puy-de-Dôme in France (Cabezuelo 2006: 110). Although they are dark, with no windows, and ideal for iron work, I did not observe any evidence (*i.e.* slag) that they could have served as smithies.

At Peña Harpada (CA-10) the quarry shares the summit of a pointed mountain with what is interpreted as medieval defensive constructions (Giles *et al.* 2011: 128-130) that could have served as a refuge for the millstone makers. The dwellings at El Torcal (MA-1) are slightly more complex (Fig. 8.11). Beside a very crude shelter with a very small door and a number of broken and unfinished millstones in the surroundings, there is a drystone enclosure measuring about 100 m2 suggesting the presence of animals. Although these shelters may be simply related to pastoral activity and have nothing to do with the quarry, they could be interpreted as a seasonal residence for the quarrymen accompanied by their animals, possibly including oxen used for millstone transport.

Fig. 8.10: Views of two different drystone dwellings at the trench quarry of El Lachar (JA-2). In the local vernacular, they are called "caracolas"*, evoking the snail shells they resemble.*

Fig. 8.11: Views of a dwelling at the quarry site of El Torcal (MA-1). The feature, strewn with finished and unfinished millstones, comprises a crude drystone hovel and a drystone enclosure: a) general view; b) detail of the hovel; c-d) views of the enclosure wall.

8.4. Transporting millstones

8.4.1. Generalities

Querns and millstones were transported by different means depending on their weight, the type of terrain and the distance to the final destination. For both short- and long-distance transport, the proximity of a road to the quarry was essential. The quern exploitation of Châbles in Switzerland, for example, actually owed its existence to the major public road a few steps from the quarry (Anderson *et al.* 2003: 185). The millstone outcrop of Villanueva de San Juan in Seville (SE- 6) illustrates the necessity of a road near millstone workings as in the words of Madoz, the site "...was not exploited for lack of a road" (Madoz 1850, Vol. 16: 207).

Small rotary querns, weighing less than 50 kg, could be lifted by one or two of the crew and placed in the bed of an animal-driven cart. Lifting larger models, weighing up to a tonne, was another story. These massive cylinders had to be guided and slid by levers along wooden tracks or stone debris slipways out of the extractive area to be loaded into waiting carts. At certain more recent and sophisticated sites, millstones were hoisted out of the quarry by cranes or capstans, features still visible at the volcanic quarries of Mayen in Germany (Harms & Mangartz 2002: 88-93) and Selbu in central Norway (Grenne *et al.* 2008: 61, fig. 25). In southern Spain, however, there is no evidence of any of these types of features.

In the study area, most sites were near major or secondary thoroughfares. These roads were often *camino reales*, maintained by the authorities, often borrowing trade routes dating from Antiquity or thoroughfares for seasonal migration of livestock.

Some smaller roads leading to millstone exploitations have left their stamp on the toponyms. As noted in a previous chapter, the site of the Albardado, for example, is accessed from the town of Belmez by a road called the *Camino de la Pedrera* (CO-10) (road of the quarry).

The possibility of damage to the millstone during transport was a constant source of worry for the millstone maker, his client and the transporter. A tale is recorded that a miller in northern Spain offered the local church a property in exchange for divine protection during the transport of his newly acquired millstone. Apparently the millstone arrived in perfect condition because the property is still today in possession of the diocese (Pascual *et al.* 2011: 248). However, uneventful transport was not always the case. Millstone accidents could explain the abandoned cylinders along the roads leading from the quarries of Alhama de Granada (GR-6) and Vélez de Benaudalla (GR-10) (Fig. 8.12).

Fig. 8.12: Examples of millstones abandoned along access roads to the quarries of a) Fuente de los Morales, near Alhama de Granada (GR-6), and b) Barranco de las Piedras, near Vélez de Benaudalla (GR-10).

8.4.2. Short-distance transport

For land transport of millstones, the most usual means was the dray or cart with a flatbed and spoked wheels. These were probably similar to the carts that served to transport building blocks illustrated in the 16th century by (pseudo) Juanelo Turriano (Fig. 8.13). From a photograph of an early 20th century block quarry cart from Extremadura (Fig. 8.14) it appears that this means of transport did not change much over time.

In the Palencia mountains as late as the 1960s, it is recorded that carts to transport millstones were well adapted for the rugged terrain and equipped with massive wooden wheels and wooden axes (Maestro 2011: 45). The author adds that to facilitate loading millstones onto the bed of the cart in inclined terrains, the cart was parked downhill with its wheels in deep holes so as to lower the level of its bed (Maestro 2011: 45).

Although mules are at times represented in old Spanish illustrations and photographs of quarry carts, most were certainly drawn by oxen, animals that were more sturdy and less excitable. In the 1502 Loja (Granada) Municipal Ordinance (see GR-2), for example, the penalty for removing millstones without the consent of the authorities was the confiscation of the oxen used to draw the millstones. In the more recent productions in the Palencia mountains, the millstone

Fig. 8.13: Drawing from the pseudo Juanelo Turriano codex (c.1595) of a quarry cart driven by mules (Archimedes Project, http://dmd.mpiwg-berlin.mpg.de/author/dmd/database/).

Fig. 8.14: Modern flatbed quarry cart from Extremadura (from Martín Galindo 2006: 849).

cutters, who also raised livestock, often used their own oxen to both drag the millstones down the mountain, as well as to pull carts (Maestro 2011: 46). An oxen-drawn millstone cart from northern Spain has recently been reconstructed in an illustration by Javier Castro (Fig. 8.15).

A second mode of millstone transport was that of wooden sledges such as a "Y" shaped sledge called *corzas* or *basna*. The millstone was attached with chains on the fork and then harnessed to oxen by means of a second chain attached to the base (Maestro 2011: 48).

Another type of simple sledge known elsewhere in Spain is the "V" shaped *rastra* (from the word meaning to "drag") (Fig. 8.16). The piercing of the eye of millstones, often carried out at Spanish quarry sites as seen by numerous abandoned examples, probably facilitated attaching the mill with ropes or chains to these types of sledges.

Millstones could be moved from their quarries by other means. A simple method identified in the Palencia mountains to haul them through difficult terrain was to "wrap" them with long branches of broom. This shrub, abundant in the area, protected the stone while it was dragged by oxen, presumably by means of the eye, to the point where it could be loaded onto a cart (Maestro 2011: 46).

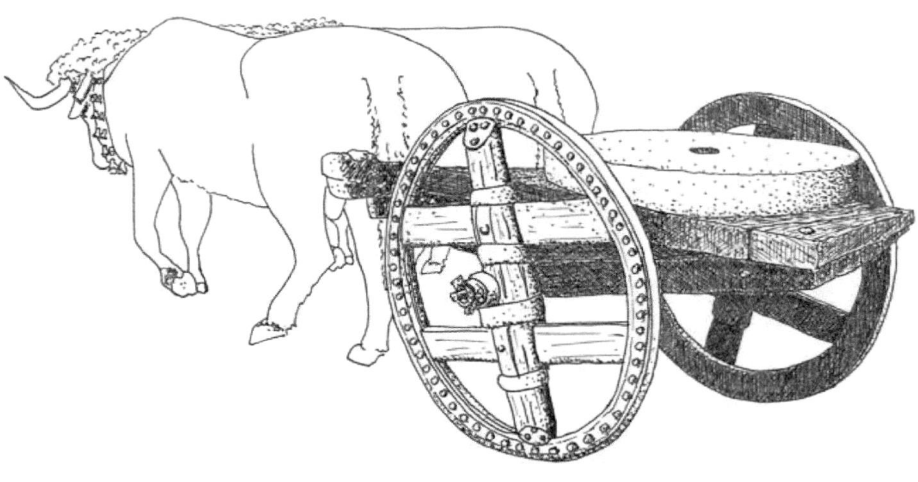

Fig. 8.15: Reconstruction of an oxen-drawn flatbed cart for the transport of millstones and other heavy products based on information from the region of Navarra (from Pascual et al. 2011: 250, fig. 25, drawing by Javier Castro).

Fig. 8.16: Examples of recent "V" shaped rastra *sledges designed to transport heavy loads (photograph to the left is from http://cuentoquenoescuento.blogspot.com.es/2012_02_01_archive.html; the photograph to the right is from http://espanolinternacional.blogspot.com.es/2008/03/el-transporte.html).*

Another means of millstone transport described by Maestro (2011: 47) consisted of staking the millstone's location with long poles and waiting until the arrival of snow. Then they were dug out of the snow and slid along the surface to town on crude snow sledges. This method, recalling that of the schist quarries of Selbu in Norway (Grenne *et al.* 2008), was certainly not applicable to millstone workings in the milder climates of southern Spain.

According to a recent quarry study in the northern autonomy of Navarra, transport of millstones in difficult terrain was carried out with *narrias*, a simple sledge with parallel blades joined by a platform of crossbeams and towed by oxen (Fig. 8.17) (Pascual *et al.* 2011: 249, fig. 23). The authors cite a notarial archive dating to 1419 recording that María Ruiz de Arregui sold some properties to a monastery but retained the right for herself and her successors to not only extract millstones, but to cut the wood necessary to transport the millstones. Two centuries later, in 1634, a document records that no wood could be cut unless needed for the transport of millstones (Pascual Mayoral *et al.* 2011: 249). Both of these documents imply, presumably, that the sledges to transport the millstones were either improvised at the extraction site or there was a need for wood to act as rails to facilitate the sliding of the sledges.

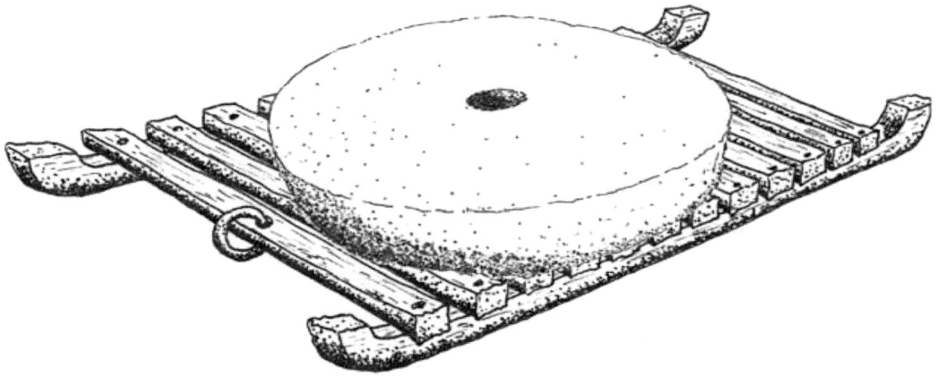

Fig. 8.17: Example of a narria *(from Pascual Mayoral et al. 2011: 249, fig. 23; drawing by Javier Castro).*

In southern Spain, there are five independent accounts that describe yet another singular means of moving millstones. This system could be labelled the "rolling stone" technique. It consisted of placing the millstone upright, slipping a beam through its eye like an axle and, while two men stabilised the ends of the axle, a third pushed it.

The "rolling stone" technique, obviously not applicable in inclined terrain, was described to me orally by "Niño" Ernesto regarding millstone transport at Moclín (GR-1a), Granada. It is also recorded in two different studies of watermills of the Alpujarra Mountains of Almería and Granada (Cara Barrionuevo *et al.* 1999: 151; Rodríguez Monteoliva 1989: 705-706). Another recording appears on the Internet regarding the quarry of Esparragal (CA-12) outside the town of Ubrique (Cádiz) (http://manuelcabelloyesperanzaizquierdo.blogspot.com.es/2010/10/como-se-puede-trasladar-una-piedra-de.html). Finally, a variant of this method is described and illustrated by Gómez Ruíz (2003: 85, fig. 7) in his study of watermills along the Odiel River of Huelva. In this last case, instead of being pushed by men, the millstone was pulled with ropes by a mule guided by a man (Fig. 8.18). This crude and seemingly risky technique, requiring that the eye be pierced, was certainly limited to very short distances and to situations where a cart was not available (Rodríguez Monteoliva 1989: 705-706).

Fig. 8.18: Reconstruction of a scene of moving a millstone by means of the "rolling stone" technique (from Gómez Ruíz 2003: 85, fig. 7).

Rutted roads are another transport feature associated with millstone quarries. This type of road consists of two parallel and equidistant tracks or grooves cut directly into bedrock that served to guide loaded carts or sledges. These features, used in tandem with ropes attached to winches or posts along steep sectors of the road facilitated the climb and descent of heavy loads. An example directly at a quarry is recorded at Claix near Angoulême (Charente) in France (Belmont *et al.* 2011: 219-220, fig. 24). In north-eastern Spain, a Roman or Medieval rutted road, identified over a distance of two kilometres, is directly related to the quarry of Cerro Redondo, Zaragoza where cylinders measuring 60 to 80 cm in diameter (possibly millstones) were scored (Cisneros *et al.* 1985: 158-159, figs. 13-14).

In southern Spain, there are no rutted roads that can be associated irrefutably with millstone workings. One such track, nonetheless, passes very near the northern entrance of the mountain pass beside the millstone workings at the Puerto de la Cadena (MU-1) (Fig. 8.19).

Fig. 8.19: Example of a rutted road cut through the slope of a conglomerate embankment. This road is beside the quern and millstone quarry of the Puerto de la Cadena (MU-1) (from http://senderosdecartagena.wordpress.com/2010/12/16/cabezo-del-puerto-de-la-cadena/).

8.4.3. Long-distance transport

8.4.3.1. Inland cart transport

Animal drawn carts were the most common means of long-distance transport since most millstones produced in Spain were destined to inland markets and their sources were not necessarily near major watercourses. Since Antiquity, the inland transport of querns and millstones certainly benefited from a wide network of trade routes whose quality is known to have fluctuated over time.

Several old written sources relate the overland transport of millstones. An example is the tale of Santiago García, who in 1766 ferried a millstone about 70 kilometres from Trébago in the Sierra del Madero (Soría) to the town of Mendavia in the south of Navarra. He was obliged to pass mountain passes of the Iberian System mountain range (up to 1000 m in height) before crossing the wide Ebro River (Pascual & García 2011: 291-292).

Inland transport was undertaken at times by the customer who retrieved the millstone himself, either with his own or a hired vehicle. At other times, it was the millstone maker who organised the stone's transport. The son of Herminio Miguel, the central character in the photograph of the *moleros* of Brañosera (*cf.* Chap. 1, Fig. 1.6), recalls that his father transported millstones to Oña in the Province of Burgos, about 80 kilometres to the south-south-east, and to Puente Almuhey in the Province of León, about 60 kilometres to the west. Sauro Miguel, to the right of the photograph, transported a millstone to Villarmoronta near Carrión de los Condes, about 60 kilometres to the south (Maestro 2011: 47). Although these may not appear to be long distances "as the crow flies", they were rough trips lasting several days through circuitous mountainous terrain.

Part of inland millstone transport was also carried out by a long-established network of professional *arrieros* or *carreteros*, transportation "guilds" that ferried a number of different products throughout the Spanish countryside by animal (mostly mule) or by cart (Espejo Lara 1985; Diago Hernando 2004: 219). A protocol in the Provincial Archives of Seville, for example, details the sale of a millstone in 1510 from the *carretero* Juan Gallego of Catedral to Alonso dé Castañeda of Huévar (Otte 1996: 58, note 24). The cart driver, in this case, seems to have also taken on the role of a millstone merchant.

Pascual Mayoral and her team, from interviews with the *arriero* Amador Gastón from Mués (Navarra), record that he hauled millstones on *narrias* (flatbed cart, from Basque) drawn by four oxen (Pascual & García 2011: 236).

A series of documents identified by Sainz Albero (2007: 176-178) in the archives of Mendieva, Navarra, indicate the distance and, on occasions, the time of inland transport from different millstone workings to the city of Mendieva. The first, dating to 1713, described the costs of acquisition and transport of a lower stone. This transport included a payment to the Mayor of Arubal (La Rioja), payment of a toll, and payment to the ferryman to cross the Ebro River. The last leg, for the people that fetched the stone, was paid with wine. A second document from 1789, notes that Antonio Elvira was paid to drive a lower stone from Arbeiza to Mendieva, a distance of about 30 kilometres.

Other documents record that Benito Balerio, in 1800, was hired to drive a stone to Logroño (La Rioja), a two-day trip corresponding to about 25 kilometres and, Fausto Durana and Isidro Robres, in 1801, were paid for a stone brought from Treviño (Burgos), about 50 kilometres away. In 1804, Santos Elvira and Benito Balerio needed three days and three sets of *caballerías* (probably mules), as well as two different carts, to convey a millstone from beyond Logrono (La Rioja), a distance of at least 30 kilometres. Finally, in 1805, the sum of 211 *reales* was paid to deliver a stone from Logroño (La Rioja) about 30 kilometres away (Sainz 2007: 176-178).

Although these different archives offer details of the price, distance and time of millstone transport, they do not record the "on the road" obstacles and dangers these men had to overcome to deliver these products.

8.4.3.2. Fluvial and maritime transport

In Antiquity, transport of querns and millstones by water was advantageous for both the millstone exporter and the boat or ship's master. Water transport gave the millstone merchant the option of wider, more distant markets, while, at the same time, afforded both pay and ballast for the vessel. Roman querns manufactured in the Cabo de Gata volcanic district (see AL-1; AL-2), for example, were a few kilometres from vessels that could have docked in the deep-water coves along the Mediterranean.

Yet no cargo of Roman rotary querns or millstones has been recorded along either the Atlantic or the Mediterranean coastline of southern Spain. In Catalonia, however, a wreck with a quern cargo was identified at Illa Pedrosa, off the coast of Girona. The site underwent its first underwater excavation in the 1960s. Its cargo comprised 130 rotary querns (60 *catilli* and 70 *metae*) originating, according to several analyses, from a quarry near the settlement of Ullastret (Girona) (Vivar 2004: 101-104). This assemblage, unfortunately never thoroughly published, is the best example of maritime transport of Roman millstones originating in quarries of the Iberian peninsula.

There are also no examples of fluvial transport in southern Spain comparable to the 3rd- century AD cargo of volcanic millstones discovered at Wanzenau (Alsace) in a silted up ancient bend of the Rhine River (Forrer 1911). It is accepted, nonetheless, that the Guadalquivir and Guadiana Rivers were navigable from the Atlantic coast into the heart of *Hispania Baetica*. It is likely, therefore, that querns and millstones from the quarries in the volcanic district of Campo de Calatrava could have been assembled at a staging point at the city of *Corduba*, as were other mineral resources, for river transport down the Guadalquivir. The Guadiana, likewise, was navigable from at least Mérida (*Colonia Iulia Augusta Emerita*) to the Atlantic coast. Once again, although likely, there is no physical evidence such as a loading bay or shipwreck that certifies transport by these waterways.

In more recent times, in the 18th century, maritime export of Montjuïc millstones from Barcelona is known to have reached as far as the Provence in France, Murcia in southern Spain, and possibly Genoa in Italy (Español 2009: 967). In southern Spain, there is no site with a volume of export comparable with that of the Montjuïc. Certain historical archives, nonetheless, point to some maritime millstone transport. For example, in 1508, four stones, probably from the site of El Torcal (MA-1), were ferried a short distance by boat along the

Mediterranean coast of Málaga by the seaman Francisco Martín from the port of Cala del Moral to Torre de Vélez (Fernández López 1982: 221, note 10). Furthermore, Prudente Arjona (citing Ponce 1981) indicates that two 19th-century Portuguese ships (*"Felicidade"* and *"O que Deus Quera"*) sailed from their home ports in Portugal to acquire shell-rich (biocalcarenite) millstones at the coastal quarry of Rota (CA-3) in Cádiz.

8.4.3.3. Rail transport

The rail network was only introduced in the south of the Iberian Peninsula during the second half of the 19th century during the Industrial Revolution. Millstone production at certain larger Spanish quarries most likely took advantage of this new means of transport. The short railway branch connecting Gerena, Seville (see SE-7), with the Aznallcóllar-Guadalquivir line, documented by an old photograph (Fig. 8.20), was built specifically to access granite construction material at Gerena's quarries (Perejil Derey 2005). Granite oil conical rollers known to have been produced at Gerena from photographs also probably benefited from this branch for regional distribution by rail (Perejil Derey 2005). It is therefore reasonable that Gerena's smaller-scale millstone workings must have also taken advantage of this new railway.

Fig. 8.20: Scene at the loading platform of the train station of Gerena (Seville) where granite blocks were loaded onto wagons. Oil rollers (and possibly millstones) probably also benefited from this means of transport (photograph from http://lafactoria-cuencaminera.blogspot.com.es/2010_12_01_archive.html).

The spread of the train network opened new markets for quarries in southern Spain. The group of five white millstones in a photograph (Fig. 8.21) at the loading platform of the station of Huelva were most likely, due to the absence of iron bands around their girth (typical of French burrstone imports), manufactured at regional quarries in Spain. It is tempting to imagine that they arrived by train in Huelva from an inland quarry either for local mills in the city or that they were in transit, waiting to be transported elsewhere by boat from the adjacent port. Another option is that they arrived in Huelva by sea and were waiting to be transported inland by rail.

In any case, while the rail helped to commercialise millstones from Spanish regional quarries in the mid 19th century, it ultimately played a crucial role in their demise because it accelerated the arrival and propagation of burrstone imports from France, the famous *meulas francesas* that were to equip a large number of the grain mills in southern Spain.

Fig. 8.21: Huelva train station towards the end of the 19th century with five bright white millstones in transit. Behind the upright example (to the left), are three others. The stone to the right, under the sitting official, appears to have a fracture along its lower rim. The absence of iron girth bands reduces the likelihood that they were composite French burrstone imports. The Huelva-Zafra rail line was built between 1867 and 1900. The station is located by the port, as seen by the boats in the background to the right behind the fence (photograph from http://lafactoria-cuencaminera.blogspot.com.es/2011/06/casas-espanolas-al-borde-de-un.html).

9. THE MEN BEHIND THE MILLSTONES

There is very little information about the individuals that worked in millstone quarries, that is, their level of expertise, their wages, the organisation of working crews, whether they exercised other professions and the impact of their work on their health. Some of the following considerations regarding these intangible aspects of millstone production derive from old written documents or archaeological excavations outside the study area. In Spain, I borrow from the writings about the sandstone workings of Montjuïc in Barcelona and the marble workings of Macael in Almería. The Montjuïc was both a producer of millstones and building blocks, whereas Macael, not a millstone quarry, was a rival to the celebrated white marble of Carrara, Italy. I have had to rely on these sources because written records specifically regarding millstone makers in southern Spain are restricted to a small number of short, fragmented vignettes.

9.1. The millstone makers

Professional millstone makers, like all stone cutters, were robust, brawny individuals. Through the physical toil and the difficult working conditions, they developed taut muscular frames and a tough mind-set. The crews of millstone makers at the Berrueco mountain in Cádiz (CA-8), according to Madoz, reserved the right to bear arms (explosives in this case) not only to level crags at the Berrueco mountain and the neighbouring town of Medina Sidonia, but also to chase down *rateros* (thieves) that occasionally turned up in the area (Madoz 1848, Vol. 11: 343).

Only a few millstone makers are known by name. Juan de Bargas, for example, was an early 17th-century stonemason who scored stones at the Albardado quarry (CO-10) near Belmez, Córdoba (González Perralbo 2008). Knowledge of the names (and even nicknames) of the two sets of fathers and sons of the Brañosera crew from the mountains of Palencia, photographed in 1933 (Fig. 9.1), is by virtue of the interviews conducted by Maestro Hernández with the millstone maker's descendants.

The photograph is also evidence that the millstone trade, like many professions of the past, was dominated by men and imparted from father to son. Furthermore, the Maestro Hernández interviews indicate that millstone making at Brañosera could be an activity secondary to their principal occupation as farmers (Maestro 2011: 48).

Fig. 9.1: Photograph dating to 1933 of moleros *(millstone makers) at the quarry of Brañosera (Palencia) in northern Spain. Some are identified by name in the study of Maestro Hernández (2011: 40): to the left are father and son Manolo and Antonio García; in the middle background only one of the men, Gerardo, is identified. Herminio Miguel is in foreground and to the right, in the background, is his son Sauro (photograph from Cuevas Ruiz 2006).*

9.2. Permanent or seasonal work

Although millstone making required a profound, intuitive knowledge of rocks and their properties, and the skills to handle specific tools and techniques developed through years' of experience, the profession was not necessarily full-time or year-long. I suspect that most quern and millstone quarries serving local and regional needs were only in operation when the climate permitted and when they received purchasing orders. A few larger production centres, serving a wider, more distant clientele, and benefitting from a constant influx of purchasing orders, were probably manned on a permanent, possibly, year-long basis.

In southern Spain during Late Prehistory and part of Protohistory, the evidence suggests that most querns originated in the form of loose surface blocks. There is little evidence of veritable bedrock quarries. The Prehistoric settlements of Los Pensadores and Las Rajas (Pinos Puente, Granada), for example, were a short walk from the presumed bedrock quarry of Zujaira (GR-12) (Anderson 2010, unpublished). Since these grinding tools were not necessarily fragile, and did not require constant replacement, it seems evident that quern making in early societies was an occasional occupation carried out parallel to domestic tasks.

For the Iron Age, the existence of monumental architecture and statuary in the Iberian Culture suggests a high degree of rock carving specialisation. Some of these stone workers certainly also extracted and fashioned rotary querns and the larger rotary pushing millstones. This does not exclude that they, like their earlier counterparts, spent most of their time tending to their fields and flocks.

In Roman times, with a massive production of volcanic rocks at sites such as Cerro de Limones (AL-1) and Hoya del Paraíso (AL-2), craftsmen could have worked exclusively (or at least for extended periods of the year) at quern production. It is also possible to imagine the separation of tasks at the quarry: specialists for extraction, specialists for fashioning the stones into shape before others transported them to the coast for sea transport. But this commitment does not exclude that the workers periodically took part in other activities.

For the Middle Ages under Islamic rule, there is no data concerning millstone makers. Outside of the sphere of Islamic rule, in present Catalonia, it is conceivable that the quarry of Montjuïc by Barcelona, owing to its excellent sandstone, as well as its potential for export by sea, provided year-round work for a number of stone cutters who could have divided their time between scoring millstones, edge rollers for the oil industry and building blocks. Montjuïc is the only documented rock quarry in Iberia in which the stonemasons were united in an official *gremio* (guild) established as early as the 13th century by the royal authorities (Capmany 1779: 256).

The notarial archives of Pozoblanco, Córdoba, indicate that Juan de Bargas was both a millstone maker and a stonemason. From the list of his possessions (mostly tools) and lack of land drawn up for his second wedding, it can be inferred that he was a stone specialist and did not participate in other activities (such as farming) (González Perralbo 2008).

The research of Maestro Hernández presents a different view of how more recent millstone makers in the early 20th century organised their work. The *moleros* of Palencia, when receiving an order of purchase, took up camp at the quarries for periods of between four to seven days (Maestro Hernández 2011: 43-44). The timing of this extra work surely depended, however, on the climate and their other duties raising animals and crops.

Related to the question of permanent or seasonal work is also the notion of itinerancy. In the Middle Ages stonemasons, like members of certain other professions, were known to be mobile, travelling from quarry to quarry or construction site to construction site. At times to make work more attractive, they were granted special exemptions from taxes imposed on local populations (Martínez Prades 1998: 11). This model, however, is less applicable to millstone workers because their main activity, millstone extraction and rough fashioning, contrary to the construction worker who moved from building site to building site, took place at the quarry. It was therefore the millstone, and not its maker, that travelled.

9.3. Master millstone makers and crews

As previously noted, the size of the crew depended in part on the product and the volume of production. At the small Roman quern quarry of Châbles in Switzerland, due to the limited amount of workspace and the homogeneity of the tool marks spread across the quarry floor, there was probably only one person scoring cylinders (Anderson *et al.* 2003: 59). The quern maker could, nonetheless, have been assisted in other tasks such as the removal of working debris. From the size of the site and the number of abandoned querns, the volume of production at Châbles was obviously far less than that of the Roman volcanic quern quarry of Cerro de Limones (AL-1), a site that could have engaged a large number of workers. Among these workers there was certainly a wide range of levels of expertise and the need for some individuals, the masters, to assert authority over others in order to maintain a rhythm and standard of work.

Fig. 9.2: Photographs of the working crews of the pit quarries of Montjuïc (Barcelona) and Gerena, Seville (SE-7). (photograph of Montjuïc from http://amf2010blog.blogspot.com.es/2012/03/gaudi-y-la-piedra-un-magico-acuerdo.html; photograph of Gerena from Acuña Carabantes, 2004).

In the Middle Ages, millstone workings probably respected a hierarchy like that of construction sites, with assistants and apprentices at each quarry led by a "master" (Martínez Prades 1998: 10). These positions could have been filled by family members. It would be the master, however, who retained contact with the client and, if he did not take part in the work directly, supervised the different activities.

For the quarry of Berrueco (CA-8) in Cádiz there is specific information spanning the first half of the 19th century regarding the total number workers. It was first the traveller Cruz y Bahamonde (1813: 91, note 1) who recorded that the site employed 50 workers. Madoz, 45 years later, recorded 23 men at the site, divided between five different workshops (Madoz 1848, Vol. 11: 343). This number would, theoretically, correspond to crews of about five men per workshop. Although Madoz and the other authors provide no information of how the crews were organised, it is difficult to conceive that they were not divided by a hierarchy based on longevity and experience.

In any case, master millstone makers and master stonemasons shared access to the affluent classes. A constant contact between the master stonemason and the noble class and clergy took place during construction of castles and churches. Similarly, the master millstone maker's contact with the upper class to replace a broken or worn out millstone, was either through the owner of a watermill, often a member of the nobility or clergy, or through the wealthy person who leased the mill. This association with the affluent class could result in an increase of the master millstone maker's social standing.

The Palencia millstone makers in the photograph worked as a team, according to the interviews conducted by Maestro Hernández (2011). In the photographs of millstone works at Montjuïc (Barcelona) and the conical oil rollers of Gerena (SE-7) (Fig. 9.2), the crews comprise between 10 and 15 men. It is not possible to know if the numbers in these last photographs are representative of the reality of millstone working crews, but they do approximate with those of the quarry of Berruecos.

There is no written evidence recording the participation of women or children in millstone making. It would, however, not be surprising that family members assisted the stonemasons, taking on auxiliary roles such as removing working debris. This might explain the presence of what appears to be a boy to the left in the background of the photograph of Gerena (Fig. 9.3).

Fig. 9.3: Scene of workers from the 1920s from a granite quarry (building blocks or paving stones) in the area of Gerena, Seville, a district known to have produced millstones. The worker to the left could be extracting a stone or metal shard from the eye of the other worker. In the background, to the left, is what appears to be a child (from Acuña Carabantes 2004).

Throughout the recent history of Spain, as other countries, children were employed in manual professions, at times under appalling working conditions. The first Spanish law prohibiting children under the age of 10 years to work in industry, workshops, metallurgy and mines, although passed in 1873, was never enforced (Tiana Ferrer 1987: 47-48). It would therefore not be surprising that children were present at certain millstone workings, in particular those situated a short distance from the permanent residence of the workers.

9.4. Earnings

We can speculate that at least since the late Middle Ages, quern or millstone makers were neither impoverished nor prosperous. A notarial protocol dating to 1619 establishing the possessions of Juan de Bargas records that his "riches" amounted to a certain sum of money (not much by the standards of the time, according to González Peralbo, 2008) and his stone working tools. This suggests that his "wealth" resided in his skills and equipment, and not in ownership of land.

In the *Survey of the Marqués de Ensenada* (1750-1754), among the professions of the residents of Abengibre (Albacete), was a *picapedrero de molinos* (millstone maker). His daily wages were six *reales*, broken down into four for his services and two for his expenses. The table of wages of other professions in the same village, gleaned from the same survey, indicates that the earnings of millstone makers surpassed that of all the other workers of manual trades (wool workers, weavers, tailors, blacksmiths and brick makers) except for that of construction workers (Table 9.1).

Table 9.1: Table of the daily wages for different professions compared to that of the millstone maker based on the responses of Abengibre (Albacete, Castilla La Mancha) to the Census of the Marqués de Ensenada (1750-1754).

Profession	Daily wage	Expenses	Total
Construction worker	*6 reales*	*2 reales*	*8 reales*
Millstone maker	***4 reales***	***2 reales***	***6 reales***
Wool worker	-	-	*5 reales*
Weaver	-	-	*4 reales*
Tailor	*2 reales*	*2 reales*	*4 reales*
Blacksmith	-	-	*4 reales*
Tailor apprentice	-	-	*1 real*
Brick maker	*2 reales*	*2 reales*	*4 reales*

Several other old texts suggest the profitable nature of millstone making. The situation of Alconera, Extremadura, cited previously in the *Royal Census* of 1791, suggests that workers earned more in the millstone and brick/tile "factories" than working their fields (http://www.nuevoportal.com/andando/pueblos/extrema/badajoz/alconera.html#historia). The Berrueco (Càdiz) (CA-8) millstone quarry in the late 18th century, according to the description of the local priest, afforded an "advantageous" daily wage to its workers (Martínez y Delgado 1875: 129).

Finally, in the Palencia mountains in the 1950s, making a millstone could mean a sum of up to 300 *pesetas* per millstone above what was earned from tending to crops and livestock. If the example of Palencia is typical of *moleros* elsewhere, their millstone earnings were an interesting supplement to what they obtained from agriculture, allowing them to live comfortably albeit modestly (Maestro Hernández 2011: 48).

9.5. Occupational hazards

Stone workers, as seen in various old photographs, were robust individuals. Through their long working hours they developed both physical strength and physical and mental endurance. Their profession, nevertheless, entailed both short- and long-term dangers.

The limbs and extremities of millstone makers were always at risk, whether when cutting or carving with a pick or with a chisel and hammer or moving a millstone. Some risks, common to most stone workers, are cited in interviews with former workers from the white marble quarries of Macael in the north of the Province of Almería (González Alcantalud 1997: 117-118). These workers state that it was the hands, eyes and back that risked the most. Back related problems were always present due to adopting unnatural (bent and arched) working positions while cutting the trenches and exerting themselves when lifting heavy materials. After years of work in both heat and cold, the skin of their hands became hard and callouses developed that remained for life. The bright white colour of the white Macael limestone reflected the rays of the sun and burned their eyes. In millstone working contexts these lesions to the eye were certainly present in the "white" limestone exploitations (reputed for white bread) spread throughout the Spanish landscape.

The González Alcantalud interviews do not mention, although it was certainly a problem, the ever-present threat to the eyes of an impact of an iron or rock shard. In the siliceous millstone quarries in France, millstone makers at times assembled crude protective eye wear out of iron wire and recycled glass (Jagailloux 2002: 158). It would not be surprising that one of the Gerena quarrymen in the photograph (*cf.* Fig. 9.3) is removing a foreign object from the eye of the other.

It is also noteworthy that dust generated by working the white Macael marble was not considered (at least at the time) to be danger to the lungs of the workers. However, most *moleros*, like quarrymen and miners, working silica-rich rocks (conglomerates, sandstone, granites, limestones), lived under the constant threat of the silent, long-term killer silicosis, a lung fibrosis provoked by the inhalation of silica dust. There are many documented cases of this ailment among the silicious French burrstone workers in France. It is said that the health of a strong man could be broken within a period of 15 years (Hockensmith 2009: 207-209). A treatise by the Bonneff brothers dating to 1908 (*Agapain* 2002: 144-150) and an article by Doctor S. Jagailloux (2002: 151-164), detail the horrors of this disease. Alain Belmont has dedicated several pages to this occupational risk in his study of the production of burrstones in France (Belmont 2006a, Vol. 2: 194-196).

These risks, although probably lower in the millstone quarries of Spain compared to the highly siliceous workings of French burrstones, were certainly present. It is noted that cutters from Macael, in the difficult years after the Spanish Civil War, feared the idea of having to immigrate and either work granite at General Franco's Mausoleum at the *Valle de los Caídos* outside Madrid, or sandstone at the quarry of Montjuïc at Barcelona. In the case of the *Valle de los Caídos*, they are said to have feared the idea of working underground, whereas at Montjuïc, they dreaded the "lung disease" (González Alcantalud 1997: 110). In Moralzarzal, a town in the Sierra de Guadarrama granite district (about 15 km from the millstone quarries of Colmenar Viejo, M-3), a monument has recently been erected to honour the legacy of the local stonemasons, who for the most part, did not perish from accidents, but from silicosis (http://www.conocermoralzarzal.es/canteria.htm).

In spite of the different occupational risks, quarry owners and administrators - if the example of Macael can be trusted - showed little concern for the workers' safety. The company only provided *botiquines*, first-aid kits or rudimentary first-aid posts and stretchers, after the cutters went on strike in 1922 (González Alcantalud 1997: 117-118). It is therefore safe to assume that the lack of concern for the safety of the white marble quarrymen by the quarry owners or administrators of Macael was probably characteristic of the whole of the stone industry, including millstone production.

10. MILLSTONE QUARRY OWNERSHIP AND CONTROL

The following notions regarding ownership and control of millstone quarries from Prehistory to the Middle Ages are based principally on the archaeological record. In the Late Middle Ages through Contemporary times, however, these notions stem for the most part from written sources. In general, and through time, it is assumed that modest quern and millstone exploitations for local consumption probably went unnoticed and were disregarded by the ruling bodies, whereas larger workings, like other potential sources of wealth (such as mines), were coveted and controlled by the ruling powers.

10.1. Prehistory to the Iron Age

Since millstone rocks in southern Spain from Prehistory to the Iron Age were readily available as surface material in a wide variety of types (granites, mica schists, sandstones, biocalcarenites, travertines, conglomerates, dacites, rhyolites, etc.), it is difficult to conceive that the quarries where they were collected could be subject to any control. These sites probably did not foster the same value bestowed on other more rare lithic resources, such as mines of flint or precious metals. This does not exclude, however, that quern and millstone production was subject to control at the level of the settlement, by certain sectors of society. Nonetheless, this subject, due to the absence of research, remains purely speculative.

10.2. Antiquity

In Antiquity a new situation can be perceived. The growing demand for volcanic querns and millstones procured well beyond the local sphere was behind the creation of a vast network of long distance trade. This would inevitably lead the ruling powers to covet authority over the quarries, as was the case of mining iron ore and precious metals. It is therefore reasonable to assume that in Antiquity the authorities maintained control of millstone quarries either by ownership, concessions or by levying taxes.

Quarries in the countryside, removed from the urban sphere, belonged most likely to the domain of a villa. The master of the villa therefore retained the right to either exploit it by his own means or lease it to a specialist. In any case, this is a possible model for the quern quarry of Châbles, Switzerland, probably under the authority of the nearby villa of Châtillon (2.5 km) (Anderson *et al.* 2003: 59). This might also be the situation for certain Roman quarries in the south of Spain near settlements. The broken, unfinished roughouts at Los Cazadores in Almería (Anderson *et al.* 2011a: 153) and Los Ceniceros, near Mazarrón, Murcia (Anderson *et al.* 2014: 118) could be indicative of workings linking these settlements to nearby quarries.

The volcanic quern and millstone quarry about 500 metres outside the Roman city of *Sisapo* (CR-1) (Almodóvar del Campo) shares the same outcrop with a building block exploitation. Judging from their proximity to the city, it is plausible that both extraction sites (if contemporary), like other exploitations in the area (notably silver and lead mines), were subject to the control of the authorities of *Sisapo*.

Roman quarries associated with urban defensive features are considered to have been managed directly by the military (Gutiérrez García-Moreno 2009: 282). Following this line of thought, an argument could be made that certain quern exploitations might have been administered by the military due to the need of an army on the march for a great number of portable handmills. Military querns with inscriptions carved along their circumference are known at Strassbourg, France, Saalburg and Xanten in Germany, and Vindolanda in England (Jodry 2011b: 88-89). Direct control by the military or not, the armies of Rome presumably had access and authority over quern production. All said, in spite of a large military presence in the Iberian Peninsula, no quern with a Roman military inscription has been identified on Spanish soil.

Independent of the large Roman volcanic productions (and to a certain extent the *ostionera* productions along the Atlantic coast in the Province of Cádiz), modest non-volcanic quarries were still in operation in Roman times, as seen through sandstone, limestone tufa and conglomerate querns brought to light in many Roman contexts. The sphere of distribution of these workings remained, nonetheless, at the local level. Querns from surface blocks in riverbeds or taluses, or hewn from sandstone outcrops, were certainly always an economical alternative to imported volcanic querns. These lower-quality exploitations, however, probably did not retain the interest of the authorities.

Fig. 10.1: General view and detail of the sector where angular building blocks were extracted from the volcanic quarry (CR-1) near the Roman city of Sisapo.

10.3. Islamic rule

The concept of the ownership of millstone workings under Islamic rule is particularly murky. In the years immediately after the Islamic conquest, properties such as mines (and presumably quarries) were only confiscated by the authorities in the regions taken by force (Serrano Ruano 2010: 193).

In southern Spain, where most of the regions were transferred to Islamic rule by means of treaties, properties presumably remained in the hands of the original owners (Serrano Ruano 2010: 193). There is no evidence, at this time, of a large, profitable network of millstone commerce comparable with the earlier Roman long-distance trade of volcanic millstones. With the abandon of the volcanic rock and the adoption of (or return to) more widely available stones, such as conglomerates, limestones and sandstones, sources of millstones would have been omnipresent. The ever-growing demand throughout the landscape for new watermills to be raised along a developing network of *acequias* (irrigation ditches) was probably met by local and regional exploitations. It is therefore plausible that millstone workings, like mines, remained under the authority of individuals or "tribal" units, and not under the control of more central authorities (Serrano Ruano 2010). Nonetheless, as with other professions, such as millers, tribute was likely to have been demanded from the millstone makers by the authorities (Lagardère 1991: 60).

10.4. Post-Islamic rule and the Modern period

In the Middle Ages, after the reconquest of present Catalonia, written documents indicate that the highest authorities were directly interested in, and presumably maintained authority over, the production of the large millstone quarries at Montjuïc Mountain just south of the city of Barcelona, an exploitation that yielded sandstone for construction and millstones since at least the early 13th century until as late as the 20th century (*cf.* chapter 1, Fig. 1.4-1.5).

An archive from 1203 records that the monarchy was awarded tribute for each millstone that exited Barcelona's port (Español 2009: 967). Gutiérrez García-Moreno, in her work on building stone quarries in north-east Catalonia, adds: "*The fact that during Mediaeval times quarrying at Montjuïc was protected by the kings gives you an idea of the importance of this industry, and in particular of millstone production*" (Gutiérrez García-Moreno 2009: 90-91). The monarchs in question, Kings Pere II, Jaume II and Pere III, maintained these interests by renewing the privileges granted to the stone cutters on three occasions (1211, 1327 and 1338).

Elsewhere in the Iberian Peninsula, the Christian reconquest left large sectors of the countryside without population. The ensuing "protection and transfer of property" was regulated through codices called *Fueros*. The *Fuero de Cuenca*, dating to the rule of King Alfonso VII in the late 12th century, sheds unequivocal light on the interest of rulers regarding millstone workings:

"All the quarries, gypsum beds, ***millstone quarries****, tile works, and also the perennial springs should be common property of* ***the council****. Whoever has a* ***millstone quarry*** *or any of these things mentioned previously on a property of his should sell it to* ***the council*** *for a double-sized property, and it becomes communal. If someone occupies it against another of the council, he should pay a hundred* aurei" (translation by J. Powers 2000: 57).

The control of these resources by the council (*consejo*), the highest unit of government under the monarchy, is not unique to Cuenca, as this *Fuero* is known to have served as a model for a series of subsequent codices in other parts of the peninsula. A later version of the *Fuero de Sepúlveda* dating to 1305, for example, yielded ownership of the natural resources to the authorities either through acquisition or by expropriation (Bermúdez Aznar 2005: 164).

The role of the church in the ownership of millstone quarries in this period is not clear. The sites of Cabra, Los Frailes (CO-1) and Monasterio de San Salvador in Guadalajara (GU-2), as their names indicate, were not far from monasteries. Although there is no evidence proving that the church owned any particular millstone production centre, it would not be surprising that some fell under monastic authority, so as to gain profit by leasing them. As is well-established, the church received or acquired numerous properties through the years. The watermills in urban Córdoba in 1236, for example, were ceded by Fernando III to the clergy shortly after the reconquest (Cordoba de la Llave 1988: 827). Property ownership by the church became in fact so extensive over the years, that the state imposed a series of *desamortisaciones*, that is, "confiscations", starting at the end of the 18th century. (This question will be elaborated further ahead).

In the south of Spain, after the fall of Islamic rule, a number of resources of raw materials, including quarries, were, following the rules of the *Fueros*, in the hands of the councils. The residents of Antequera in 1501, for example, were outraged with the *Malagueños*, from the coastal city 30 kilometres to the south, who intruded into their territory and took millstones without paying tribute. This rift, presumably involving the site of El Torcal (MA-1), ended in a legal bout that reached the highest court of the day, that of the Catholic Monarchs (Ferdinand and Isabella) in Granada (Fernández López 1982: 222-223; AML, Leg. 49, p. 9. 1 cuartilla + 1).

An unquestionable example of the control of the authorities over millstone workings is seen through the ordinance of 1502 in the Municipality of Loja (Granada), served in the form of a *pregón* (oral proclamation). The regulation threatened harsh punishment for any unauthorised extraction, that is, not only a fine and the confiscation of the millstone, but seizure of the oxen used for its transport (Pregón 1502, enero, 10. Loja, Pregón de la ordenanzas de Loja sobre no sacar piedras de molino. AML, Leg. 49, p. 9. 1 cuartilla + 1; http://www.teresadecastro.com/Fuentes/DocMunic/DOCLOJA.6.htm).

At Llerena (BA-2) in Extremadura, a similar municipal ordinance dating to 1566 prohibited "foreigners" from extracting millstones from the local quarry, unless they, like the locals, paid the official tariff (http://manuelmaldonadofernandez3.blogspot.com.es/2010_04_01_archive.html).

Even the highest authority of the peninsula presumably showed an interest in rock quarries. King Philip II, towards the middle of the 16th century, acquired the limestone exploitation of Colmenar de Oreja (M-2), reputed for its excellent, hard white rock. The intention behind this purchase was to control the blocks destined for royal constructions, such as the Royal Palace of Aranjuez and other buildings in Madrid. Yet it is perfectly conceivable that if millstone workings were in operation at Colmenar de Oreja at that time, as they are known to have been two centuries later, they also would have fallen under the monarch's authority.

Yet another example dates to 1745, when the municipal authorities of Rota, based on historical archives, reinforced the prohibition of millstone extraction along the coast unless the millstone makers paid a tax (José Antonio Martínez, Chronicler of Rota, pers. comm.).

A different angle to the question of millstone quarry ownership has been noted by Pilar Pascual and Pedro García in their research in La Rioja and parts of Castilla y León. They have pointed out that some municipality borders seem to be deliberately traced to split up millstone workings (Pascual & García 2011: 291). Some of these divisions could have been intentional attempts by the authorities to establish an equitable division of the workings to avoid disputes between the different local authorities (Basterra 2003: 249-251; Maestro 2011: 24-28). In any case, this is the case, based on contracts drawn up in 1706 and 1744 (and others possibly since the 15th century), between different *consejos* in the Palencia mountains that sought to regulate the size, quality and quantity of millstones and avoid "excesses that would damage their commercialisation" (Basterra 2003: 250).

No such divisions or contracts between municipalities are known in southern Spain. Several workings are, nonetheless, found along the boundaries of present municipalities, such as Castillo de Locubín (J-1); Loja, La Merced (GR-4); Miraflores de la Sierra (M-5) and Bohanal de Ibor (CC-4). It is obvious that drawing up contracts to avoid disputes between municipalities requires that the rock outcrop be extensive, as is the case of the puddingstone in Palencia. It is therefore unlikely that production at Castillo de Locubín (J-1) and Loja, La Merced (GR- 4), relatively small sites, was significant enough to influence municipal boundaries. The only case in southern Spain with a situation similar to that of Palencia is the granite outcrop at *Los Molares* (HU-8) in the Province of Huelva, a site that stretches over the Municipalities of Almonaster la Real and Santa Ana la Real. But there is no evidence this division was deliberate.

Not all millstone production was exclusively in the hands of the authorities in Late Medieval times. In 1419, in the Province of Navarra, a document records the sale by María Ruiz de Arregui of some of her mountain properties to the Monastery of Roncesvalles. The sale was concluded on the condition that she and her successors retain the right not only to exploit millstones, but to take as much wood as needed to be used during the transport of the millstones (presumably to build sledges). Of interest is that documents show that this accord was still in effect two centuries later, as it was renewed in 1634 (Pascual *et al.* 2011: 249). It is unknown, however, whether similar accords, permitting an unlimited access to a resource without possessing the land, existed in the south of Spain.

10.5. Contemporary period

The example of LLera (BA-3) in the Province of Badajoz (Extremadura) throws light on a different angle of the question of quarry ownership. The Royal Census of Extremadura of 1791 records that the town did not possess any particular mineral worth noting, with the exception of rock outcrops for lime kilns and millstones (specifically for millstones destined for flour mills). The census goes on to state that these rocks are at the disposal of whoever needs them ("... *de las que se haze uso quien las nezesita*", Caso Amador 2008: 133) suggesting that the quarry was not regulated. Yet the text does not reveal whether the authorities levied a tax for their exploitation.

One of the rare cases identifying millstone quarry ownership in southern Spain is noted in the early 19th century by Cruz y Bahamonde. This distinguished traveller recorded, in his writings about commerce in the Province of Cádiz, that the Berrueco quarry (CA-8) was the property of the Duke of Medina Sidonia and that the aristocrat was awarded 18 millstones a year in exchange for the concession (Cruz y Bahamonde 1813: 91, note 1).

The family of the Duke of Medina Sidonia has a long history of ownership of mills and quarries. At least one watermill (Molino de La Teja) in Vejer de la Frontera, about 20 kilometres from Medina Sidonia, was their property from the beginning of the 16th century (http://www.patrimoniojandalitoral.es/f_vai_8.htm). In addition, archives dating to the 19th century indicate that the family also possessed a windmill in Vejer de la Frontera (Molino de Santa Inés) equipped with Berrueco millstones (http://www.wikanda.es/wiki/Molinos_de_Viento_de_Vejer_de_la_Frontera). The periodic replacement of millstones, especially limestone models that were prone to wear and had a short life, could explain why the noble family demanded the lease be settled in millstones instead of currency. This would have been an advantageous business for the Duke of Medina Sidonia since at this time, before the influx of French siliceous stones, the Berrueco stones were reputed to be the best in the region.

The ownership of siliceous millstone production in the Brie region of northern France, notably at la Ferté-sous-Jouarre, resulted in a rare social phenomenon labelled "*siliçocratie*", that is, a "rule" over the years by a series of millstone merchant families that amassed huge fortunes by exporting millstones throughout Europe and beyond (Belmont 2006a, Vol. 2: 251). A similar situation, to a lesser degree, is documented at the quarry district of Selbu, Norway, where Frederik Birch, in the middle of the 19th century, acquired many sectors of the vast quarries and introduced new more efficient extractive equipment and techniques, resulting in a lucrative business. Birch even went so far as to exhibit a Selbu millstone in the Universal Exposition of Paris in 1855 and 1878, as can be seen by diplomas displayed on the walls in the Selbu Museum.

This phenomenon cannot be applied to Spain. Although some larger exploitations, like the Montjuïc in Barcelona, El Berrueco in Cádiz (CA-8) or the Cantera de los Frailes in Cabra (CO-1), certainly afforded their owners a certain amount of prosperity, there is no evidence that these sites afforded a comparable massive wealth and social prestige as the "*siliçocratie*" sites in the Brie region of France (Belmont 2006a, Vol. 2: 251).

As stated previously, it is well-established that the church accumulated numerous properties throughout the Middle Ages and Modern period. Judging from their ownership of many watermills, it would not be surprising that some millstone workings also fell under their control. It is unknown, however, whether any of these hypothetical quarries were affected by the *desamortisaciones* (confiscations), cited previously, of church properties that began in 1798, and continued periodically into the early 20th century. One of the more notable series of confiscations took place in 1855 and is known as the *Desamortisación de Madoz*. This Minister of Revenue for Spain was the same Madoz who a few years earlier penned the *Diccionario geográfico-estadístico-histórico de España* in 16 volumes (1845-1850), the most important source for the identification of millstone quarries. However, without more research, it is impossible to say whether any of the sites recorded in the pages of Madoz' geographical dictionary passed from church to private hands.

10.6. Millstone quarry concessions

Throughout this chapter I have alluded to millstone makers leasing, in the form of concessions, sectors of quarries from the quarry owners. This notion is difficult to confirm through observations in the field. At the Mont Vouan (Haute Savoie) in France, some side-by-side underground exploitations separated by thin partitions that were not sturdy enough to serve as pillars (to prevent collapse), have be interpreted as evidence of individual concessions leased to different work crews (Alain Belmont, pers. comm.).

The only site in the study area that shows any evidence of compartmental organisation that could correspond to different concessions is that of Cantera de los Frailes, Cabra (CO-1) in the Province of Córdoba, dated to Contemporary times (Fig. 10.2). From the aerial view it is possible to make out a series of internal compartments following a roughly orthogonal pattern of quarry faces, corresponding, possibly, to limits between concessions.

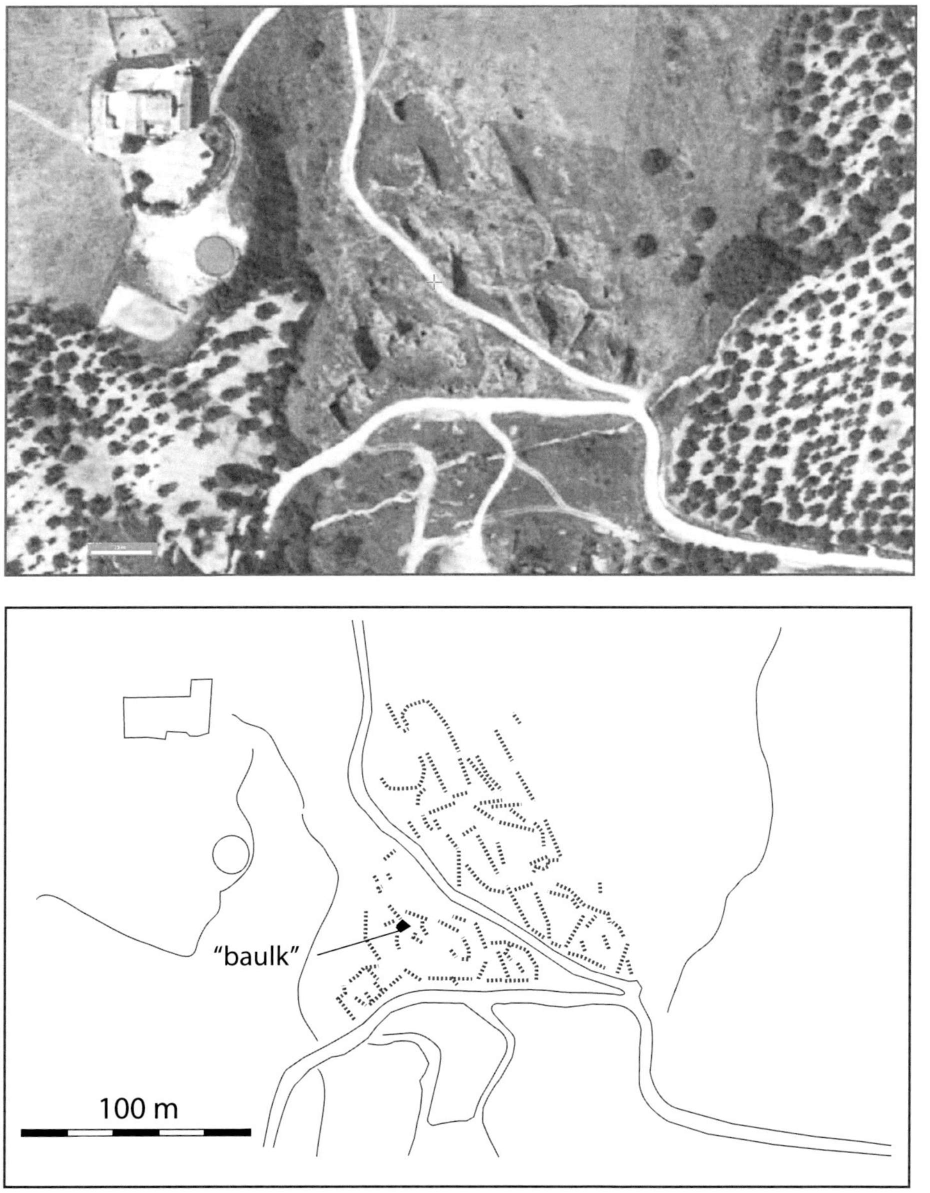

Fig. 10.2: Top: aerial view of the millstone quarry of Cantera de los Frailes, Cabra, Córdoba, The linear quarry faces seem to line up into compartments that could correspond to concessions (SIGPAC). Bottom: interpretation of the different quarry faces and the position of the "baulk".

In the south-western sector there is a massive vertical rectangular "baulk" (Fig. 10.3). This type of feature has been observed at open-air construction block quarries such as the Roman sites of *Sisapo* (see CR-1) and El Mèdol in Tarragona (Gutiérrez García-Moreno 2009: 156-157) and is difficult to interpret. The famous *aguja* (needle) of El Mèdol reaches a height of 20 metres after recent excavations along its base (A. Gutiérrez García-Moreno, pers. comm.). These rectangular or square features might not only correspond to boundaries between concessions, but as bearings to measure the depth of quarry floors to control the volume of extracted rock.

The Cantera de los Frailes is one of the larger millstone workings in the south of Spain. It is known to have exported its millstones as far as Málaga (85 km) (Ezquerra del Bayo 1856: 385) and Córdoba (60 km) (Montero 2008). Because of its size, it would not be surprising that a number of millstone makers would have desired a concession at the site.

The situation of the quarry of Fuentelárbol in the Municipality of Quintana Redonda (Soria) in the north of Spain is quite unique from the point of view of granting millstone working concessions. On the outskirts of the town is a long, curved alignment comprising 255 millstones (Fig. 10.4) originating from a vast limestone quarry a short distance away. The closest parallels to this feature, although anachronistic, are prehistoric megalithic alignments. Although the date of the curious alignment is not recorded, the reference to the "excellent millstones" by Madoz (1847, Vol. 8: 216) places production before the middle of the 19th century.

Fig. 10.3: Views of the Cantera de los Frailes. Top: views of the vertical "baulk" from the south and the east. Bottom: details of rectangular "compartments" that may have corresponded to concessions attributed to different millstone making crews.

Folklore has it that this long, unique alignment of millstones was raised as a tribute to the municipal authorities. The story narrates that "young people who married and decided to stay in the town had the right to exploit a parcel of the quarry without charge on the condition that the first millstone they produced be placed in the alignment" (Pascual & García 2011: 286). This was indeed a singular way to be granted a millstone concession and pay tribute to the local authorities.

Fig. 10.4: Alignment of 255 millstones in the Fuentelárbol (Quintana Redonda, Soria). Oral tradition has it that these millstones were placed in this feature as a tribute to the authorities (photograph by Pilar Pascual and Pedro García).

11. MILLSTONE QUARRY CHRONOLOGY

11.1. Introduction

Millstone quarries, on the whole, were places of work and not permanent residences. In archaeological terms, these sites, as opposed to settlements, are usually devoid of domestic artefacts, notably pottery, that allow the trained eye to date them. This is corroborated by the few millstone quarry excavations undertaken in Europe, notably in France and Switzerland, where the number of artefacts collected in stratified quarry backfill is minimal. Millstone quarries are therefore, even in the best conditions, difficult and delicate to date.

To complicate matters, quarries (or quarry districts) can endure for centuries, even millennia. There is evidence that the *grès de la Serre* in the French Jura was worked from Late Prehistory to Contemporary times (Jaccottey 2011). The *grès coquillier* of the Swiss Plateau, likewise, was exploited from the Late Iron Age until Contemporary times (Anderson *et al.* 2003). The white limestone of Claix (Charentes) and the hard arkose of Mont Vouan (Haute Savoie), were worked for millstones from (at least) the Early Middle Ages through the 19th century (Belmont & Anderson 2010; Belmont *et al.* 2011). Hence, a millstone quarry can have a number of different overlapping working phases spanning a long period of time.

From my research in southern Spain, where no quarry has been the object of an archaeological excavation, the key to attain a more reliable date for a site is to assemble the largest number of chronological indicators, as illustrated in Figure 11.1.

The principal chronological indicators to date these sites, the subject of the first section of this chapter, are old written sources and, to a certain extent, the typology of the millstones abandoned at the workings. Yet, since these products were not finished, they rarely bear the complex driving and tentering typological features (*e.g.* handle and rynd cuttings) that help date them. It is therefore the morphometrics (dimensions) of the abandoned roughouts or blanks (positives) or extraction hollows (negatives) that serve as chronological indicators.

Rock type is also an important element for dating. It is becoming more and more evident that certain rocks tended to be more pervasive in certain periods. Other indicators are *mola*-related place names, certain extraction techniques, the volume and nature of the products (homogeneous or heterogeneous), stratigraphy and the geographical proximity to a dated feature. Oral information also serves to date the more recent production phases. Yet, even in the best of cases, with a combination of indicators, millstone quarries often defy conclusive dating, and can, at best, only be placed in broad chronological niches.

The second section of this chapter focuses on applying the different chronological indicators to the millstone workings identified in the south of Spain. This research has yielded three major chronological assemblages described in detail. This chapter then concludes by touching on the question of the longevity of certain workings.

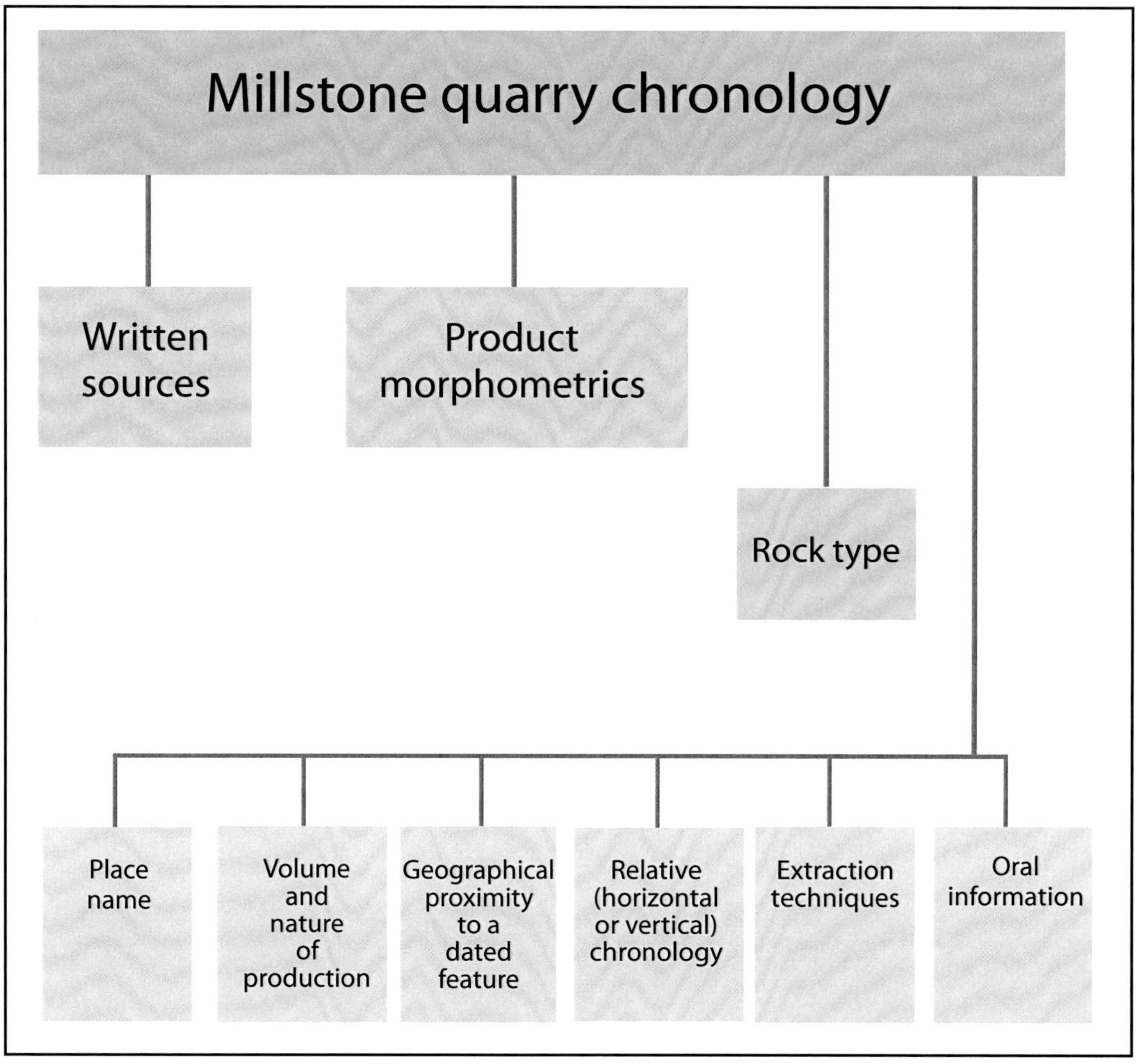

Fig. 11.1: Organisational chart presenting the hierarchy of the principal chronological indicators of millstone quarries.

11.2. Chronological indicators

11.2.1. Written sources

Old written sources such as Royal Censuses, notarial protocols and historical, geographical and geological texts, are extremely useful to date phases of millstone workings (Fig. 11: 2). Many of these texts were penned in the middle of the 19th century and reflect millstone production roughly in the first half of the 19th century, before the advent of the Spanish Industrial

Revolution that facilitated the arrival on a large-scale of imported French burrstones. Without the information from the 19th century geographical dictionaries of Pascual Madoz (1845-1850), and, to a lesser extent, Sebastián Miñano (1826-1829), the number of sites in the study area would be rather slim and their dating far from conclusive. In fact, sites pinpointed by these tow sources (61) make up more than half of the total identified in southern Spain.

Historical records in the form of notarial protocols or municipal ordinances are extremely valuable for identifying millstone workings from the end of the Middle Ages through Modern times. Some date as far back as the late 15th century. It must be noted that I did not sift through these old records myself, but relied heavily on the work of other historians, notably the medievalist Ricardo Córdoba de la Llave, who uncovered through his study of very early watermills a number of references to quarries in the archives in the Province of Córdoba (1988; 2003; 2011).

Written sources, besides their role as chronological markers, at times also offer additional valuable information related to milling terminology, the grinding properties of millstones and their transport, facets that I have examined in previous chapters. These documents, it must be noted, can reflect the personal bias or intention of their author. As mentioned in a previous chapter, it is possible that Madoz had certain political intentions when recording the presence of millstone quarries in particular geographical localities. It was Madoz himself, shortly after the publication of his *Geographical Dictionary* in 16 volumes (1845-1850), who headed the commission of one of the *desamortizaciones* (confiscations) organised by the State that resulted in the forceful change of hands of many properties (often neglected) owned by the Church. This possible turnover of millstone quarries from the Church to the State, however, remains unclear and deserves future study.

In spite of their obvious chronological value, written sources are not conclusive to establishing the total phasing of a quarry. They in fact only offer a truncated view of a quarry life-span since they reflect only a specific phase of a the total production and do not cast light on earlier or later working phases. A rare exception is the reference by the geologist Ezquerra del Bayo, who, writing about the Cantera de los Frailes, Cabra (CO-1), mentions that production had existed already for a "long time" (Ezquerra del Bayo 1856: 385). It is nonetheless impossible to gauge if the term "long time" is a question of years, decades or centuries.

The case of Moclín (GR-1) in the Province of Granada is a good example to illustrate the truncated view of the chronology of a millstone quarry gleaned from a written text. If I were to rely only on Madoz, the Moclín workings would date exclusively to about the first half of the 19th century. The presence of a series of smaller quern and millstone models in certain sectors take the site back at least to Modern and possibly Medieval times. Oral information from a local *Moclineño* also places workings in the early 20th century, a half century later than the Madoz reference.

A more precise idea of the total "life" of a site is attained when it is described by several independent written sources. El Torcal (MA-1) and Loja (GR-2/3), for example, are each mentioned in historical archives toward the turn of the 16th century (shortly after the fall of Islamic rule), and then, once again, in geographical texts of the 18th and 19th century. It can therefore be surmised that millstone production at these two sites carried on, possibly uninterrupted, for at least four centuries.

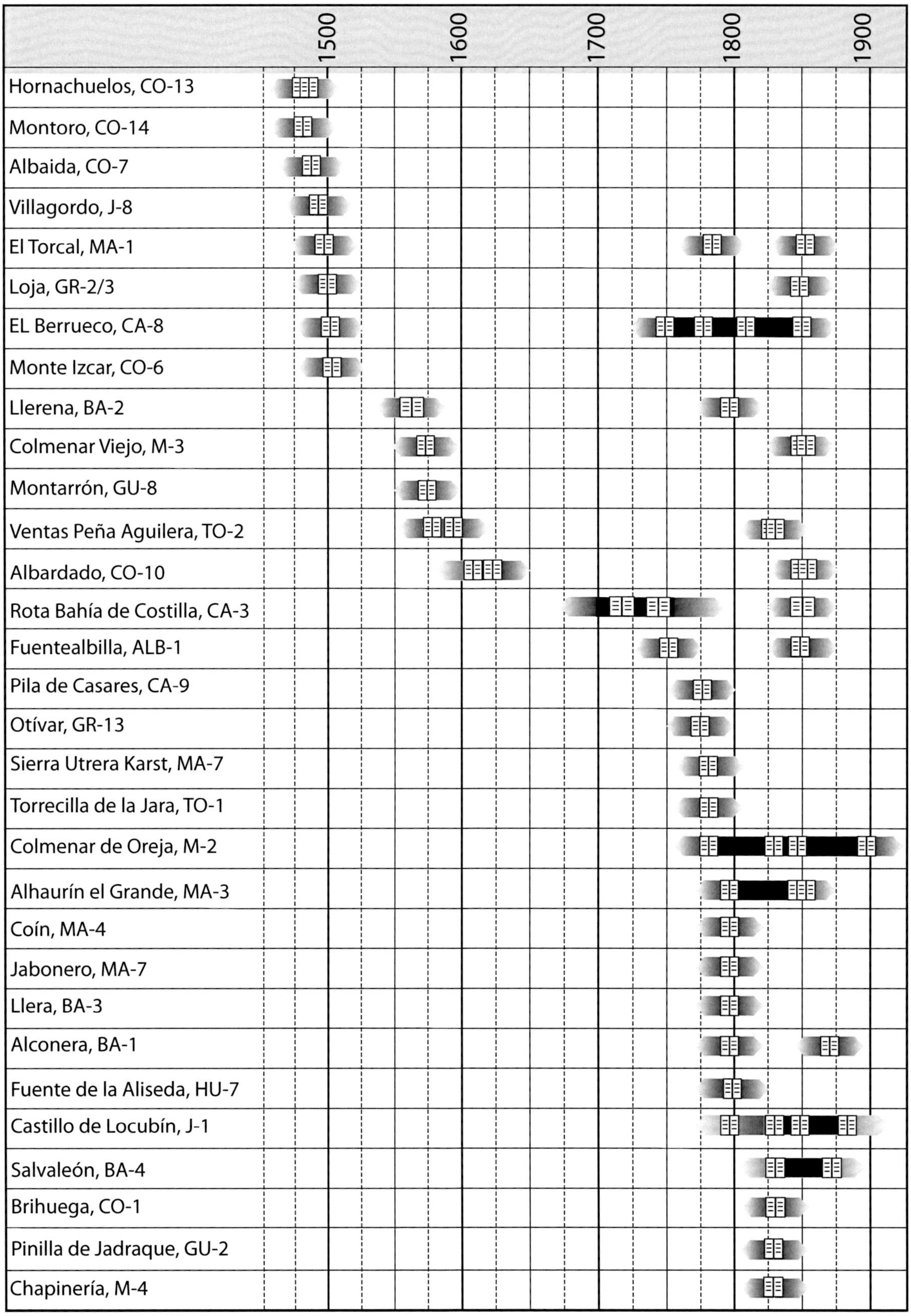

Fig. 11.2: Chronological chart of millstone quarries mentioned in old written sources between the end of the 14th century and the beginning of the 20th century. The symbol in the centre of the page represents the year of publication of the document. So as to reflect an estimated period of production, the arrows with grey gradients indicate (arbitrarily) 25 years before and after the year of the document. Some of the sites are recorded in more than one reference resulting in a prolonged period of exploitation.

Fig. 11.2: (continued from previous page).

11.2.2. Morphometric indicators (millstone diameters)

The term "typology" in the case of millstones comprises three aspects: 1) form, 2) dimensions and 3) fittings. The problem with typology, in the context of millstone quarry dating, is that the artefacts under study, cylindrical blanks and roughouts, are unfinished, hence devoid of aspect 3, their fittings. These fittings, comprising cuttings for the systems of traction and tentering (handles, levers, rigs, rynds, spindles, crosspieces), are crucial elements in establishing the chronology of millstones. It is therefore rare that typology can be used to establish the chronology of a quarry.

In these cases one must rely on less precise morphometric aspects of millstones. At true extractive quarries (MQ-2a) devoid of roughouts (often pillaged by collectors), the morphometrics of the cylinders can be reconstructed by measuring the diameter and height of the extraction hollow and then subtracting the width of the circular trench.

To establish the general morphometric indicators of chronology presented schematically in Figure 11.7, three analyses were carried out and presented in detail below. Two of these studies, with reference to Antiquity and the Middle Ages, comprise measurements taken of the diameter of roughouts and extraction hollows at extraction sites in the Province of Almería. The third, covering the period ranging from the Late Middle Ages to Contemporary times, is based on written sources.

Analysis 1

The diameter of 69 querns was measured at the volcanic quarry of Cerro de Limones (AL-1). The line graph plotting the measurements reveals diameters ranging between 36 and 42 cm, with a peak at 40 cm (Fig. 11.3). These measurements can be correlated with a high degree of precision with models from Roman contexts found in museums. Cerro de Limones is also a rare example of a site with abandoned querns bearing chrono-typological features, such as radial slot cuttings on upper stones and central protuberances for "*sombrero*" lower stones.

Fig. 11.3: Analysis 1. Line graph indicating the diameters of 69 roughouts and blanks measured at the Roman quern quarry of Cerro de Limones (AL-1). The graph shows a clear dominance of the diameters between 36 and 42 cm with a peak at 40 cm.

Fig. 11.4: Measuring extraction hollows at the Rambla Honda (AL-3a) quarry (photograph by Francisco Martínez).

Analysis 2

The second analysis consists of data gleaned from measurements taken at the pebble conglomerate Rambla Honda quarry (AL-3) in collaboration with Francisco Martínez and Antonio Gallegos (Fig. 11.4). Here, since no roughouts were conserved, the diameters of 151 cylinders were extrapolated based on the measurement of the extraction hollows. These results, published in transactions of the Rome Colloquium (Martínez López *et al.* 2011), reveal four basic millstone diameter groups (Fig. 11.5).

The first group (a), from 40 to 50 cm, corresponds to a small assemblage of rotary hand querns. The second (b), from 80 to 95 cm, presents a peak at 90 cm. The third (c), from 1.05 to 1.20 m, is more numerous (about one third of the total) with a peak at 1.15 m. The last group, comprising about 20 extractions (d) measuring from 1.25 to 1.40 m in diameter, is the largest assemblage.

From the chronological standpoint, the dating of the site is not precise (no roughouts hence no typological features) and I am in no position to discern if the quarry had a prolonged life or not, or if all the extractions were contemporary. The following chronological notions can, nonetheless, be gleaned.

The few hand quern extractions in group (a), from 40 and 50 cm, are on average larger than the typical Roman querns at the Cerro de Limones (AL-1) of analysis 1. This size, coupled with their thin, discoidal form, point to a post-Roman date. Conglomerate querns of these sizes, for example, are reported at the Cabezo del Moro, a medieval site in the Province of Murcia (Gutiérrez Lloret 1996: 205-207). Furthermore, there are no examples of Roman querns in museum depositories hewn from this type of pebble conglomerate. An additional factor suggesting a post-Roman date is the absence, among group (b), of thick extractions that could correspond to Roman cylindrical mills, a type also unknown in this rock. The presence of a large number of discoidal extractions in groups (b) and especially (c), different in size from any known Roman mill, combined with the absence of a "*molares*" related place name, and no mention of the site in any historical written record, lead us to believe that at least the core of this site, *i.e.* extraction groups (a), (b) and (c), dates to the Middle Ages.

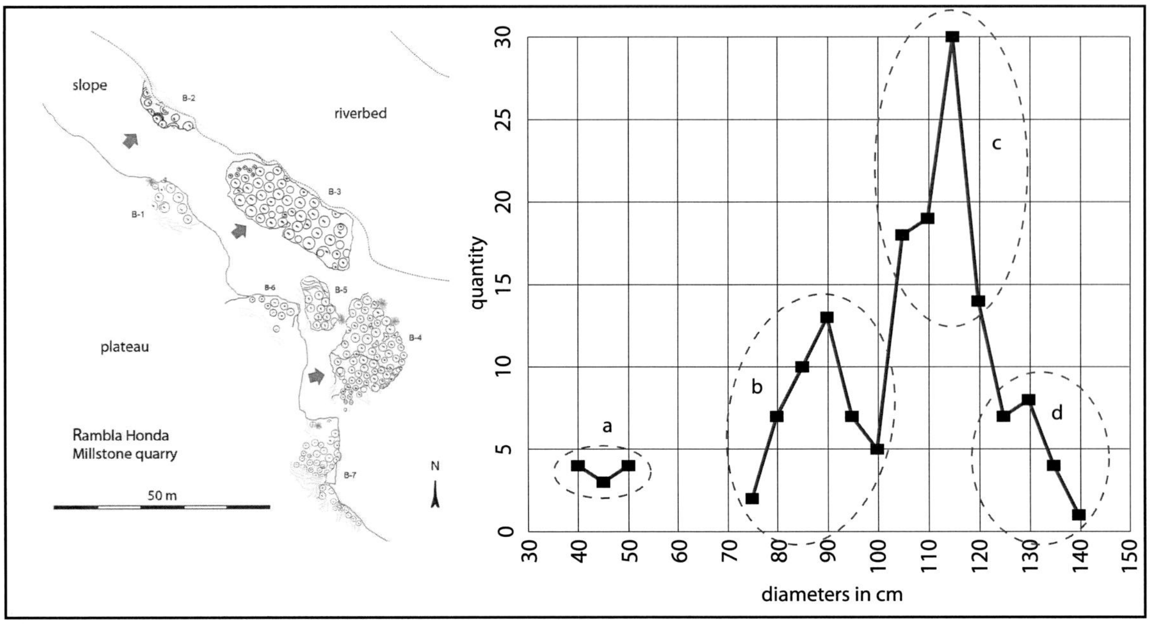

Fig. 11.5: Analysis 2. Line graph representing the diameters of 151 quern and millstone hollows in cm at the Rambla Honda quarry (from Martínez et al. *2011). Four diameter groups stand out: a) from 40 to 50; b) from 80 to 95; c) from 105 to 120; d) a final group, between 130 and 140, is poorly represented.*

Analysis 3

The results of the third analysis, regarding the diameters of millstones from the Late Middle Ages to the Contemporary period, are based on historical written documents. I have compiled a list of cases where the date and the diameter of the product is recorded (Table 11.1). For example, a millstone extracted from the quarry of Hornachuelos (CO-13) measuring a diameter of eight *palmos* (palms, *c.* 125 cm) is recorded in a historical archive dating to 1481. Likewise, a notarial protocol dated to 1606 records millstones from the Albardado quarry (CO- 10) measuring six *cuartas*. The conversion of these early systems of measurement to the metric system, based on the Madrazo table, indicates that the millstones measured, respectively, 1.68 and 1.26 m (Madrazo 1984: 226).

By correlating the dates and diameters on a scatter graph (Fig. 11.6), the tendency of millstones, starting with a single case in the 12th century, seems to increase progressively over time, attaining between 1.40 and 1.70 m in the 16th century. At this time, the trend tends to reverse itself culminating in the 19th century with millstones ranging between 77 and 99 cm (Vallejo 1833). On the surface, this appears to be incoherent, especially when comparing these smaller diameters with the last phase of Contemporary extractions of Zone III of the French millstone quarry of Claix (Charente) that measure above 1.70 m in diameter. The size of the smallest group of Spanish mills could, however, be explained by the fact that they were scored at Colmenar de Oreja (M-2) and not necessarily meant to be driven by the powerful waterwheels, but by horses, mules or donkeys in the *tahonas* of the city of Madrid.

The low number of examples renders the graph devoid of real statistical value and shows that dating millstones based solely on the size of the diameter can lead to confusion. The 80 to 90 cm range, for example, is shared by large Iron Age mills, by Roman ring-mills and watermills, as well as by medieval mills of different type. Hence, the indicator of diameter is not always valid and must be used in tandem with other indicators such as cylinder thickness and rock-type.

Table 11.1: List of written sources (historical archives or molinological publications) that correlate a date with the diameter (or range of diameters) of millstones.

Site code	Site or area	Type mill	Date	Ø (cm)	Unit of measure	Reference
-	Guadalbullón, Jaén	Watermill	1114	106	*4,5 empans*	Lagardère 1991: 109, note 105
-	Alcocer, Guadalajara	Watermill	1339	126	6 *palmos*	Martín Prieto 2006: 838
CO-13	Hornachuelos, Córdoba	Watermill	1481	168	8 *palmos*	Córdoba de la Llave 2003: 305-306, note 26
J-9	Villargorda, Jaén	Watermill	1499	147	7 *palmos*	Córdoba de la Llave 2003: 306
CO-6	Izcar, Córdoba	Watermill	1557	140	6 *cuartas y media*	Córdoba de la Llave 2011: 106
CO-10	Albardado, Córdoba	Watermill	1606	126	6 *cuartas*	González Peralbo 2008, website
-	Puente Genil, Córdoba	Watermill	18th-19th c.	140	-	Esojo Aguilar 2011: 13
-	Murcia region	Watermill	19th c.	130	-	Cara Barrionuevo 1999: 155
-	Contraviesa, Almería	Watermill	19th c.	90	-	Cara Barrionuevo 1999: 155
M-2	Colmenar Oreja, Madrid	Tahona	1833	77-92	1 *vara*	Vallejo 1833: 387
-	Alpujarra Mts., Almería	Watermill	19th-20th c.	115-120	-	Cara Barrionuevo 1999: 155

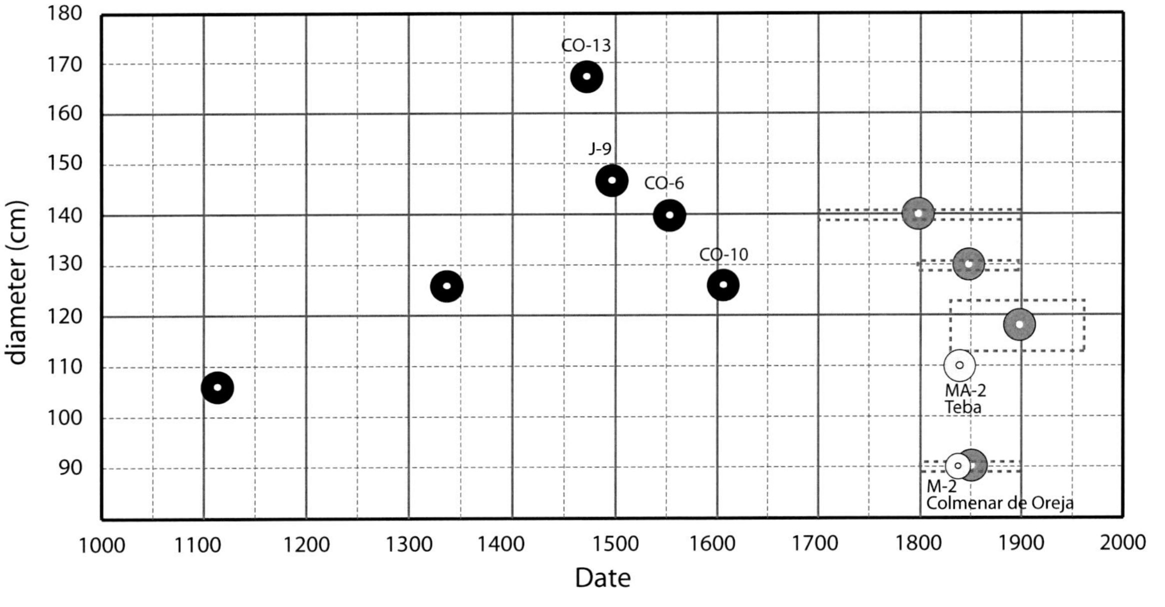

Fig. 11.6: Analysis 3. Scatter chart illustrating the diameters of millstones recorded in written sources from the 12th to 19th centuries. The dashed lines indicate chronological (horizontal) and diameter (vertical) ranges when the data of the written source is approximate. The quarry sites, notably Teba (MA-2), are indicated by their code. The small circle is the quarry of Colmenar de Oreja (M-2), known for its production of small animal-driven millstones measuring from 77 to 92 cm.

The quarry of Teba (MA-2), Málaga, is particularly interesting from the standpoint of combining the diameter indicator with chronology. From the publication of three written sources (*Diccionario Geográfico Universal*, 1833: 608; Madoz 1849, Vol. 14: 752; Espinosa 1848: 265), production took place roughly in the first half of the 19th century. From field work observations, the millstones were homogeneous, measuring systematically 1.10 m in diameter (Fig. 11.7). This diameter follows the tendency based on written sources seen in Table 11.1 and Fig. 11.6 suggesting that "small" extractions, between 1.00 and 1.20 m, were being scored as late as the early 19th century.

A series of other sites such as Vélez de Benaudalla (GR-10), Peña Harpada (CA-10) and El Hacho (SE-3), all cited by Madoz (1845-1850), have a number of cylinders measuring between 1.00 and 1.30 m, confirming that this diameter was common for the middle of the 19th century.

Fig. 11.7: The dozen abandoned millstones from the site of Teba (MA-2), dated by written sources to the first half of the 19th century, measure an average 1.10 m in diameter.

From measurements collected at quarry sites, museum depositories and the little existing millstone literature, I have arrived at the following general morphometric indicators that serve to establish chronological tendencies.

Four basic chrono-morphometric categories (Fig. 11.8)

1- Extractions ranging in diameter from 35 to 42 cm most often correspond to Iron Age and Roman rotary querns. Iron Age models are usually smaller than their Roman counterparts. Querns from the Middle Ages, although generally larger than Iron Age and Roman models, also fall at times into this diameter range. They differ from the earlier models in that they are discoidal and proportionally thinner.

2- Thin, discoidal quern extractions 50 cm in diameter date, with a high degree of confidence, to the Middle Ages. Thicker models of this diameter (and up to 60 cm) are consistent with Contemporary animal fodder rotary querns.

3- The 60 to 90 cm diameter range is the most complex and covers all of the chronological periods. In general, Iron Age and Roman "drum" extractions in this range are ring-mills, usually as thick as they are wide. The tendency is for the Roman models to be slightly larger than the Iron Age models. Lower, discoidal models of this diameter are assumed to be either animal-driven or water-driven millstones ranging from the Middle Ages to Contemporary times. Therefore, the 60 to 90 cm range, without other chronological indicators, is not valid for dating millstone workings.

4- Millstones with a diameter of 1.00 m to 1.50 m, can date from at least the Late Middle Ages to Contemporary times. This wide range, although not very useful for dating, does permit discarding early periods such as Antiquity and, presumably, the Early Middle Ages.

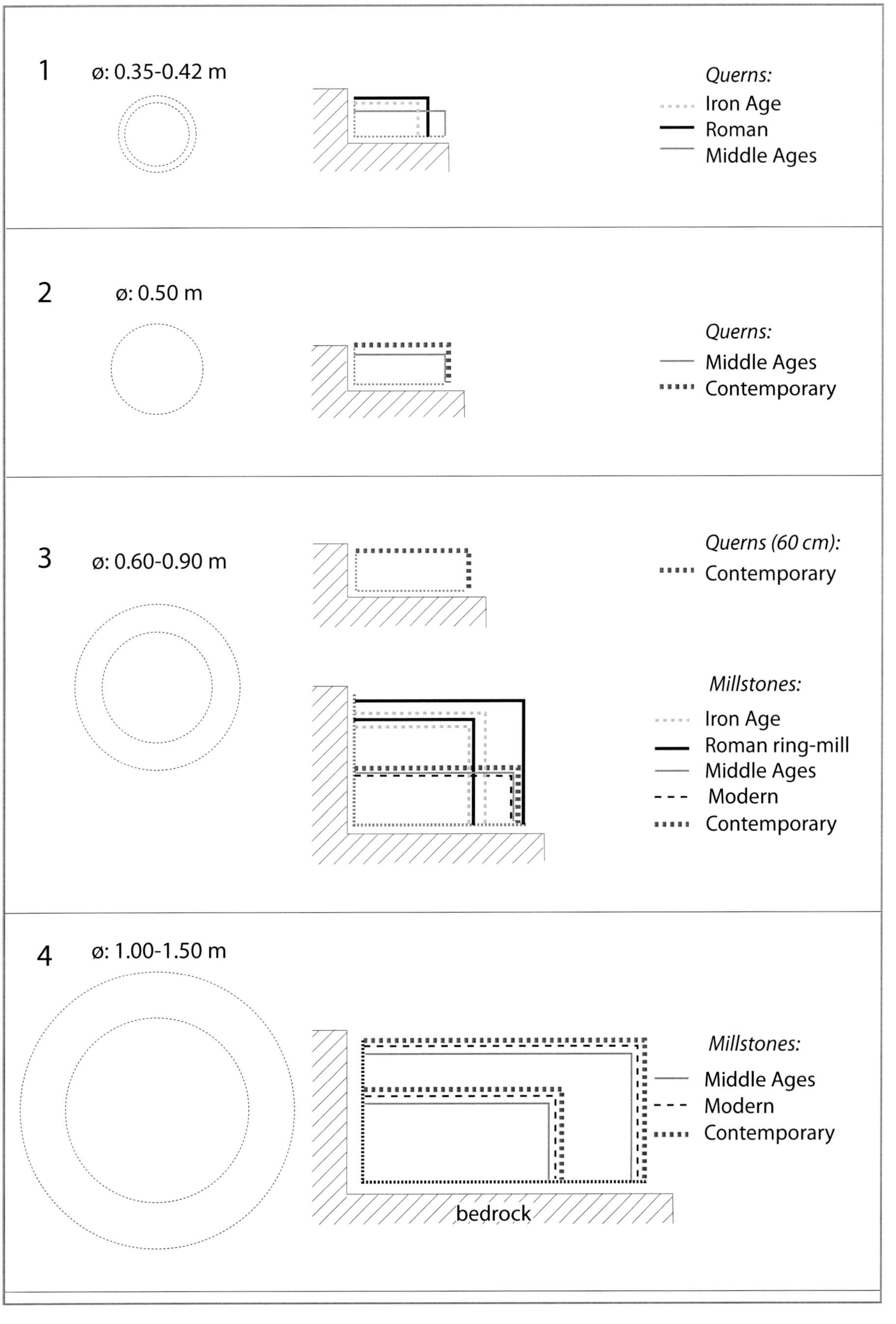

Fig. 11.8: Schema illustrating the four major categories of quern and millstone extractions (both cylinders and hollows) based on approximate proportions of diameter and thickness. The diameter of products overlap in the different chronological periods noted beside each drawing. This overlapping, especially in the case of category 3 (ranging from 0.60 to 0.90 m), complicates the dating process. In this case in particular, other indicators are needed to establish quarry chronology.

11.2.3. Rock type indicator

The penchant for hard, porous volcanic rocks during the Roman period is well-established. Basalts, andesites, and rhyolites were worked, for example, in the Eifel of Germany (Harms & Mangartz 2002), in the Vulsini (Orvieto) and the Veneto districts on the Italian mainland, as well as the islands of Sardinia and Sicily (Antonelli & Lazarini 2010). In France, volcanic rocks were exploited in the Massif Central and in the surroundings of Agde. Closer to the study area is the Farkhana outcrop in northern Morocco and the Olot-Garrotxa in Catalonia. This last district is known for its quern production from the Late Iron Age throughout Roman times (Peacock 1980; Williams-Thorpe & Thorpe 1987; Williams-Thorpe 1988; Portillo 2006).

The vast Campo de Calatrava in the Province of Ciudad Real and the south-eastern Spanish volcanic district (stretching from the Cabo de Gata in Almería to areas of Murcia and even a small area in the Province of Albacete) are the two volcanic districts in the study area of southern Spain. As a result of targeted research over the last few years, these two volcanic districts, heretofore undocumented as millstone districts, have joined the register of Roman volcanic millstone productions spread throughout Europe and the Mediterranean Basin.

Since these rocks were so highly prized in Roman times and were abandoned for most of the Middle Ages, the probability is very high that a volcanic millstone dates to Roman times.

Other rock types can also play the role of chronological indicators. Coarse conglomerates with large rounded clasts and hard white limestones, for example, are unknown in Roman quern and millstone productions. Fine white limestones in the south of Spain also tend to be associated with Medieval to Contemporary productions, as are the cases of Moclín (GR-1a) and Berruecos (CA-8).

11.2.4. Other chronological indicators

Other indicators, although not as revealing as written sources, morphological aspects and rock types, can provide notions as to the dating and life-span of a millstone quarry.

11.2.4.1. *Mola*-based place names

The toponym *molares* and other derivatives of the Latin *mola*, in the study area, in particular in the south of Spain, due to their Latin origin, probably postdate the Islamic domination and reflect the renaming of certain tracts of land after the Reconquest. In this sense, they can potentially serve as chronological indicators, placing them, roughly, at the transition from the Late Middle Ages to Contemporary times. However, as stressed previously, since an outcrop can be exploited over a long period, a *mola*-based toponym cannot exclude millstone workings previous to the Reconquest.

It is reasonable, nonetheless, to suppose that older sites that had already abandoned millstone production before the Reconquest, would not have been renamed with a *mola*-based name unless millstone workings were renewed. Consequently, the absence of a *mola* place name can be an indicator of production in a distant past. The sites interpreted as dating to the Islamic domination, notably the Rambla Honda (AL- 3a), Zagra (GR-5) and Almadén de la Plata (SE-1), are all devoid of *mola*-based place names.

11.2.4.2. Geographical proximity to a dated feature

The proximity of a millstone quarry to a well-dated feature can also serve to link it, at times, to a chronological period. The best example is the *Sisapo*-Castillejo volcanic quarry (CR-1) dated to the Roman period because of its nearness to the Roman city of *Sisapo* (Fig. 11.9).

Likewise, assemblages of broken or unfinished querns and millstones in dated contexts, such as the group of volcanic roughouts at the Roman settlements of Oreto y Zuqueca (Granátula de Calatrava, Ciudad Real) and Los Ceniceros (Mazarrón, Murcia), link nearby volcanic quarries to Roman times. This is the case of the supposed workings at Cerro Columba (CR-6) and Cerro de Cabezo de Oliva (MU-2). In similar fashion, the Prehistoric saddle querns of the settlements of Los Pensadores and Las Rajas Prehistoric date the nearby sandstone quarry of Zujaira (GR-12). However, as mentioned before, this chronological indicator cannot stand alone and must be accompanied by other indicators.

Fig. 11.9: Top: aerial view (SIGPAC) of the volcanic domes where Romans quarried querns and millstones (CR- 1). The proximity (500 m) of the quarry to the Roman city of Sisapo *suggests a chronological link (SIGPAC); bottom: view from the south-east of the Roman city and the quarry (photograph by Tor Grenne).*

11.2.4.3. Relative chronology and vertical stratigraphy

In the controlled context of a modern archaeological excavation of a millstone quarry one can establish stratigraphic relations between different features such as backfills, quarry faces and quarry floors. Furthermore, one can at times collect artefacts in secure stratigraphical contexts that furnish key clues to dating a site. Surface surveys of millstones quarries, on the contrary, do not benefit from these dating techniques.

During the excavation of a settlement or cemetery, the trained eye can observe how individual features, such as pits or tombs, can be cut by other pits or tombs, establishing a relative chronology between the different features that serve, with other factors, to determine the general phasing of a site. In millstone and construction quarries these types of relationships between individual extractions rarely exist. Individual stone extractions, although contiguous, rarely overlap because when choosing the location of a cylinder, a millstone maker sought out a homogeneous sector of rock and invariably avoided the cuttings of earlier cylinders. Therefore, cylindrical hollows, although contiguous, rarely come into direct contact with each other, making it difficult to establish chronological relations.

There are at times, nonetheless, exceptions to this rule that serve for relative chronology. Smaller querns or millstones, were sometimes cut directly into the heart of larger abandoned cylinders, as is the case at Zone I of the excavation of Claix (Charente) (Belmont *et al.* 2011: 206, fig. 5). From the chronological standpoint, it is obvious that the smaller millstone was scored after the aborted attempt to extract a larger millstone.

In the context of millstone research this direct example of relative chronology is very interesting. A rotary quern cut from the heart of a large, abandoned millstone suggests quern workings dating to the Middle Ages. In any case it excludes a Roman date for the quern because large discoidal millstones over one metre in diameter are unknown in Roman times. A medieval date for the workings of Zone 1 at Claix, was confirmed by the thermoluminescence analysis of a potsherd (Belmont *et al.* 2011: 205). In southern Spain, there is an identical case at the Puerto de la Cadena (MU-1) (*cf.* Fig. 11.24) that points to workings from the Middle Ages.

Regarding vertical stratigraphy, an example of its use for dating a millstone quarry is illustrated by the relationship between a sector of the Montesa quarry (V-2) and a fortification perched directly above it (*cf.* Fig. 11.25). The fortification certainly antedates the quarry, not only because its foundation cuts through the top of the tubular extractions, but for reasons of the castle's stability, it would not have been wise to score millstones within such a close range of the foundations of the fortification.

11.2.4.4. Extraction techniques indicator

Although pick work evolved little over time, detailed study of extraction techniques in the framework of a controlled archaeological excavation can, at times, provide evidence of relative chronology. For example, the chronology between Zones I and II of the site of Claix (Charente) was evidenced, among other things, by subtle differences in splitting techniques. The querns of Zone I, interpreted as the older workings, were split by wedges lodged in large, single, trapezoidal holes, whereas those of the later Zone II were split by wedges in triangular cuttings (Belmont *et al.* 2011: 205-207).

Nonetheless, extraction techniques in the south of Spain are rarely visible in surface surveys. This indicator is therefore only valid in very specific cases where the tool marks are still visible, such as the medieval workings of Zagra (GR-5) and Almaden de la Plata (SE-1a/b).

11.2.4.5. Oral information

Oral information or history, especially the memory of direct descendents of millstone makers, can be very useful to date at least the last phases of workings of certain sites, as in the case of Moclín (GR-1), Benaocaz (CA-12) and Playa de Agua Dulce (CA-4). In the case of Moclín, I was informed by two older residents that millstone scoring was still taking place in their youth, placing work toward the 1950s. In the case of Benaocaz, the oral history, recorded in an

Internet blog, describes the "rolling stone" transport of a millstone by a local resident. Finally, José Antonio Martínez, the chronicler of the city of Rota, reports that an older resident of the city recalls that his father, a miller, had a millstone scored at the nearby Playa de Aguadulce. All of these cases indicate that production was still active in the first half of the 20th century.

Likewise, as noted previously, the dating of the last phases of the workings in the mountains of Palencia in northern Spain is based on interviews by Maestro Hernández (2011) of the descendents of the *moleros* that appear in a black and white photograph dating to 1933 (see chapter 1).

11.3. The chronology of millstone quarries in southern Spain

11.3.1. Introduction: The chronological assemblages

To illustrate the dating of the millstone quarries in southern Spain, I have divided the different workings into three chronological assemblages, represented in three tables, based on their earliest reputed phase of production. The description of each assemblage, summarised below, is developed further in the coming pages.

Chronological assemblage 1 *(Table 11.2)*

Chronological assemblage 1 presents the sequence of quarries with the oldest phases of production. These are for the most part Prehistoric, Iron Age, Roman and Medieval productions of saddle querns, rotary querns, ring-mills and small millstones. Some of these workings, for example Rota (CA-3), Moclín (GR-1) and Castillo de Locubín (J-1), also have recent chronological phases, identified by texts or oral tradition. The indicator of the rock type, especially in the case of Roman sites (volcanic rocks), plays an important role in establishing chronology.

Chronological assemblage 2a and 2b *(Table 11.3a and Table 11.3b)*

This chronological assemblage is established exclusively from written sources and is illustrated in two tables separated by a hiatus of about 30 years. The sequence is founded on the earliest known extractive phase of the sites. Instead of using the less accurate, traditional chronological units, the tables are divided into more precise columns by centuries.

The first assemblage (2a) (Table 11.3a), divided into four subgroups (a, b, c, d), presents the sites with phases dated by sources from 1481 to 1794 and corresponds broadly to the Modern period. Assemblage 2b, although the logical prolongation of the first, is marked by the transition to Contemporary times marked by the French Revolution. Most of these quarries are cited in the dictionaries of Miñano (1826-1829) and Madoz (1845-1850). Here are also the few sites from the end of the 19th or beginning of the 20th century, dated by written or oral sources.

Chronological assemblage 3 *(Table 11.4)*

This last chronological assemblage records the approximate phasing of the remaining millstone quarries that do not benefit from any means of dating except for the morphometrics of their extractions and extraction hollows. From observations in the field (the first nine sites on the list) and from photographs gleaned from hiking itineraries or historical Internet websites, it appears that production at these sites consisted of cylinders over 1.00 m in diameter. Based on this dimension, these exploitations therefore date from anytime between the Middle Ages and Contemporary times. Older working phases at these sites cannot be excluded as the sites have not been the object of a field survey. Here, due to the lack of chronological precision, I am constrained to apply broad chronological units as in the case of the first assemblage.

The tables present the symbols indicated below. The different patterns for the symbols, the same as those used in the geological chapter, indicate the petrography.

The explanations of the chronological indicators for each site are given in the following pages. Those of assemblage 1 are much more complex and require much more ink than the more straightforward assemblages 2 and 3.

As future millstone quarry research progresses, the chronology of many of these sites will be refined, completed and corrected. One of the objectives of this study is to pave the way for this future work.

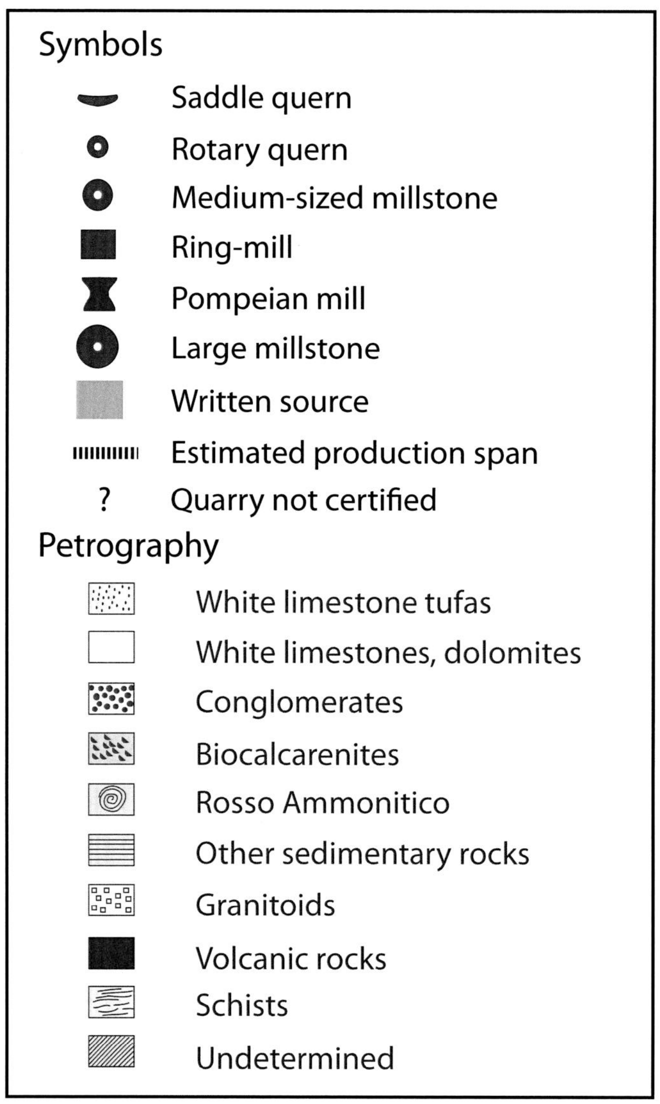

Symbols used in the different chronological tables.

11.3.2. Chronological assemblage 1

This assemblage comprises 31 sites. Its indicators suggest early phases of work (Table 11.2).

11.3.2.1. Prehistoric and Early Protohistoric quarries

Saddle querns recovered at settlements date the three Prehistoric or Early Protohistoric quarries identified in the study area. Of the three, only El Barronal (AL-10), in the volcanic fields of Cabo de Gata, reputedly has extraction features in the form of small pits that correspond, presumably, to diggings to detach blocks from the summit of the columnar jointing. The chronology of these workings is corroborated by surface finds (pottery and stone artefacts), dated approximately from the Chalcolithic and the Bronze Age (Carrion *et al.* 1993).

The second site, Cabezo de la Oliva (MU-2), also exploiting a volcanic rock (lamproite), was identified by an undated saddle quern surface find in the 1970s by Saturnino Agüera (Agüera Martínez *et al.* 1999). Unfortunately, besides the typology of the quern, there are no other chronological indicators to date this site that saw use for rotary querns in Roman times (see below).

The presumed yellow sandstone quarry of Zujaira, Granada (GR-12) is dated by saddle querns (and other lithic material of identical rock) collected at two nearby Prehistoric settlements. A first Late Neolithic extraction phase of the quarry is corroborated by saddle querns associated with coarse globular vessels and flint at the settlement of Los Pensadores (Fig. 11.10). A second later Chalcolithic phase is based on saddle querns associated with large, low carinated vessels and minute, barbed, flint arrowheads at Las Rajas (Anderson 2010, unpublished). This extraction site, as discussed below, is also presumably the source of Roman rotary querns. Recent block workings, notably for the local church, has probably destroyed all or part of the older extraction sectors.

Fig. 11.10: Example of a complete saddle quern recovered in the backfill of a silo at the Late Neolithic settlement of Los Pensadores. The source of the yellow sandstone is presumably the nearby outcrop of Zujaira (GR-12), a site that was exploited for building material in recent times..

11.3.2.2. Late Iron Age rotary quern and millstone quarries

Although not identified in the field, the chronology of the exploitation of limestone tufa (called travertine) quarries in the surroundings of Almedinilla (CO-15) is established with a high degree of precision towards the middle of the 2nd century BC based on the findings of rotary querns and larger millstones at the stratigraphically secure settlement of Cerro de la Cruz. It was, in fact, a Roman military incursion that brought about the settlement's sudden and violent end

Table 11.2: Chronological table of the earlier quern and millstone quarries. This chart illustrates, for the most part, sites that are not recorded in written texts. The exploitations of Rota (CA-3), Moclín (GR-1) and Castillo de Locubín (J-1), identified by recent written or oral sources, appear in this chart because of their old saddle quern or rotary quern extraction phases.

Code	Site	Surveyed	Neolithic Chalcolithic Bronze Age	Late Iron Age	Roman	Middle Ages	Modern	Contemporary
AL-10	El Barronal	no						
MU-2	Mazarrón	no	?		?			
GR-12	Zujaira	yes	?		?			
CO-15	Almedinilla district	no		?	?			
AL-1	Cerro de Limones	yes						
AL-2	Hoya del Paraiso	yes						
CR-1	Sisapo	yes						
CR-2	Las Herrerías	yes						
CR-6	Cerro Columba	no			?			
AB-2	Cancarix	yes			?			?
CA-1	Trafalgar	yes						
CA-3	Playa de Costilla, Rota	yes						
CA-2	Pta. Camarinal/Pta. Paloma	yes			?			
MU-4	Fortuna	no			?			
BA-6	Mérida district	no			?			
SE-7	Gerena District	no			?			?
CO-11	Piedras Moleras	no			?	?		
CC-3	Villar de Plasencia 2	no						
A-2	Barranco Molinos, Ibi	no						
GR-5	Zagra	yes						
SE-1a	Almadén de la Plata	yes						
SE-1b	Almadén de la Plata	yes						
SE-2	El Pedroso	yes						
GR-4a	La Merced 1	yes						
MU-1	Puerto de la Cadena	yes						
AL-3a	Rambla Honda	yes						
AL-3b	Los Leonardos	yes						
GR-4b	La Merced 2	yes						
V-2	Montesa, La Mola	no						
GR-9	Playa de Carchuna	yes						
HU-8	Almonaster, Los Molares	no				?		
AL-8	Cerro el Chispas	yes						
J-1	Castillo de Locubín	yes				?		
GR-1a	Moclín	yes						
CO-12	El Patriarca	yes						

Fig. 11.11: The limestone tufa (travertine) Late Iron Age production is dated by a series of millstones discovered in situ *at the well-dated Iberian Culture settlement on Cerro de la Cruz, by Almedinilla (photograph left: http://www.rutasconhistoria.es/loc/poblado-ibero-el-cerro-de-la-cruz); photograph right by T. Anderson).*

(Quesada *et al.* 2010). Due to this destruction of the site by fire, several millstones remained *in situ*, some on their podiums (Fig. 11.11). The characteristic handle cuttings and lugs of the upper stones correspond to driving mechanisms known at other well-dated Iron Age sites in the north-east of the Iberian Peninsula (*e.g.* Alonso *et al.* 2011). This rock type, as seen by other Iron Age examples in southern Spain, is also an indicator of this period.

11.3.2.3. Roman quern and millstone quarries

Dating rotary quern and millstone quarries from Roman times relies first and foremost on the comparison of the abandoned querns at the quarry with examples in museum collections. The large number of finds in museum collections, comprising rotary querns, ring-mills and a few Pompeian mills, constitutes a reference group large enough to date a number of quarries broadly to the Roman period. However, advancing a finer dating for these workings at this stage is not possible due to the absence of well-defined chronological millstone assemblages.

A second general chronological indicator for the Roman period is their propensity to exploit hard, vesicular volcanic rocks. In southern Spain these rocks comprise basalts, andesites, rhyolites and lamproites.

Fig. 11.12: Unfinished querns at the Cerro de Limones dacite quarry. The start of a radial cutting and "sombrero" protuberance are examples of typological indicators of the Roman period.

Roman volcanic quern and millstone quarries

The quern quarries of Cerro de Limones (AL-1) and Hoya del Paraíso (AL-2) in the Cabo de Gata, on the southern edge of the south-eastern Spanish volcanic district, are dated to Roman times by the typology and morphometrics of the numerous abandoned blanks and broken roughouts at the site (Anderson *et al.* 2011a: 156-157). The first is a dacite exploitation while the second, based for the moment only on macroscopic analysis, falls broadly into the dacite-rhyolite unit. A typological feature linking production from the quarry of Cerro de Limones with examples in museums is an upper stone with an abandoned radial handle slot cutting (Fig. 11.12a). But it is the many discarded *"sombrero"* type lower stones with their central protuberance (Fig. 11.12b) that indicate a connection between these workings and the querns from Roman contexts stored in different museums throughout south-eastern Spain.

It is reasonable to suppose that the volcanic (lamproite) outcrop at Cabezo de la Oliva (MU-2) near Mazarrón, Murcia (cited above as a potential source of Prehistoric querns), included a Roman phase. This is based on several unfinished or broken lamproite querns recovered at the nearby Roman villa of Los Ceniceros, notably a broken *"sombrero"* lower stone (Fig. 11.13) that is typologically equivalent to the Cabo de Gata production.

The volcanic quarry of *Sisapo* (CR-1) (Anderson *et al.* 2011a: 159-161) in the volcanic district of Campo de Calatrava in central Spain, is dated to Roman times by the indicator of petrography and the sizes of the extractions (40 and 80-90 cm) that correspond respectively to Roman rotary querns and cylindrical mills (Anderson *et al.* 2011a). Unfortunately, no discarded roughouts were visible during my visit. Nonetheless, in photographs published in an archaeological study of the city of *Sisapo* (Fernández Ochoa *et al.* 2002: 121), there are examples of both abandoned rotary querns and cylindrical mills, notably an aborted ring-mill upper stone (Fig. 11.14). The proximity (500 m) of this volcanic exploitation to the Roman city is also a chronological indicator.

Fig. 11.13: The "sombrero" *type lower stone from the Roman villa of Los Ceniceros, Mazarrón (Murcia), dates to Roman times. The source is presumably the Cabezo de Oliva, the nearest lamproite outcrop (3 km away).*

Fig. 11.14: Abandoned cylindrical drum measuring about 80 cm in diameter from the quarry of Sisapo. *The circular cavity on the upper face suggests that it is an aborted Roman ring-mill upper stone (photograph from Fernández Ochoa* et al. *2002: 121, fig. 6: 4).*

The Las Herrerías quarry near Bolaños de Calatrava (CR-2), also in the Calatrava volcanic district, exploited a nephelinite olivine or olivine melilitite lava. The original extraction areas have been destroyed by the modern exploitation. Roman workings are certified by ten abandoned roughouts, miraculously set aside by the present day quarry workers (Anderson *et al.* 2014: 120) (Fig. 11.15). All of them bear morphometric features characteristic of mills identified on Roman settlements throughout southern Spain. Among the lot there is a fine example of an aborted "*Volubilis*" ring-*catillus* (no. 10).

The Cerro Columba (CR-6) in Granatula de Calatrava (Anderson *et al.* 2011a: 159-161), another a volcanic dome in the Calatrava district on the banks of the Jabalón River, is presumed to be the source of the construction material of a Roman bridge (Alañon 1982: 230), now under the waters of the Jabalón Dam. Roman quern production is inferred from roughouts and blanks measuring between 38 and 40 cm in diameter at the nearby settlement of Oreto y Zuqueca (Fig. 11.16). These roughouts are associated with the site's older Roman levels or recycled as construction material in subsequent phases dating to the Middle Ages.

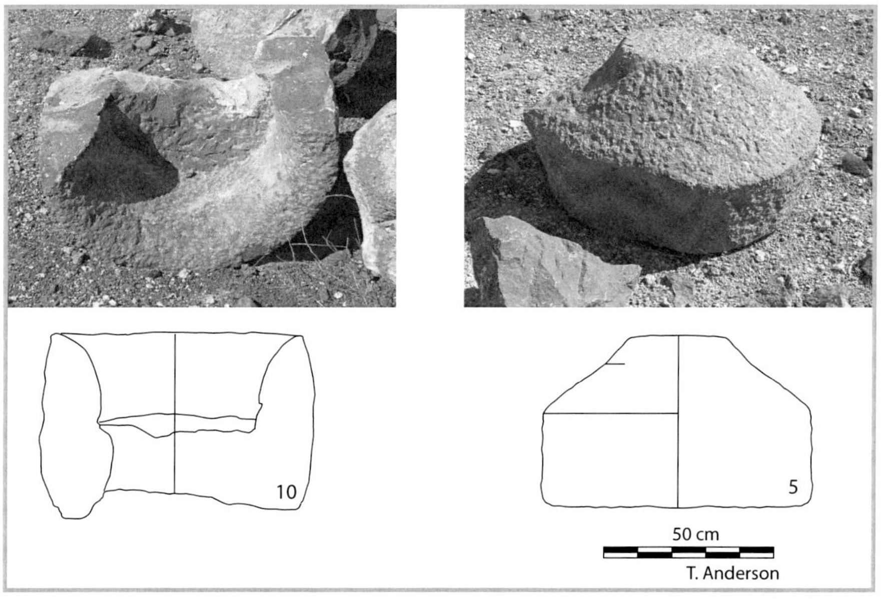

Fig. 11.15: Examples of unfinished millstones from the volcanic quarry of Las Herrerías (CR-2) near Bolaños de Calatrava. No. 10 is a broken ring-catillus and no. 5 is an unfinished conical lower stone. These types are characteristic of the Roman period (sketches by T. Anderson).

Fig. 11.16: Examples of unfinished or broken volcanic rotary querns at the Roman and Medieval settlement of Oreto y Zuqueca, presumably from the quarry of Cerro Columba (CR-6), a nearby volcanic dome. The diameters of the half dozen unfinished querns, ranging from 38 to 40 cm, point to a Roman production.

The Pitón de Cancarix (AB-2), another impressive volcanic (lamproite) outcrop in the Sierra de las Cabras (Albacete) of the south-eastern volcanic province, is presumably the source of Roman rotary querns discovered at the nearby Roman settlements of La Horca and El Tolmo in the Minatda-Agramón Valley (Jordan *et al.* 1984: 222-223, 227). It is also the reputed source of modern rotary querns found in local farmhouses (Jordan 1997: 17). The prominent columnar jointing of this outcrop, and its naturally detached columns scattered over its southern slope, could have been worked for querns. Yet, my recent visit to the southern sector of the site with Tom Heldal (NGU) only confirmed modern (19th or 20th century?) workings.

Roman millstone quarries exploiting other rocks

Roman biocalcarenite millstone quarries: In the south-western sector of the study area, far from the volcanic districts, rotary querns were scored from of shell-rich, highly porous and abrasive, biocalcarenite outcrops. Two of these *ostionera* sites have been identified at Trafalgar (CA- 1) (Anderson 2011a: 231-232, fig. 2-3) and Rota (CA-3). Their quern extractions are roughly dated to Roman times by comparison with querns from Roman contexts at settlements such as *Baelo Claudia*. The study of the *Baelo Claudia* querns indicates a range of diameter between 24 to 43 cm (Fig. 11.17), perfectly compatible with the morphology of extraction hollows of these quarries.

Other examples of Roman *ostionera* querns compatible with the extraction hollows of coastline quarries are known at other Roman settlements near the Straits of Gibraltar, notably at *Carteia* (Salvador Bravo, pers. comm.) and *Iulia Traducta* (Vargas & Bernal 2009: 179: fig. 24; Domínguez & Bernal 2011: 454, fig. 4). The eleven *Iulia Traducta* querns are dated more precisely to the 5th-6th centuries AD and interpreted as having served to grind fish flour.

*Fig. 11.17: Examples of biocalcarenite (*ostionera*) rotary querns from the Roman city of* Baelo Claudia*. The study of the complete querns indicates a wide range of diameters, as seen in the line graph. These sizes are compatible with the extraction hollows of the quarries located on the coast of Cádiz.*

Roman *ostionera* millstone production was not limited to querns, as evidenced by an unfinished "hour-glass" Pompeian *catillus* recovered during early excavations at *Baelo Claudia* (Fig. 11.18). This unquestionably Roman millstone, probably a copy of an imported volcanic model, is an isolated case and cannot be dated precisely. It is nonetheless, for the moment, the only confirmed example of a Pompeian *catillus* produced on Iberian soil, as pointed out in the colloquium of Lons-le-Saunier (November 2011) (Anderson *et al.* in press). It most likely represents the recycling of a construction element (column drum?) scored from a nearby Roman quarry either at Punta Camarinal (CA-2a) or Punta Paloma (CA-2b).

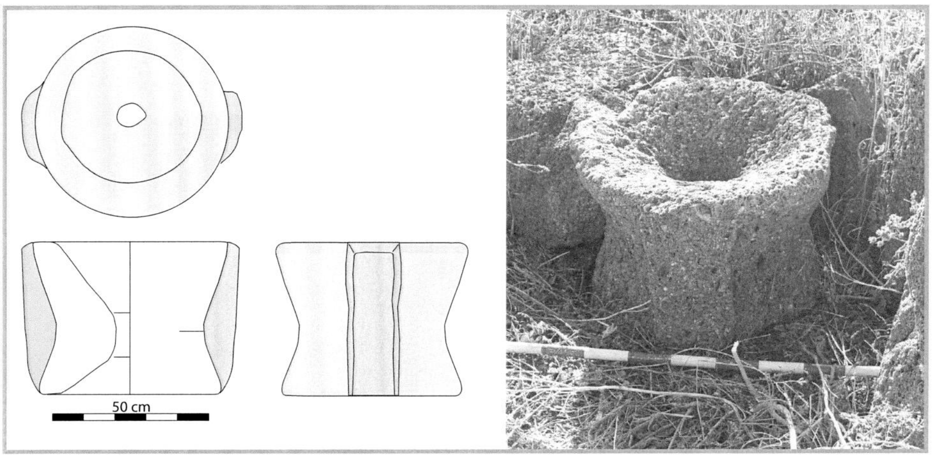

Fig. 11.18: Unfinished biocalcarenite Pompeian upper stone. It was probably was recycled from building material, such as a column (photograph and drawing by T. Anderson).

Roman limestone tufa (travertine) millstone quarries: Limestone tufa workings at Almedinilla, Córdoba (CO-15), known since the Iron Age, did not die out after the Roman conquest. The evidence that production continued in Roman times is based on the typology of a few rotary querns in the collection of the Museum of Almedinilla (Fig. 11.19). One upper stone has grain receptacles and radial slot cuttings that are identical to those of volcanic models produced in Roman quarries in the Cabo de Gata. A second, with a rectangular lateral handle hole and rynd cuttings on its upper surface, betrays another Roman tradition specific to the south of Spain. It must be noted that no quern workings of this rock has been identified in the field.

Roman calcareous sandstone millstone quarries: To conclude the sector on the means of dating the different Roman quarries, there are other sites exploiting sedimentary rocks that have not be confirmed in field work. The first is inferred from a single, unfinished ring-mill 60 cm in diameter illustrated in the article by Matilla in the Contemporary oil roller quarry of Fortuna (MU- 4), Murcia (Matilla 2001: 273, foto 7). This solitary example, typical of the Roman period, appears to be a sandstone or fine conglomerate and could correspond, like the Contemporary oil rollers from the site, to an exploitation of surface blocks. A second fine conglomerate ring-mill, typically Roman, is stored in the depository of the Museum of Murcia (no. 31, Anderson *et al.* 2011b: 16).

The second quarry is inferred from a yellow sandstone quern fragment recycled as construction material in the wall of the Roman villa of El Tesorrillo, Granada (excavated by I. Rodríguez). This piece, securely dated to Roman times, probably originated, as the saddle querns from the Prehistoric sites of Los Pensadores and Las Rajas (see Prehistoric quarries), from the nearby outcrop of Zujaira (GR-12). These examples, like the limestone tufa querns of Almedinilla, point to local Roman quern workings that were an alternative to the volcanic imports.

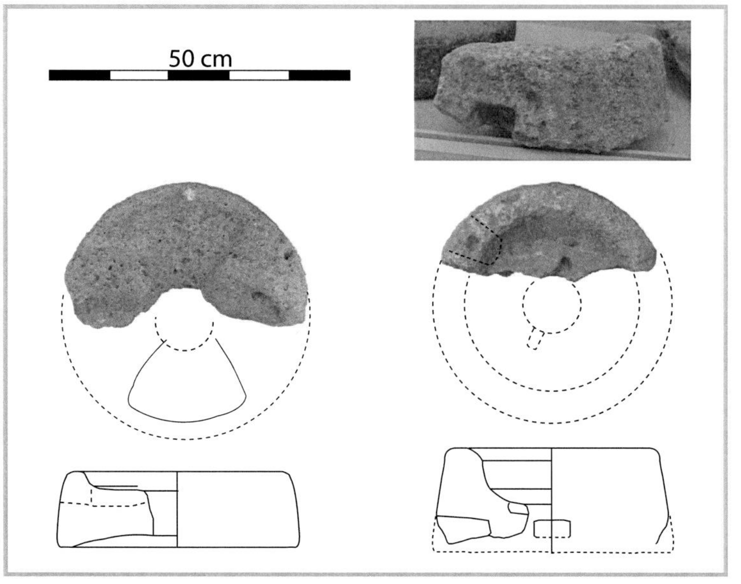

Fig. 11.19: Roman limestone tufa querns in the Museum of Almedinilla (photos and drawings by T. Anderson).

Roman granite millstone quarries: For logistical reasons, I have spent little time in the granite production districts to the north of the Guadalquivir River Basin and have not identified any Roman millstone exploitations in the field. All the same, there is sufficient evidence from examples stored in museum collections or mentioned in publications, to validate that granite, because of its hardness and accessibility (in the form of surface boulders or bedrock outcrops), was exploited for querns and millstones on a large-scale in Roman times.

Production centres around Mérida (BA-6), for example, must have abounded, as seen by the many granite rotary querns and ring-mills in the depository of the Museo Nacional de Arte Romano and at the *lapidarium* of the Alcazaba. Because of storage conditions, unfortunately, I had only limited access to this collection. All said, the basic construction material of Roman Mérida was granite as seen by its most imposing monuments. An important source of granite blocks was the huge "pit" quarry just outside Mérida that later became the Proserpina Dam. Other nearby quarries supplying granite for the Temple of Diana are Cuarto de la Charca, El Hinojal and Royanejos (Álvarez 1991: 86). It is reasonable to suppose that any of these workings could have also supplied millstones.

The granite districts of Gerena (SE-7) and Piedras Molares (CO-11) by Villanueva de Córdoba present similar situations. Each of these districts shares a long tradition of granite production and their museums store both Roman and Medieval rotary querns.

11.3.2.4. Undated rotary quern quarries

The rotary quern quarries of the Barranco de los Molinos at Ibi (A-2) in Alicante and Villar de Plasencia 2 (CC-2) in Cáceres, sites that I have not been able to visit personally, are awkward to date and are not included in the table of assemblage 1. These two true extractive exploitations, with their small, relatively thick extractions (Fig. 11.20), could date anywhere from Late Iron Age to the Middle Ages. Although, a later date from the Middle Ages can probably be dismissed due to the diameter/thickness proportion, the rock type of Ibi is better suited to the Middle Ages. The search for other chronological indicators for these sites (for example, similar rotary querns in local museums or nearby settlements) is essential to narrow down the date of these workings.

Fig. 11.20: Example of a small unfinished quern extraction at the quern quarry of Ibi (A-2). The site is difficult to date and does not appear in any chronological assemblage. Although the proportion of the cylinder's diameter to its width suggests a Roman or even Iron Age date, its rock type is more characteristic of the Middle Ages (photograph by Ágata Marquiegui).

11.3.2.5. Middle Age rotary quern and millstone quarries

The dating of quern and millstone quarries to the Middle Ages, a span of about a millennium, is particularly complicated. The problem is the practical absence of (or absence of access to) written sources and other chronological indicators, as well as the scarcity of securely excavated archaeological settlement complexes and published millstone assemblages. It is, even so, obvious that the Middle Ages inherited a large array of milling traditions from Roman times, from domestic rotary querns, to animal-driven mills, to industrial watermills, and this variation seems to be reflected in the wide variety of millstone diameters and petrography.

One of the few factors that can be used to date these workings is their difference with earlier Roman workings. Hence flat, discoidal, non-volcanic rock quern extractions, measuring about 50 cm in diameter, is one of the few characteristics pointing to a date from the Middle Ages. Typological factors of abandoned querns (in particular handle holes) cannot be used because these quarries, on the whole, have left nothing behind but extraction hollows. To complicate establishing their chronology, most of these sites (except the homogeneous Zagra, GR-5) are characterised by a variety of extraction sizes, possibly from different periods. This is the case, for example, at Rambla Honda (AL-3) and Puerto de la Cadena (MU- 1).

As to publications of millstones from the Middle Ages in southern Spain, the most complete assemblage is the dozen illustrated querns in the doctoral dissertation of Encarnación Motos Guirao from the settlement of El Cerro de Castillón in Montefrío (Granada) dated between the 7th and the 12th centuries (Motos Guirao 1987: 465-467, 473-474, 479-480, PL. I-IV). I have recently had access to this assemblage. Of note is the dominance of limestones and sandstones. A second publication includes drawings of two rotary querns from the settlement of Cabezo del Moro in Murcia (Gutiérrez Lloret 1996: 205-207). Other information on querns has been gleaned from querns exhibited in museums and depositories (notably the Museums of Siyâsa and Murcia). Yet, these querns devoid of context only serve to establish general dating guidelines. Publications including the larger medieval animal-driven or water-driven millstones is, to my knowledge, non-existent. Furthermore, these larger models are also absent from museum collections, complicating the task of dating quarries.

Outside the study area, the work of Joaquín Sánchez on the Balearic Island of Minorca has brought to light a number of quern quarries with extraction hollows reportedly dating from the 10th-13th centuries (Sánchez Navarro 2011). Another interesting reference to a few rotary querns from the end of the 9th to the middle of the 10th century is an article about a shipwreck off the coast of France. According to the authors, the pottery at this wreck, as well as the querns, probably originated in southern Spain at Pechina (Almería) (Joncheray & Sénac 1995: 30-32).

Samuel Longepierre, in his recent thesis regarding southern France, has also illustrated a series of rotary querns similar to those observed in southern Spain, dating from the 7th to the 9th centuries. Although discoidal like the Andalusian models, these querns show different driving fittings (Longepierre 2011: figures 358-361). Even as far as Stavanger, Norway, rotary querns of this period are very thin and reveal flat working surfaces (Dahlin & Anderson 2014a).

Thus, to recapitulate, the main indicators to date Middle Age quarries are morphometric, that is, the presence of quern extractions measuring 50 cm and/or the presence of millstone extractions between 70 and 90 cm in diameter. Other "negative" evidence suggesting a Medieval date is the absence of written sources, the absence of characteristic place names and the shunning of volcanic rocks.

The coarse sandstone site of Zagra (GR-5), a true extractive quarry, falls into this chronological range due to extraction hollows corresponding to discoidal cylinders about 50 cm in diameter (Fig. 11.21a). A date of the Middle Ages is reinforced by the study of the neighbouring cemetery just steps away from the millstone workings, suggesting that the rock-cut burial chambers date to the 6th or 7th century (end of Visigothic domination) (Jiménez Puertas 2002: 226). The few surface potsherds, however, point to a date from the 10th to the beginning of the 12th century (Jiménez Puertas 2002: 226). Since the dating of the pottery does not concur with that of the cemetery, it is plausible that it was related to the quarry, placing the quern workings during the Islamic rule.

It is, in fact, difficult to conceive that both features at Zagra, one dedicated to the realm of the sacred and the afterlife and the other a millstone workshop, saw use simultaneously. The single, aborted millstone measuring 1.20 m, toward the periphery of the site, reveals that the site was frequented briefly at a later stage (Middle Ages or Contemporary times). Production of larger millstones, however, never progressed at this later stage.

Fig. 11.21: View of the quern quarry of Zagra (GR-5) with extractions corresponding to discoidal querns about 50 cm in diameter suggesting a date from the Middle Ages. On the picture to the right, one can perceive (with difficulty) the splitting marks.

Another detail related to the chronology of Zagra is the splitting tool marks. Most are weathered and no longer visible. Yet, in a few cases, it is possible to perceive regularly spaced indentions along the periphery of the cylinder that could point to chiselling, either for direct splitting or to carve holes for small iron wedges or pegs (Fig. 11.21b). In any case, these marks recall the techniques identified by Joaquín Sánchez at the millstone workings that he dates to the Islamic period on the Island of Minorca (Sánchez 2011).

The conglomerate exploitation along the Calzadilla riverbed at Almadén de la Plata (SE-1a/b) is another example of a rotary quern quarry dated to the Middle Ages. The discoidal extractions 50 cm in diameter (Fig. 11:22a) are identical to those of Zagra (GR-5). The splitting technique, particularly well preserved, is also analogous to that of Zagra. Furthermore, the lower sectors of the site (SE-1b), along the Calzadilla riverbed, show a number of extractions measuring a diameter of 80 cm (Fig. 11:22b). These larger extractions, most likely contemporary with the rotary quern hollows, were probably destined to be medieval animal- or water-driven mills.

Two other small quern extraction sites, La Merced 1 (GR-4a) and El Pedroso (SE-2), can be dated broadly to the Middle Ages. The chronology of the first, with discoidal quern extractions of 50 cm, seems relatively secure based on the morphometrics of the hollows. Although the extractions of the second are smaller, around 40 cm, they are too thin to be older than the Middle Ages.

Rambla Honda (AL-3a) and Puerto de la Cadena (MU-1), two large extensive "puddingstone" quarries, produced a wide variety of millstones. Each of these sites reveal rotary quern extractions measuring from 40 to 50 cm, "medium-sized" extractions between 70 and 90 cm, and larger models over 1.00 m in diameter (up to 1.40 m). The rotary quern workings at these two sites correspond to thin, discoidal models that bear no resemblance to Roman models. The medium-sized extractions from 70 to 80 cm are also discoidal and cannot be mistaken for the thicker "drums" destined for Roman ring-mills (Fig. 11.23). The even larger millstones, a size unknown in Roman times, could date anywhere from the Middle Ages to Contemporary times. Hence all of these factors preclude a Roman dating.

As noted previously, the Puerto de la Cadena (MU-1) quarry includes a rotary quern hollow cutting carved directly into the heart of an unfinished millstone three times its size (Fig. 11.24). This detail reveals that rotary querns were being produced at this site simultaneously or after

Fig. 11.22: Details of the quarry of Almadén de la Plata (SE-1a/b). a) Discoidal extractions corresponding to querns measuring 50 cm in diameter with peripheral chisel or small wedge or peg holes. b) Larger extractions measuring 80 cm in diameter for animal- or water-driven mills.

large millstones. Based on this feature, as well as the presence of extractions from 70 to 80 cm and the absence of a *mola*-based place name, the Puerto de la Cadena (MU-1), and its counterpart, the Rambla Honda (AL-3a), can be dated presumably to the Middle Ages.

To this group can also be added the quarry of Los Leonardos (AL-3b) and the subterranean quarry of La Merced 2 (GR-4b). Although devoid of rotary quern workings, each reveals "medium-sized" discoidal extractions, as well as later undated workings.

The quarry of Montesa, La Mola (V-2) is interesting from the chronological standpoint. Its place name *La Mola*, probably deriving from *Molares* (millstone quarry), suggests a date after the Christian Reconquest, an event, in this region, dating to the beginning of the 13th century. The upper layers of the high vertical extraction tubes of one of its oriental sectors are cut (according to Juan García Cerdá) by the foundations of a fortification perched on the top of the hill. This would indicate that the fortification was built after the abandonment of this working area (Fig. 11.25), probably towards the 14th century during the region's re-population after Islamic rule (Guichard 2001: 248). In any case, it is doubtful, for reasons of stability of the fortification, that extraction would have been allowed along this quarry face if its foundations were already in place. This stratigraphical relation, therefore suggests millstone extraction sometime before the 14th century.

Fig. 11.23: Medium-sized examples of the unfinished extractions of 70 and 80 cm in diameter at the Puerto de la Cadena (MU-1). This model does not correspond to earlier Roman ring-mills and must therefore date to the Middle Ages.

Fig. 11.24: Rare example of a small rotary quern extraction in the heart of a large abandoned millstone at the quarry of Puerto de la Cadena (MU-1). This feature indicates the relative chronology of the two extractions and establishes that querns were produced after large millstones, suggesting a date in the Middle Ages (from: http://senderosdecartagena.wordpress.com/2010/12/16/cabezo-del-puerto-de-la-cadena/).

Montesa, La Mola (V-2) merits a detailed study that would certainly provide surprises. It is possible to conceive that the extraction of at least some of the rotary querns and medium-sized millstones might, in spite of the Latin *mola*-based place name, have taken place during earlier Islamic rule. The larger extractions, surpassing a metre in diameter, are probably more recent, from the Late Middle Ages or Modern period. It is therefore reasonable that these later extractions explain the post-Islamic *mola* place name.

Another quarry in the study area that combines three rotary quern extractions (*c.* 40 cm in diameter) with a dozen large hollows corresponding to millstones 1.50 m in diameter is the modest site of Playa de Carchuna (GR-9) (Fig. 11.26). Although the quern hollows correspond to the diameter of Roman quern extractions, the rock, a coarse puddingstone with large rounded clasts, is not known to have been sought out in Roman times in this area. This, once again, points to a date from the Middle Ages for the querns and possibly the larger extractions.

Fig. 11.25: View of a vertical tubular face at the quarry of Montesa (V-2) at the base of the Montesa castle. The upper extractions, at the fortification's foundations, were disturbed by (and therefore antedate) the castle's construction. This therefore places the quarry probably under the period of Islamic rule (photograph by Juan García Cerdá).

According to the archaeological inventory of Huelva, the granite quarry of Los Molares (HU-8), near Almonaster la Real has extractions ranging from 27 cm to 1.30 m. The larger extractions are compatible with Medieval or later exploitations. The problem of the smaller extractions is that 27 cm is generally too small even for the oldest rotary querns of the Iron Age. Hence these undated small extractions deserve revision as they could correspond to a different product.

Several sites dated from the early 19th century in the dictionary of Madoz, have, along with their large extractions, a small number of rotary quern extractions suggesting earlier workings. Cerro el Chispas (AL-8) has a couple of small puddingstone hollows corresponding to rotary querns about 40 cm in diameter and Castillo de Locubín (J-1) has a quern blank measuring 50 cm in diameter. These sites therefore could potentially have early phases dating to the Middle Ages. At Moclín (GR-1a) a phase from the Middle Ages seems apparent, not only by the handful of quern extractions, but by the medium-sized hollows tantamount to millstones of 80 cm.

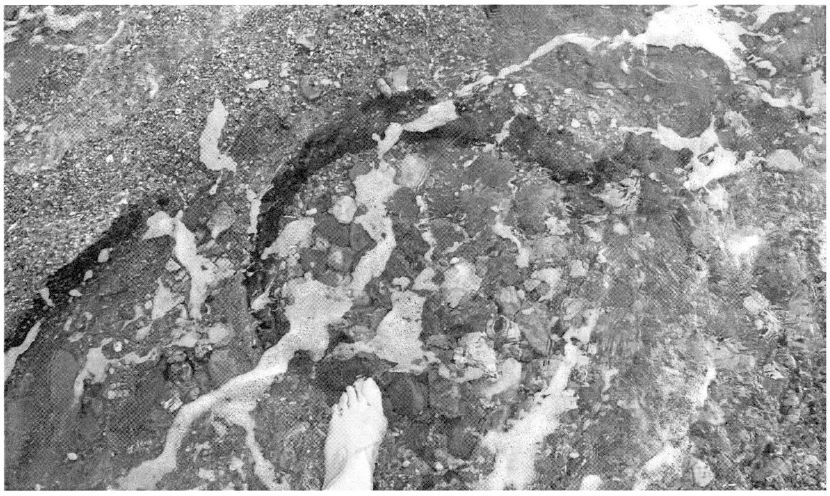

Fig. 11.26: View of a quern extraction measuring 40 cm in diameter at the coastal quarry of Carchuna (GR- 9). The size of the extraction coincides with that of Roman or Medieval times. The rock type, in turn, a sort of puddingstone, is generally associated with workings dating from the Middle Ages to Contemporary times.

To conclude the first assemblage of millstone quarries is the site of El Patriarca (CO-12), on the outskirts of the Córdoba, a city known for its watermills. The site comprises hollows indicating millstones from 0.90 to 1.45 m in diameter. Altamirano and Antón (2012: 339) propose a date from either Roman or Medieval times. The sizes of these discoidal millstones, however, do not concur with Roman models. They do, nonetheless, fit the model of Middle Age millstones, in particular, the smaller examples. It is interesting to note that in spite of the quarry's proximity to the city of Córdoba, the workings are not mentioned in the historical archives identified by the medievalist Córdoba de la Llave. This absence could indicate that the Patriarca site has phases that precede the Christian Reconquest.

11.3.3. Chronological assemblages 2a and 2b

11.3.3.1. Assemblage 2a: Quarries dated from 1481 to 1794 (Table 11.3a)

Chronological assemblage 2a comprises 25 millstone quarries with working phases certified by written sources roughly spanning the Modern period. The oldest of these references is a historical archive from 1481 and the most recent is a treatise from 1794. The lowercase letters preceding each group refer to the letters in the first column of Table 11.3a.

a) The millstone quarries from 1481-1509

This assemblage begins with seven sites dated in historical archives over a thirty year period from 1481 to 1509, corresponding to the transition of the Late Middle Ages to the outset of Modern times. The three from the Province of Córdoba (Hornachuelos (CO-13), Montoro (CO-14), Albaida (CO-7), as well as that of Villagordo from Jaén (J-8)), are related to watermills in Córdoba and were identified in research undertaken by the medievalist Ricardo Córdoba de la Llave. There is no evidence that millstone production continued at any of these four sites beyond the date of the written document. For the last three sites (El Torcal (MA-2), Loja (GR-2/3) and El Berrueco (CA-8)), however, there is textual proof of exploitation well into the 18th and 19th centuries.

b) The millstone quarries from 1557 - 1606

The next group comprises six sites spanning half a century in the Modern period dated by historical archives from 1557 to 1606. All of them, with the exception of Montarrón (GU-8), also have written evidence of later workings in the 18th and 19th centuries.

c) The millstone quarries from 1750 - 1754

The site of Fuentealbilla (AB-1), recorded in the Survey of the Marqués de Ensenada (1750 - 1754), is the only quarry that spans the end of Modern times through to the middle of the 18th century. A second site alluded to in the Marqués de Ensenada Survey is that of El Berrueco (CA-8). El Berrueco, however, was, according to an archive in Vejer de la Frontera, already in operation in 1509 and continued to be exploited well into the 19th century.

d) The millstone quarries from 1781 - 1794

This group comprises 14 quarries dating to the end of the Modern and beginning of the Contemporary period. Three are known through archives to have had earlier working phases: El Torcal (MA-2), El Berrueco (CA-8) and Llerena (BA-2). About half are recorded in official surveys. The Pila de Cásares (CA-9), described in detail by the priest of Medina Sidonia Martínez y Delgado (1875), is probably the first, along with that of El Berrueco (CA-8), to be mentioned in a written source other than that of a notarial archive, municipal ordinance or survey response. The exact phasing of this quarry is not known because the text was published in 1875, about a century after the death of the priest. In any case, this source marks the advent of a new medium of recording millstone quarries: the geographical or historical treatise or dictionary.

11.3.3.2. Assemblage 2b: Quarries of the Contemporary period (Table 11.3b)

Assemblage 2b, the extension of assemblage 2a, comprises a total of 49 millstone quarries, yet only 37 appear on this table. The others have earlier phases of production and appear in assemblages 1 or 2a. The letters preceding each group refer to the letters in the first column of Table 11.3b.

e) The "Miñano" millstone quarries from 1826 - 1829

Miñano records production at Chapinería (M-4), Pinilla de Jadraque (GU-2) and Villar de Plasencia 1 (CC-1). Two other sites cited in his dictionary are Castillo de Locubín (J-1) and Ventas con Peña Aguilera (TO-2). These are known to have had earlier working phases and appear respectively in assemblages 1 and 2a.

f) The millstone quarries from publications of 1833

Salvaléon (BA-4), Brihuega (GU-1) and Teba (MA-2) are sites that are dated by three different sources published in 1833. They mark the transition between the sites indicated by Miñano and the large number identified a dozen years later by Madoz.

g) The "Madoz" millstone quarries from1845 - 1850

This assemblage comprises 24 sites identified by Madoz in his geographical dictionary published between 1845 and 1850. It must be noted that three other sites recorded by Madoz appear previously in assemblage 1 (Cerro Chispas (AL-8); Castillo de Locubín (J-1); and

Table 11.3a: Chronological assemblage 2a corresponds to sites gleaned from written sources published between 1481 and 1794, broadly from the Middle Ages to the end of the Modern period. About a third reveal subsequent working phases identified by later written sources.

	Code	Site	Surveyed
a	CO-13	Hornachuelos	no
a	CO-14	Montoro	no
a	CO-7	Albaida	yes
a	J-8	Villagordo	no
a	MA-1	El Torcal	yes
a	GR-2/3	Loja	yes
a	CA-8	EL Berrueco, Cádiz	yes
b	CO-6	Monte Izcar	yes
b	BA-2	Llerena	no
b	M-3	Colmenar Viejo	no
b	GU-8	Montarrón	no
b	TO-2	Ventas Peña Aguilera	no
b	CO-10	Albardado	no
c	AB-1	Fuentealbilla	no
d	MA-7	Sierra Utrera Karst	no
d	GR-13	Otívar	no
d	CA-9	Pila de Casares	yes
d	TO-1	Torrecilla de la Jara	no
d	M-2	Colmenar de Oreja	no
d	MA-3	Alhaurín el Grande	no
d	MA-4	Coín, Sierra Gorda	no
d	MA-7	Jabonero	no
d	BA-3	Llera	no
d	BA-1	Alconera	no
d	HU-7	Fuente de la Aliseda	no

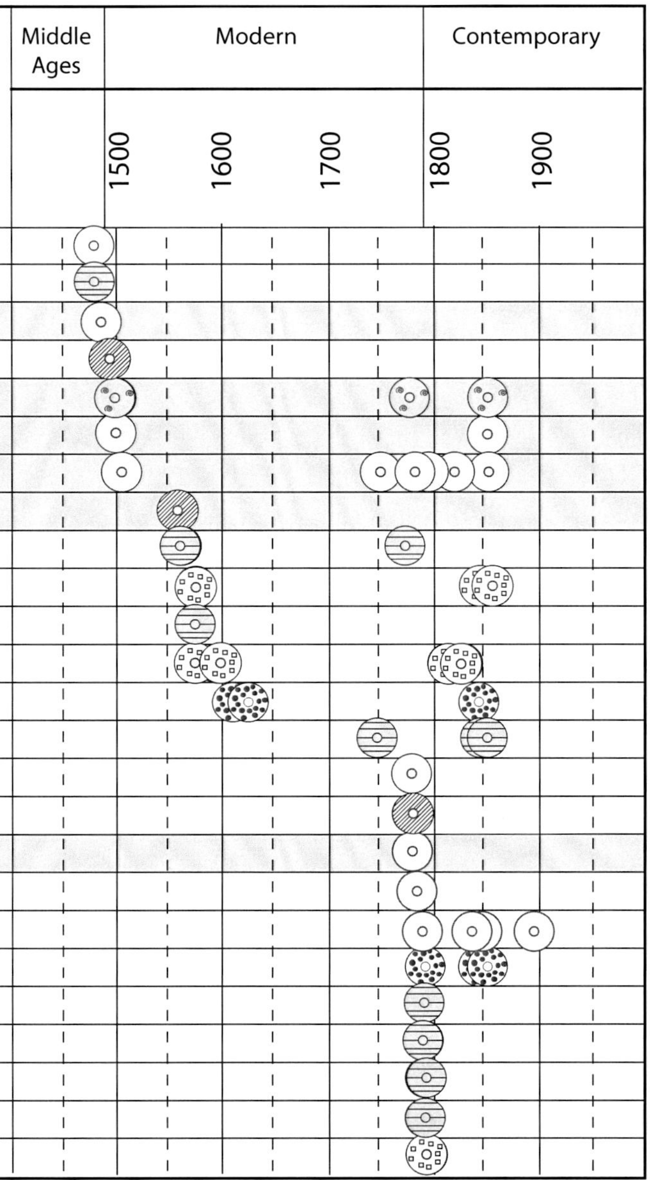

Moclín, (GR- 1a)) because of the presumed older rotary quern or small millstone workings. Likewise, five sites (Loja (GR-2/3); Berrueco (CA-8), Colmenar Viejo (M-3), Colmenar de Oreja (M-2), Alhaurin el Grande (MA-3)) cited by Madoz have older phases of work documented by written sources that appear previously in assemblage 2a (*cf.* Table 11.3a).

h) The millstone quarries from the mid-19th to early 20th century

Los Frailes by Cabra (CO-1) and Tobes (GU- 5) are the two sites cited in geological texts from the late 19th to early 20th century. The fact that the Cabra quarry, cited in other mid-19th century texts, escaped Madoz and his collaborators is surprising due to its size and reputation for top quality millstones.

Table 11.3b: Chronological assemblage 2b. This table corresponds to millstone quarries cited in publications from 1826 to the early 20th century. The list begins with three sites mentioned by (e) Miñano (1826-1829). Most of the workings, however, are recorded by (g) Madoz (1845-1850). The last five (i), known to have been active only in the early 20th century, are identified by oral sources.

	Code	Site	Surveyed	Middle Ages	Modern 1500	1600	1700	Contemporary 1800	1900
e	M-4	Chapinería	no						
	GU-2	Pinilla de Jadraque	no						
	CC-1	Villars de Plasencia 1	no						
f	BA-4	Salvaleón	no						
	GU-1	Brihuega	no						
	MA-2	Teba	yes						
g	CC-4	Bohonal de Ibor	no						
	CA-10	Peña Harpada	yes						
	SE-3	El Hacho	yes						
	GR-6	Alhama de Granada	yes						
	GR-10	Véñez de Benaudalla	yes						
	GR-11	Caniles	yes						
	HU-6	Zalamea la Real	no						
	CC-6	Logrosán	no						
	CR-3	Chillón	no						
	SE-5	Alanís	no						
	CO-9	Los Arenales	no						
	J-7	Linares District	no						
	AL-9	Vera	no						
	J-5	Andújar, Los Morales	no						
	J-6	Andújar, El Pedroso	no						
	J-3	Huelma-Solera	no						
	CO-16	Minas de Espiel District	no						
	CO-17	Fuente Obejuna Jud. Dist.	no						
	SE-6	Villanueva de San Juan	no						
	GU-6	Corduente	no						
	GU-7	Castilnuevo	no						
	B-5	Jerez de los Caballeros	no						
	MA-6	Alozaina	no						
	MA-5	Guaro	no						
h	CO-1	Cabra, Los Frailes	yes						
	GU-5	Tobes	no						
i	CA-12	Benaocaz	no						
	CA-4	Playa Aguadulce	yes						
	CA-5	Roa Martín	yes						
	CA-13	Guadalquitón	no						
	HU-5	Cerro del Águila	no						

i) The millstone quarries of the 20th century

The quarries of Benaocaz (CA-12), Playa Aguadulce (CA-4), Roa Martín (CA-5), Guadalquitón (CA-13) and Cerro del Águila (HU-5) are identified orally by individuals or descendants of individuals that personally witnessed millstone production. Although there is no evidence of earlier working at these sites, this could change with future research.

The quarry of Moclín (GR-1) can also be added to the list of sites identified by oral tradition. However, Moclín is consigned to a previous assemblage due to references in old texts and other indicators.

These last quarries coincide with the arrival on a large-scale of French siliceous millstones, an event that marks the end of the long tradition of millstone production in the south of the Iberian Peninsula.

11.3.4. Assemblage 3: Undated quarries (Table 11.4)

This last assemblage comprises 37 undated millstone quarries identified for the most part in hiking itineraries or historical websites on the Internet. None, to my knowledge, is cited in an old text, and none, at least of those that I have been able to survey in the field, reveals any type of extraction feature indicative of older work phases (for example rotary querns). I gather from the photographs and brief descriptions that the products of these sites measure over 1.00 m in diameter and therefore fall broadly into the long chronological range between the Middle Ages and Contemporary times. In this table, as in the case of Table 11.2, the columns represent broad chronological units. The chronological haziness of these workings will only be clarified by future research.

11.4. Millstone quarry longevity

As has been noted previously in this study, production at certain millstone quarries or districts endured a very long time, even centuries. The *grès coquillier* in Switzerland, the *grès de la Serre* in France and the volcanic rocks around Mayen in Germany, for example, are known to have been exploited continuously for thousands of years. In the Iberian Peninsula, according to written and photographic records, the Montjuïc by Barcelona was exploited continuously from as early as the 13th (Gutiérrez García-Moreno 2009: 91) to the 20th century. Among the millstone workings in the south of Spain, there is no evidence suggesting any production with a life-span comparable to that of the Montjuïc. There is nonetheless evidence that suggests a longevity at certain sites spanning several centuries. Some of these seemingly long and uninterrupted workings, notably Moclín (GR-1) and Berrueco (CA-8), are among the predominant millstone quarries in southern Spain.

Two of the three reputed prehistoric sites producing saddle quern quarries, Zujaira (GR-12) and Cabezo de Olivo (MU-2), for example, appear to have also supplied rotary querns centuries later in Roman times. There is no evidence, however, of a continuity of work between Prehistory and Roman times. Production in these two different periods was probably a coincidence.

Table 11.4: Chronological assemblage 3 lists 37 quarries that cannot be dated precisely. From the size of their extractions, over 1.00 m in diameter, they could date anywhere from the Middle Ages to Contemporary period. At the sites surveyed in the field (in grey), there is no indication of workings preceding the Middle Ages. The sites highlighted in grey were surveyed in the field.

Code	Site	Surveyed	Neolithic Chalcolithic Bronze Age	Late Iron Age	Roman	Middle Ages	Modern	Contemporary
GR-8	Guajar Faragüit	yes						
GR-7	Padul, Los Guillares	yes						
GR-4c	La Merced 3	yes						
J-2	El Lachar	yes						
CA-6	Chipiona	yes						
CO-3	Carcabuey, Cudillas	yes						
CR-5	Granátula de Calatrava	yes						
CO-2	Cabra, Cortaores	yes						
AL-4	Guardias Viejas	yes						
CO-5	Baena, Molino de la Piedra	yes						
AL-5	Barranco Baena	no						
AL-6	Barranco Palancón	no						
AL-7	Los Loberos	no						
HU-2	El Prao de Abad	no						
HU-3	Las Malenas	no						
HU-4	La Obra Pía	no						
HU-1	El Campillo	no						
J-4	Cambil, Arbuniel	no						
CA-11	Salto de la Mora	no						
CO-4	Vega de los Molares	no						
GR-14	Ugíjar	yes						
CS-3	Borriol	no						
CS-2	Lucena del Cid	no						
MU-3	Cantera de los Porches	no						
CR-4	Pedrizas de Piédrola	no						
CC-5	Guijo de Galisteo	no						
CC-1	Plasencia	no						
CS-1	Soneja	no						
GU-3	Ruguilla, Cifuentes	no						
GU-4	Sigüenza, La Cuerda	no						
GU-9	Cobeta	no						
CU-1	Portilla, Los Molares	no						
V-1	Canals, Les Moles	no						
A-1	Sierra del Molar	no						
M-1	El Berrueco, Madrid	no						
M-5	Miraflores de la Sierra	no						

Concerning Roman sites, the wide variety of products (querns, millstones, building blocks) from the volcanic outcrop beside the Roman city of *Sisapo* (CR-1) suggest a certain, albeit poorly defined, longevity. Yet the longevity of other Roman workings such as Cerro de Limones (AL-2) and Hoya del Paraíso (AL-2) is not assured. The fact that they only produced a single, standardised product could be indicative of a short production, not more than a generation or two.

The case of the duration of Moclín (GR-1a), estimated at centuries, is distinctive. Its longevity is not evidenced by written sources (the oldest text is that of Madoz, 1845-1850), but by the number and variety of its extractions, from rotary handquerns 40 cm in diameter to larger millstones about 1 m in diameter. Once again, the massive quantity of available white limestone and the site's proximity to a town (a few steps away) facilitating infrastructure (lodgings, smithy, transport) favoured a long and continuous exploitation. A similar situation would be expected at other workings with extensive outcrops beside towns such as Alconera (BA-1), Montesa (V-2) and Rota (CA-3).

The large puddingstone production of Puerto de la Cadena (MU-1), with its mixture of rotary querns and millstones, probably endured a long time. The combination of the rock, a type that was sought out in the Middle Ages, with the site's position along a main thoroughfare, suggests a long period of exploitation. Longevity can be excluded, however, at the puddingstone workings of Carchuna (GR-9) and Cerro El Chispas (AL-8), sites also combining quern and millstone workings, simply due the extremely modest number of extractions.

An uninterrupted production at a number of sites in southern Spain spanning three to four centuries is interpreted from a series of written sources (Fig. 11.27). Scoring limestone at El Berrueco (CA-8) for at least four centuries is interpreted from archives from 1509, 1750, *c.* 1775, 1813 and 1846 (see catalogue for sources). This long production is perfectly compatible with the mountain's enormous mass and its complex infrastructure, notably the written reference to dwellings by Madoz. In spite of the lack of physical evidence, it cannot be excluded that millstone production at El Berrueco preceded the 16th century as hypothetical earlier phases of extraction would probably have been erased by more recent millstone workings.

In the case of Loja (GR-2/3), the evidence of continuous work for almost four centuries is less certain. There are two written sources separated by about 350 years (1502 and 1847). The extensive source of rock (limestone) in the surroundings of the city known for its many watermills would, nonetheless, have favoured uninterrupted workings. Once again, this cannot be proven by the modest workings identified in the field.

Continuous exploitation for at least three centuries of the granite outcrops is possible to imagine at the Ventas con Peña Aguilera (TO-2) due to four different texts (1576, 1587, 1821 and 1828). Furthermore, in the 16th century, the granite workings of Ventas con Peña Aguilera had the reputation (certainly inflated) of being the best millstone quarry in Spain (Viñas & Paz 1951: 216).

Other sites evidenced by written sources pointing to uninterrupted work spanning three centuries are El Torcal (MA-1) (1500 and 1789), Llerena (BA-2) (1566 and 1791), Albardado (CO-10) (1606 and 1846) and Colmenar Viejo (M-3) (1574 and 1847).

There were certainly other quarries in southern Spain among those identified in this study that saw long and continuous millstone activity. Their identification, however, requires future research both in the field and in archives.

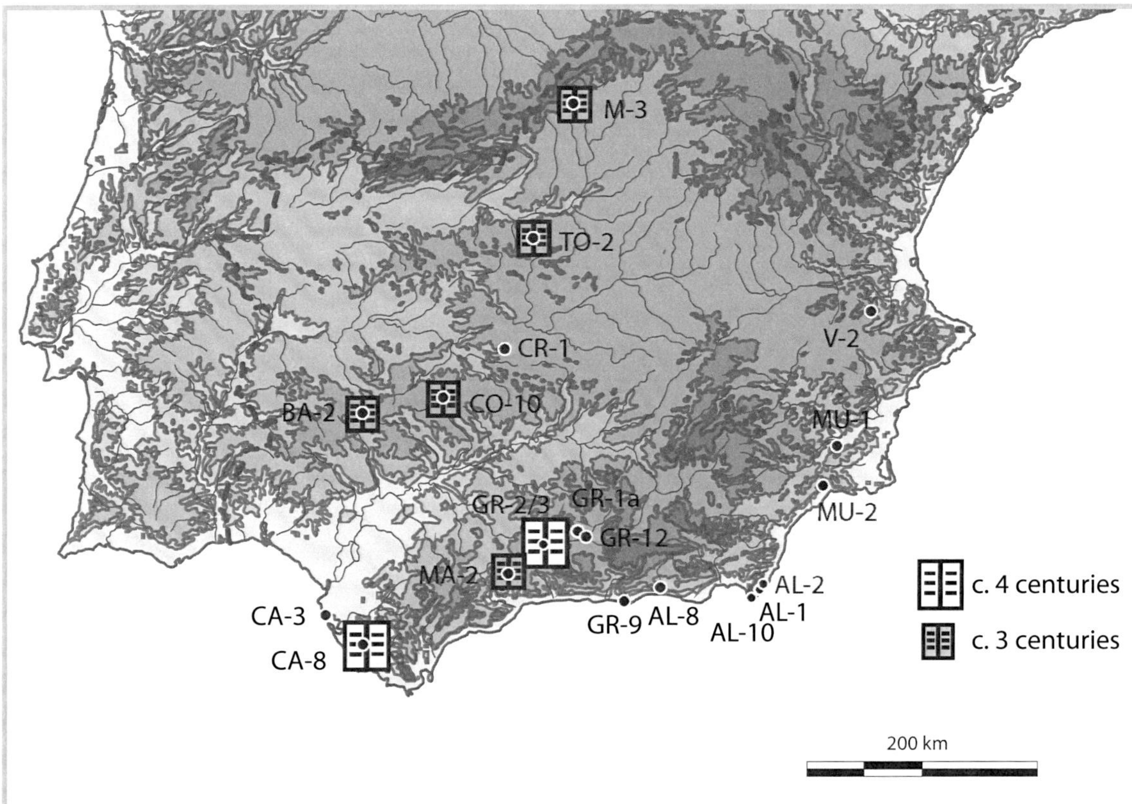

Fig. 11.27: Map of the sites mentioned in section 11.4. The symbols indicate workings with a probable continuity of production of three or four centuries based on written sources.

12. FROM QUARRY TO MILL: MILLSTONE DISTRIBUTION

12.1. Generalities

Quern and millstone productions, depending on their spheres of commercial distribution, are generally divided into three categories: local, regional and long-distance. Through time, the distances that define the boundaries between each of these categories probably fluctuated depending on the nature of the product, its size and weight, and the means and conditions of transport. It is awkward, for example, to compare the transport of a small saddle quern hewn from a small surface boulder collected in riverbed near the settlement with the transport in an ox-driven cart over a distance of 100 kilometres of a millstone measuring 1.30 m in diameter weighing almost a tonne and produced in an extractive quarry. Therefore, the definitions proposed to distinguish the categories of the commercial reach of quarries, as well as the physical distances, are hypothetical and must be considered in their chronological contexts.

The setback to the diachronic approach of the study of millstone distribution in southern Spain, as elsewhere, is the imbalance of evidence from period to period. There is very little data, for example, from Prehistory and Protohistory, whereas for Antiquity, the number of querns and millstones in museum depositories is large enough to establish reliable distribution maps. For the Middle Ages, especially during the Islamic domination, the data, once again, are almost nonexistent. In the Modern period the evidence is also very limited, in spite of a growing number of references to millstone quarries in historical archives that foreshadow an even higher number of written accounts, hence more data, in Contemporary times.

Category 1: Local millstone quarries

Local millstone quarries are defined as small exploitations producing a limited number of stones destined to nearby settlements or mills. Work at these sites, undertaken by small teams, was intermittent and depended on specific needs or purchase orders. In Prehistory and Protohistory, these sites were probably inside the "site catchment" sphere of several kilometres, corresponding to a walking distance of a few hours from the settlement. In later periods, the distance from the mill to the quarry could be greater, about 20 kilometres, roughly the distance that could be travelled by a cart in one or two days (*cf.* Sainz 2007: 176-178). The relationship between the client and the millstone maker in these cases was most likely direct, without the intercession of a merchant. Likewise, the transport of the product was mostly undertaken by the millstone maker or the client. Although the transport could at times be complicated by difficult terrain and weather conditions, it did not entail more than one or two days.

Category 2: Regional millstone quarries

Regional millstone quarries usually exploited massive rock outcrops and produced many more millstones than local quarries. In early times, from the technical point of view, they differed little from local extractive quarries except for the volume and the reach of their production. In more recent times, they were worked either year-round or seasonally by specialists who required the consent of the outcrop's owner. Sectors of these quarries were leased, sometimes in exchange for an established number of millstones a year. Crews, usually led by a master (sometimes the family head), maintained contact with both the quarry owner and the client. Production followed a sustained rhythm to satisfy specific purchase orders. These productions also had the option of storing millstones that could be subsequently commercialised directly to clients or by means of millstone merchants. Regional quarries required a more complex infrastructure. In the case of permanent or seasonal presence, this implied room and board for the workmen, and at times, their families. The families also could participate in menial aspects of the production process (removing debris, for example) and in other domestic tasks. Tool maintenance required access to a smithy. Stonemasons at times possessed the skills of blacksmiths and could maintain their own tools. Regional sites supplied millstones farther away than their smaller local counterparts, over a distance of roughly 20 to 50 kilometres (at times up to 80 km). This transport required more sophisticated logistics and, at times, the intervention of transport specialists (*arrieros*, *carreteros* or *barqueros*). Depending on the distance and the natural obstacles (mountains, rivers), transport could take several days. These factors, in addition to toll fees, raised transport costs.

Category 3: Long-distance millstone quarries

Long-distance millstone quarries possessed all the characteristics of the regional quarry. The difference is that their production attained an even larger scale, requiring not only a far greater source of rock, but more workers and a more sophisticated infrastructure. Their millstones had to benefit from an excellent reputation, so as to draw the attention of distant clients. From the logistical standpoint, these workings required more complex means of transport and a higher developed network of trade. This resulted in a higher cost of the product. Millstones from these sites could pass through the hands of a number of individuals from their source to their final destination. This process began with the millstone maker and continued through the merchant and the professional transporter. During shipment, the product could transit through various trading posts. This appears to be the case of the 130 volcanic querns in the Roman shipwreck of Illa Pedrosa, originally from quarries around Ullastret (Girona), that were stored in and redistributed from the ancient port of Empuries, Girona, before their last ill-fated journey by sea (Vivar 2004: 108).

In more recent times, specialised millstone merchants, like the millstone makers themselves, made a living by commercialising these products. The degree of distribution and distances these products could travel in Contemporary times, before the huge influx into Spain of French siliceous stones, is revealed in a Royal Decree of the 21st of May, 1829, that prohibited Marcelino Franco, a merchant of Pontevedra, Galicia, from importing Portuguese millstones due to the possibility of acquiring Spanish millstones in the Provinces of Álava (Basque Country) and Granada or other quarries in faraway Andalusia (Ferrer 1830: 219-220). Hence, from early times in southern Spain, there was the large divide, sometimes measured in hundreds of kilometres, between the quarry and the millstone maker and the final client.

12.2. Quern distribution in Prehistory and Early Protohistory

There are three presumed saddle quern quarries in the study area (Fig. 12.1). None are completely certain, and each would benefit from further fieldwork and petrographical analyses.

The reputed volcanic quarry of Mazarrón, Murcia (MU-2) was destroyed in the 1970s (Saturnino Agüera, pers. comm.) and does not provide data as to distribution. Zujaira (GR-12), on the other hand, supplied querns (and other lithic material) to the settlements of Los Pensadores (Late Neolithic) and Las Rajas (Chalcolithic) located, respectively, 200 m and two kilometres away (Fig. 12.2). It is not possible to gauge if this is simply a local exploitation, or if it supplied more distant settlements. El Barronal (AL-10), identified by Francisco Carrión at the south-eastern point of Cabo de Gata, produced volcanic saddle querns among other artefacts (Carrión *et al.* 1993). Querns from El Barronal (or from another unidentified site in the Cabo de Gata), are known at the famous settlement of Los Millares, 50 kilometres away (*cf.* Fig. 12.1) (Manuel Ramos, pers. comm.), indicating that certain volcanic rocks circulated beyond the local sphere as far back as in Late Prehistory. These volcanic models were probably perceived as a higher quality alternative to the more common garnet-mica schist querns, the dominate rock of the Los Millares assemblage from surface material collected in local riverbeds and ravines.

Current knowledge of quern quarries in Late Prehistory and Protohistory is far too incomplete to advance a coherent view of saddle quern distribution. Yet it is perfectly conceivable from the little evidence that these querns were circulating at the regional level, in the same manner as other products such as stone axes and flint.

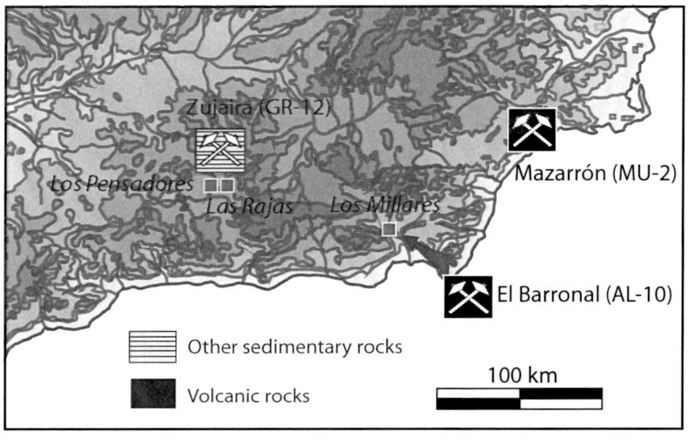

Fig. 12.1: Location of three presumed Prehistoric saddle quern quarries and the settlements they supplied. El Barronal (AL-10) is reported to have supplied querns to Los Millares, about 50 kilometres away. Zujaira (GR-12) supplied the nearby settlements of Los Pensadores and Las Rajas (see detailed map below).

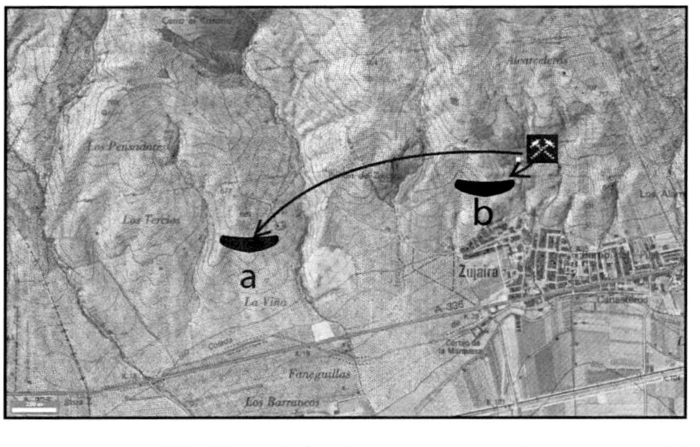

Fig. 12.2: Detail of the Zujaira quarry (GR-12) and the Prehistoric settlements of a) Los Pensadores and b) Las Rajas. The sites are respectively at a distance of 1.5 kilometres and 200 metres.

12.3. Iron Age quern and millstone distribution

The question of Iron Age rotary quern distribution in southern Spain was briefly presented at the Bergen colloquium in 2011 (Anderson *et al.* 2014). Since then, I have had access to more Iron Age rotary querns and pushing mills that permit me to expand the perspective of distribution in the Iron Age beyond the local sphere, to include both regional and some long-distance imports. In any event, research on this period is still very incomplete in southern Spain. For example, nothing is known of the Iron Age millstone production in the volcanic districts of Campo de Calatrava or Cabo de Gata, areas that would be exploited for millstones on a large-scale in later Roman times.

Most of the Iron Age rotary querns and pushing mills in the study area (Fig. 12.3) are poorly dated surface finds from *oppida*. The Zujaira pair, in a private collection, are reputed to come from the *oppidum* of *Ilurco* at the Cerro de los Infantes, Granada (Fig. 12.4). Another upper stone in the Museum of Alcalá la Real is, presumably, from La Mesa at the Aldea de la Ribera Alta, Jaén, also a reputed Iberian *oppidum* (Carlos Calvo, pers. comm.) (*cf.* Chap. 3, Fig. 3.4, c). The millstone group in the Jódar Museum is from the *oppidum* of Cerro Castillejo, Jaén (López Marcos 1995). The few finds at Baza, Granada, are from the *Basti* complex (Andrés Adroher, pers. comm.), a site known for the celebrated Dama de Baza statue brought to light in its necropolis in the 1970s.

The inventory is completed by a very modest number of *in situ* examples. The first, a compete pushing mill, is from the "Celtic" hilltop *oppidum* of Castrejón de Capote, Higuera la Real, Badajoz (Berrocal Rángel 1989). The others are either rotary querns or pushing mills from the Cerro de la Cruz, Almedinilla, Córdoba (Quesada *et al.* 2010), a hilltop settlement with an important occupation dating to the middle of the 2nd century BC (Quesada *et al.* 2010; Quesada *et al.* 2014).

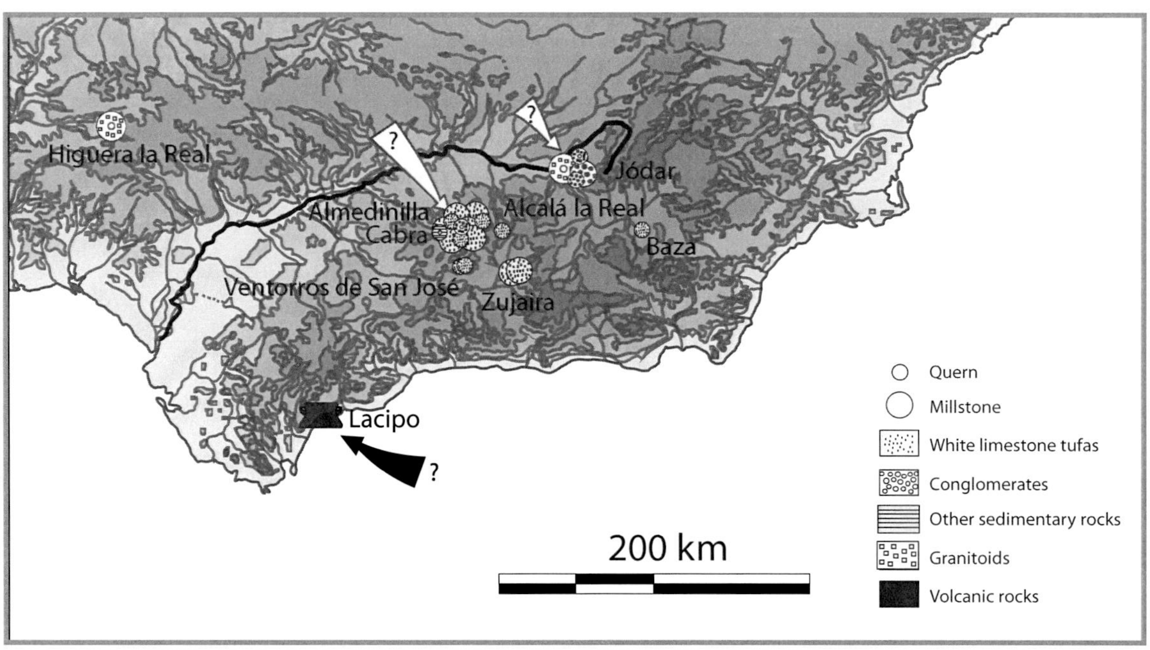

Fig. 12.3: Distribution map of querns and millstones in the Late Iron Age. There is only one supposed tufa millstone quarry in the study area at Almedinilla (CO-15), Córdoba, that supplied the local sites. The granite querns of Cerro de la Cruz and Jódar are exogenous since the closest outcrops are respectively c. 70 and c. 40 kilometres away (see arrows). The volcanic biconical millstone of Lacipo is an example of a pre-Roman long-distance import from an unknown volcanic source somewhere in the Mediterranean Basin.

As to petrography and provenancing, the *in situ* mill at the of Castrejón de Capote (Badajoz), slightly over 50 cm in diameter, was equipped with lateral inverted keyhole cuttings and set on a stone-built podium (Berrocal Rángel 1989: 258; 284, fig. 26). Its rock, granite, is local or regional, characteristic of the Ossa Morena and South Portuguese geological zones.

Most of the rotary querns and pushing mills (*cf.* Chap. 3, Fig. 3.13-3.14) from Cerro de la Cruz, on the contrary, were scored from a highly porous limestone tufa, or "*travertino*" (Quesada *et al.* 2014), a type of rock with outcrops that dot southern Spain. Although no millstone quarry has been certified in the field, limestone tufa is found in the riverbed of the Almedinilla River (CO-15), a short distance from the Cerro de la Cruz, suggesting local workings. The recent discovery of a Roman limestone tufa ashlar quarry in a valley near Almedinilla (Muñiz *et al.* 2012: 165) could be a clue to the location of earlier millstone workings. It is therefore safe to assume that the Cerro de la Cruz limestone tufa millstones reflect local work.

An upper quern stone at Baza is another example of limestone tufa. Its source is not known. It could be local as a limestone tufa outcrop is known near the Baños de Zujar, 10 kilometres to the north of Baza (Geological map IGME, 972).

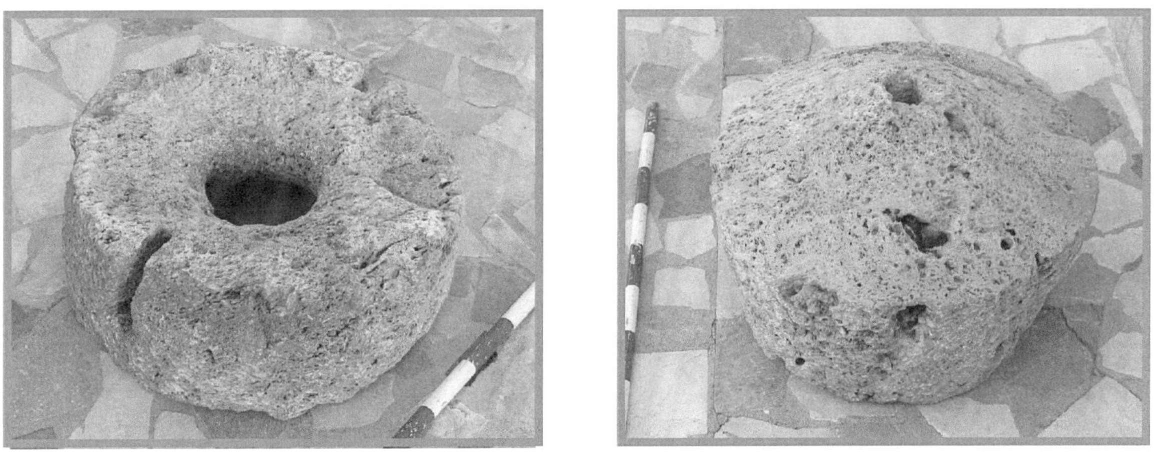

Fig. 12.4: Limestone tufa millstone pair from a private collection in Zujaira, Pinos Puente, Granada (Ø: 57 cm). The upper stone is equipped with opposite vertical slits. This pushing mill reputedly comes from the nearby Iberian Culture settlement of Ilurco, *at the Cerro de los Infantes.*

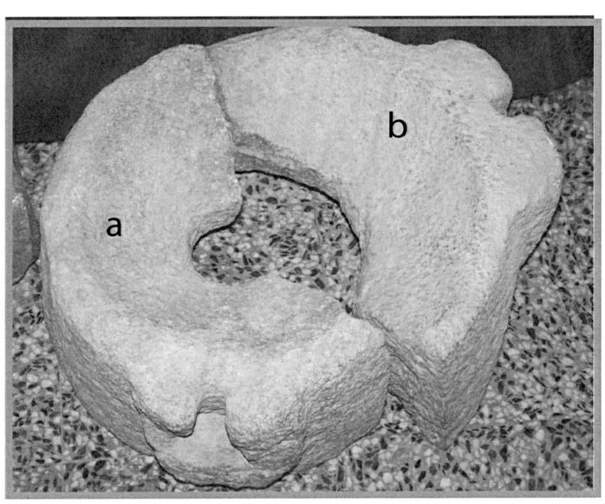

Fig. 12.5: Quern and millstone from Jódar, Jaén. Rotary quern a) is of conglomerate (Ø: 40 cm); the millstone to the right b) is of granite (Ø: 52 cm). Both have protruding handle lugs. The lug of the quern (a) is equipped with an "inverted keyhole" fitting (photograph by Ildefonso Alcalá, Museum of Jódar).

The source of the conglomerate rotary quern of Jódar (Fig. 12.5a), another type of rock exploited in the Iron Age, is not known. It probably reflects either the exploitation of a surface block or an angular block detached from bedrock.

The granite rotary quern from the Cerro de la Cruz (*cf.* Chap. 3, Fig. 3.4, e-f) and the granite pushing upper stone from Jódar (Fig. 12.5b), by contrast, are exogenous, as their find spots are in regions devoid of granite. The source of the Cerro de la Cruz upper stone is most likely somewhere to the north, beyond the Guadalquivir River Basin, in the Central Iberian granite district (*cf.* Chap. 4, Zone D) at least 70 kilometres away. The nearest potential granite outcrop to Jódar is in the surroundings of Linares (J-7), about 40 kilometres to the north-west. Linares is in fact an area known for millstone production in Contemporary times (Madoz 1847, Vol. 10: 290; Lozano Muñoz 1867: 17). These two quern stones reveal that granite in the Iron Age was an alternative to local sedimentary rocks, and that it was circulating beyond the local sphere.

The *Lacipo* volcanic "hourglass" *catillus* (Fig. 12.6), with no known typological parallel in or outside the Iberian Peninsula, is a rare example of a long-distance Iron Age millstone import. This upper stone, stored in the Museum of Málaga, is a surface find from the ancient city of *Lacipo* (Casares, Málaga), a site with strong Punic links dating to the 3rd or 2nd century BC (inv. A/CE 2585; José Suárez, pers. comm.; Puertas 1982: 88, fig. 50). Its section is biconical with an upper "hopper" about half the size of the lower. Its grinding surface features parallel furrows. Its driving fittings are particularly unique and revealing. It was originally equipped with opposite "ribbon" handles (Fig. 12.6.1), a type of handle that has, to my knowledge, no parallel. After one of these fragile handles broke, it was replaced by an "inverted keyhole" cutting (Fig. 12.6.2) that served to lodge a rynd-handle crosspiece stretching from cutting 2 to the ribbon handle 1. Apparently, the combination of ribbon handle 1 and cutting 2 was not an efficient driving system so the fitting was replaced by a crosspiece set in two smaller "inverted keyhole" cuttings 3a and 3b carved on the opposite sides (Fig. 12.6.3a-b).

The closest counterpart to the Lacipo *catillus* is the Morgantina mill upper stone, a type common to Sicily and identified at the shipwreck of El Sec off the coast of Mallorca as early as the 3rd (White 1963: 205) or 4th century BC (Williams & Thorpe 1990). Morgantina models share similar petrography, dimensions and asymmetric hoppers. But the comparison stops there, especially considering the driving fittings. The original ribbon handles near the upper rim of the *Lacipo* upper stone are far from the massive square lateral lugs along the base of the Morgantina type.

All said, the repairs of the handles of the *Lacipo* millstone are compelling. "Inverted keyhole" cuttings, as noted previously (*cf.* Chap. 3.3.1), are characteristic of rotary querns and pushing millstones of the Late Iron Age of the Iberian Peninsula, as seen from the examples from Cerro de la Cruz (Córdoba), Jódar (Jaén), Numantia (Soria) (Fig. 12.7) and Castrejón de Capote (Berrocal Rángel 1989: 284, fig. 26.1). This feature therefore places this volcanic import firmly in pre-Roman times.

Yet the origin of the *Lacipo* upper stone is not known. There is no evidence of early, pre-Roman rotary biconical millstone production in the three volcanic districts of the Iberian Peninsula. This stone must therefore be a long-distance import brought by sea traders from one of a number of volcanic districts elsewhere in the Mediterranean. This is therefore a rare example of the use of volcanic material in the Iron Age in southern Spain and a precursor of the Roman Pompeian volcanic imports that would arrive a century or two later.

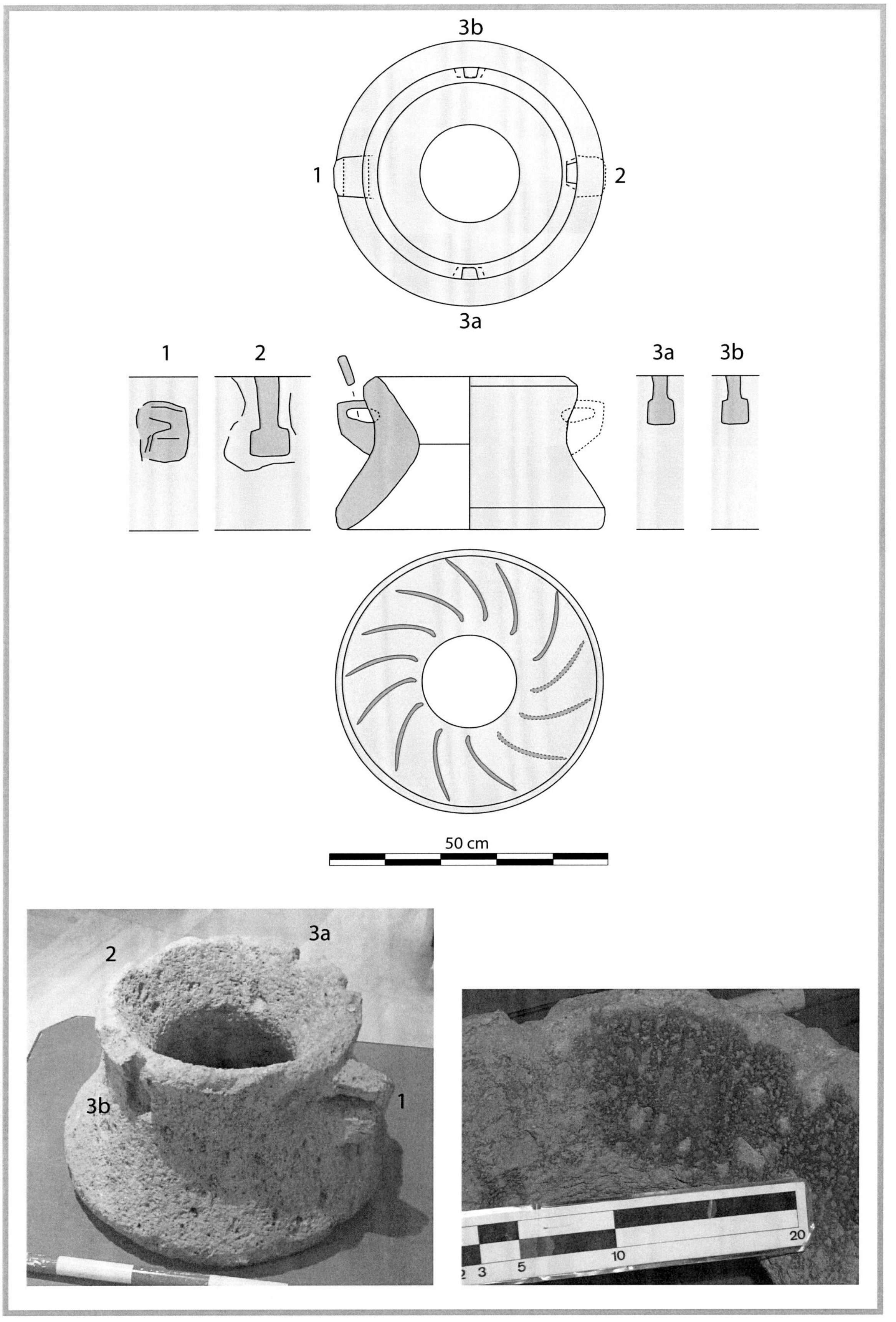

Fig. 12.6: The volcanic biconical, furrowed, upper stone is a surface find from the ancient city of Lacipo, *Málaga. Its original system of 1) ribbon handles was replaced successively by 2) and 3a-b) "inverted keyhole" cuttings, typical of the Late Iron Age in the Iberian Peninsula. The dark vesicular stone is a long-distance import from an unknown source (drawing by T. Anderson modified from the original drawing in Puertas 1982: 88, fig. 50; photographs by T. Anderson).*

Research in Catalonia and Valencia on the question of distribution of Iron Age mills is far more developed than that of southern Spain (Alonso 1999; Asensio *et al.* 2001; Portillo 2006). The model of quern and millstone procurement developed in Catalonia, for example, is probably applicable to that of southern Spain. To simplify, most millstones are either local or regional (Portillo 2006; 26, 459) and long-distance imports are rare.

For example, among the 23 rotary querns at Alorda Park, Tarragona (Barcelona), from the middle of the 5th to the 3rd century BC, there are local bioclastic limestones that were probably scored at the celebrated El Mèdol quarry about 20 kilometres away. Otherwise, the Alorda Park assemblage shows no sign of regional or long-distance imports (Asensio *et al*. 2001: 58). A bioclastic limestone lower stone from Els Vilars (Arbeca, Lleida), in eastern Catalonia, dating to the middle of the 4th century BC, is also possibly from El Mèdol, about 60 kilometres to the south-east. This example, unique at Els Vilars, could fall into the category of a regional import.

A volcanic lower stone at the settlement of Ca n'Olivé (Cerdanyola de Vallès, Barcelona), from the middle of the 5th century BC, is about 100 kilometres from its most likely source in the Olot-Garroxta volcanic district (Alonso 1999: 262, fig. 171). This is probably, at least to date, the earliest Iberian manifestation of volcanic rotary quern workings and a rare example of a long-distance import in the Iron Age.

A major difference between southern Spain and Catalonia, at least at this stage of research, is the widespread presence in Catalonia of volcanic mills in later Iron Age contexts. This is especially the case of the Indigècia region with 48% of its grinding stones hewn from basalt (Portillo 2006: 26). This high percentage is not surprising since the mills come from settlements that are very near the Olot-Garrotxa volcanic province.

Fig. 12.7: Querns and millstones from the celebrated Iberian Culture settlement of Numantia, Soria. Most are local sandstones or conglomerates. There is no indication of long-distance imports. Both b) larger millstones, as seen in the last row to the left in photograph a), and c) rotary handquerns are equipped with characteristic Iron Age lateral "inverted keyhole" cuttings. Less frequent driving fittings are opposite lugs with vertical handle slots and vertical slits (cf. Checa et al. *1999: 65, fig. 2).*

North-central Spain is represented by the large collection of conglomerate and sandstones from the celebrated site of *Numantia* (Soria) (Fig. 12.7). These are reported to be of local origin, collected from a radius of a few kilometres (Checa *et al.* 1999: 64). Their exact source, however, is not known. From what I have been able to observe, due to their hemispherical sections, they probably were the result of working small surface boulders. In any case, among the querns and millstones exhibited at the site, there is no evidence of granite or volcanic materials indicative of long-distance imports.

Summing up, most Iron Age querns and millstones throughout Iberia appear to be local or regional products. The granites at Cerro de la Cruz and Jódar and the bioclastic limestone of Arbeca reflect regional trade between at least 40 and 70 kilometres. Some could potentially come from farther and even qualify as long-distance trade. This would not be surprising since in the La Tène Culture of Central Europe, reddish breccia querns from Schweigmatt in the German Blackforest, for example, are known at Bern, Switzerland, about 100 kilometres from their source (Joos 1975). Furthermore, reddish "rhyolithtuff" querns in Germany were travelling from their presumed source in the Odenwald Mountains up to 120 kilometres along the Neckar River (Lehmkuhl 2011: 82) and a puddingstone quern recovered at Rowbury in southern England crossed the English channel from Normandy (Peacock & Cutler 2011: 77). These Iron Age examples of longer distance millstone trade, including the Lacipo and Ca n'Olivé examples, appear to be, from the current evidence, more of an exception than a rule. Their trade in no way reflects the vast network of long-distance millstone imports that would subsequently be introduced after the Roman conquest.

12.4. Roman millstone distribution

Compared with previous periods (and to a certain extent later periods), the circulation of millstones in southern Spain, as elsewhere in Europe, reached a peak in the Roman period. An essential factor for this expansion was the massive introduction of volcanic stones from quarries in the volcanic fields of Campo de Calatrava and the S-E Spanish volcanic district spread over the Provinces of Almería, Murcia and Albacete, as well as from more distant Mediterranean volcanic sources. These centres provided high-quality products over both short and long distances, profiting from long-established trade routes for traditional products such as olive oil, metals, cereals and fish derivatives.

The number of Roman quarries identified (or inferred) in the field, as well as the number of Roman millstones in museums, allows drawing up individual distribution maps according to mill types.

The first (Fig. 12.8) indicates the distribution of rotary querns. These can be divided into three main zones where three different rock types prevail. The eastern sector was dominated by volcanic material that travelled far from their source. The southern sector is dominated by the shell-rich biocalcarenites hailing from quarries known along the Atlantic coast in the Province of Cádiz. The western sector was dominated by granitoids. It is of interest to note that the southern biocalcarenite and western granite sectors remained "faithful" to their traditional local or regional rocks and appear to have "resisted" the arrival of volcanic rocks. The second and third maps (Fig. 12.14 and 12.20), presenting the spread of the larger Pompeian mills and ring-mills, are less conclusive. Although they are indicative of short, medium and long-distance imports, it is obvious that more research needs to be focused on these mills.

12.4.1. Roman rotary quern distribution

The general spread of rotary hand querns throughout southern Spain can be divided into three sectors corresponding roughly to three different rock types.

12.4.1.1. The eastern volcanic rotary quern sector

The main spread of volcanic rotary querns (*cf.* Fig. 12.8) covers the eastern half of the study area coinciding with the two volcanic districts (S-E Spanish volcanic province and the Campos de Calatrava district), as well as the area in between the two districts. Isolated cases are known as far as *Emerita Augusta* (Mérida) and Carmona (Seville) to the west. To the south the are also isolated examples at *Baelo Claudia* (Cádiz) on the Atlantic coast and Almuñecar (Granada) on the Mediterranean coast.

The quarries of Cerro de Limones (AL-1) and Hoya del Paraíso (AL-2), in the southern sector of the S-E volcanic province, exclusively produced rotary querns. They obviously supplied the local and regional settlements in the area, as seen by examples at several settlements in the Province of Almería. The proximity of these quarries to the coastline, four and two kilometres respectively, would have permitted them to benefit from maritime trade to access markets up and down the Mediterranean. But these more distant international markets are hypothetical and are not evidenced, at least for the moment, by geological analyses.

The lamproite productions in the northern part of the S-E volcanic district (*e.g.* Mazarrón, Murcia), could have also benefited from sea transport and the nearby port of *Carthago Nova* (Cartagena). In addition, as noted in a brief geological study of a millstone assemblage from Baza (Molina & Cultrone 2012: 39), stones from these still unidentified quarries could have circulated by land westward through the Almanzora Valley, a long-established trade route connecting the coast with the interior trade arteries passing through the centres of Guadix and Baza (Granada). However, this remains to be confirmed because most of the lamproite outcrops are still unexplored and have not been of the object of geological analyses.

The quarries of the Calatrava volcanic district, far from the coast in the heart of the Iberian Massif, traded their rotary querns and millstones by means of inland routes crisscrossing the centre of the Iberian peninsula. The workings of *Sisapo* (CR-1), Las Herrerías (CR-2) and Cerro Columba (CR-6) by *Oretum* (Ciudad Real) are all beside major ancient thoroughfares. This explains the distribution of volcanic querns throughout the Province of Córdoba (Fig. 12.9a) and as far west as the Roman city of *Augusta Emerita* (Mérida, Extremadura). To attain Mérida they could, theoretically, have been ferried down the Guadiano River on barges. From Roman *Corduba,* they could have been transported along the Guadalquivir River to reach Carmona (Fig. 12.9b), and farther down to the Atlantic coast to Cádiz. Once again, there is much work to be done to establish both the typology and spatial distribution of the rotary querns originating in the Campo de Calatrava.

The few volcanic rotary querns that attained *Baelo Claudia* and *Sexi* (ancient Almuñecar) were probably brought by sea from quarries in the Cabo de Gata or in Murcia (Fig. 12.9c-d). Other volcanic districts in the Mediterranean, notably in North Africa at Farkhana or Olot-Garrotxa in Gerona, Catalonia, can be ruled out as sources because they are not known to have produced *catilli* with radial cuttings or "*sombrero" metae.*

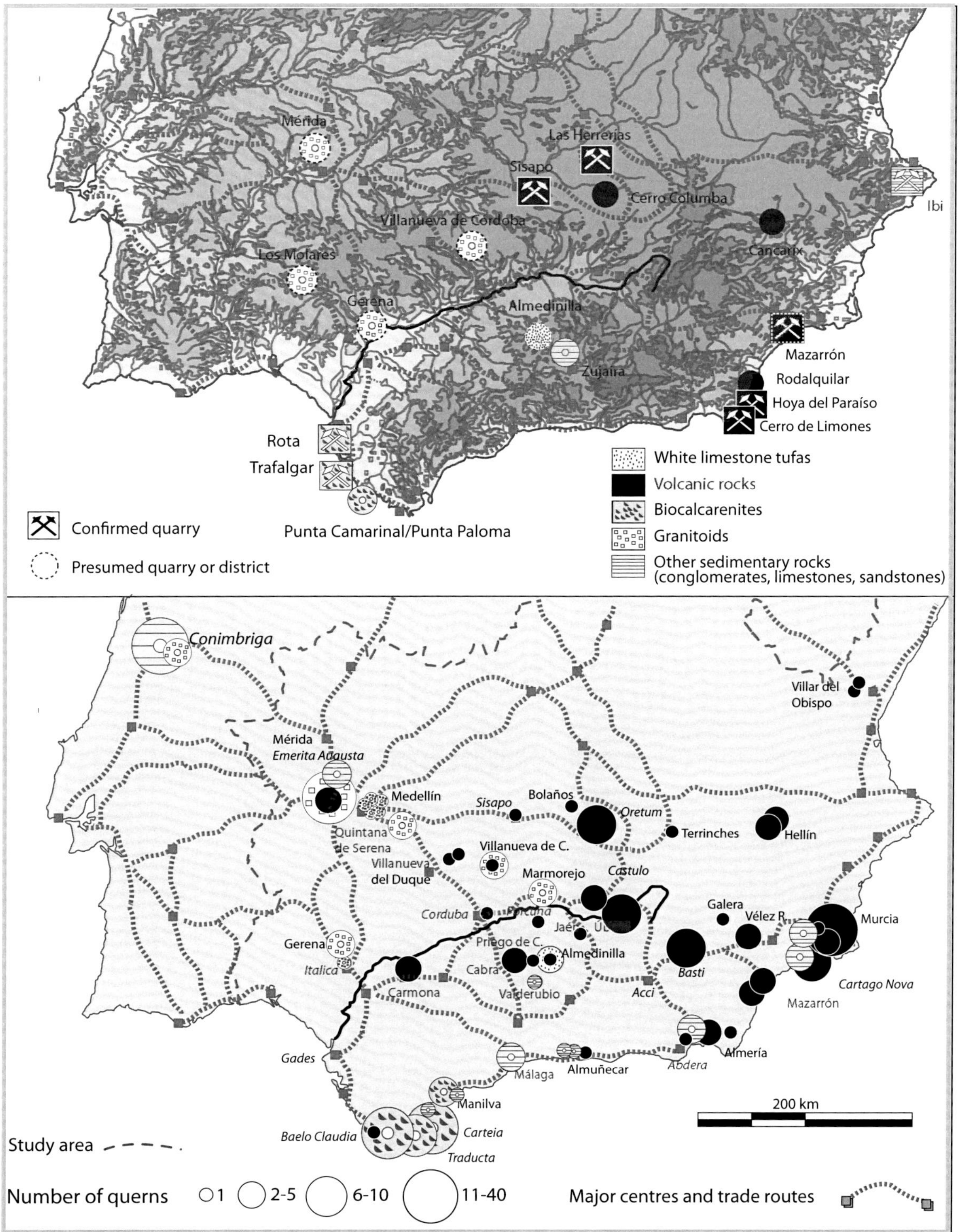

Fig. 12.8: Spread of Roman rotary querns by rock type and quantity. The main concentration of volcanic querns is in the eastern part of the study area, nearer their sources. The southern sector, along the Straits of Gibraltar, is dominated by biocalcarenites, whereas the western sector is dominated by granitoids (drawing by T. Anderson). Querns cited in publications: ***Albacete****: Hellín (Jordan* et al. *1984; Jordan & Matilla 1995);* ***Cádiz****: Iulia Traducta (Vargas & Bernal 2009);* ***Ciudad Real****: Terrinches (Benítez* et al. *2011);* ***Córdoba****: Villanueva del Duque (Garcia Romero 2002), Villanueva de Córdoba (Palomo & Fernández 2007);* ***Extremadura****: Medillín (Haba Quirós 1998), Quintana de Serena (Léon & Carmona 2006);* ***Jaén****: El Centenillo (Gutiérrez Soler & Corpas 2011);* ***Málaga****: Estepona (Soto Jiménez 1977: 64);* ***Murcia****: García Blánquez* et al. *1990;* ***Sevilla****: Itálica (Berrocal Rángel 2007: 281);* ***Valencia****: Villar del Arzobispo, (Pérez Mínguez 2006). Museum depositories and private collections (unpublished):* ***Almería****: Almería; Isla del Moro; Vélez Rubio;* ***Cádiz****: Baelo Claudia, Carteia;* ***Ciudad Real****: Bolaños de Calatrava; Oretum;* ***Córdoba****: Cabra; Córdoba, Almendinilla, Priego de Córdoba, Villanueva de Córdoba;* ***Extremadura****: Mérida,* ***Granada****: Almuñecar; Baza; Galera;* ***Jaén****: Linares; Porcuna; Úbeba;* ***Málaga****: Málaga; Manilva;* ***Murcia****: Murcia, Siyâsa; Cartagena Museum; ARQVA* ***Seville****: Carmona.*

There is a potential overlap of volcanic rotary querns from the two different Spanish volcanic sources as seen by examples in the Province of Jaén at Ubéda and *Castulo* (Linares), situated about halfway between the Campo de Calatrava and the S-E volcanics in the Province of Murcia (*cf.* Fig. 12.8). Petrographical analyses could certainly clarify the source of these querns.

All said, volcanic rotary quern quarries not only supplied stones to the local and regional clients, but also exported their products hundreds of kilometres by both land and by water. This is not surprising as this distribution model has been established elsewhere in the Roman Empire for other volcanic productions. Querns travelled down the Rhine several hundred kilometres from the Eifel volcanic district in Germany to settlements in northern Switzerland (Anderson *et al.* 2003: 66-67). Likewise, the *Baelo Claudia* volcanic quern (Fig. 12.9c), for example, is more than about 330 kilometres (as the crow flies) from its source either in the Cabo de Gata (Almería) or the Campo de Calatrava (Ciudad Real). It is possible to envisage that this quern, like others, traded hands on several occasions between the quarry and its final place of use.

The identification of a number of volcanic production centres, coupled with the identification of volcanic rotary querns in their final place of use, as seen below (*cf.* Table 12.1), provides statistical evidence to quantify the distances that help define local, regional and long-distance distribution in Roman times.

Fig. 12.9: Examples of Roman volcanic rotary quern catilli *with radial cuttings to lodge rynd-handle crosspieces. Their find spots are far from volcanic sources : a) Priego de Córdoba; b) Carmona, Seville; c)* Baelo Claudia, *Cádiz; d) Almuñecar, Granada.*

12.4.1.2. The southern biocalcarenite rotary quern sector

In the southern-most area of the Iberian Peninsula, Roman millstone makers exploiting a yellowish, shell-rich biocalcarenite (*piedra ostionera*) along the Bay of Cádiz at Trafalgar (CA-1) and Rota (CA-3) supplied rotary querns to cities and settlements in the Province of Cádiz along the Atlantic coast such as *Baelo Claudia* and *Carteia* (Salvador Bravo, pers. comm.) and *Iulia Traducta* (Vargas & Bernal 2009: 179: fig. 24; Domínguez & Bernal 2011: 454, fig. 4). These shell-rich rotary querns are known as far as Manilva (Málaga), 25 kilometres north of Gibraltar on the Mediterranean coast.

Among the 50 to 60 *ostionera* known rotary querns in the area, there is only one volcanic import (the *Baelo Claudia* example cited previously) (*cf.* Fig. 12.8). The proportionally few volcanic rotary querns indicate that the locals remained faithful to their traditional rock. The reason for this "faithfulness" remains obscure. The local stone may have been better suited for grinding local products, in particular products derived from fish.

The spread of biocalcarenite rotary querns recalls the distribution of *grès coquillier* rotary querns in Roman Switzerland. The spread of the Swiss shell-rich stone, by far the dominant grinding stone of the Swiss Plateau, was "between" two concentrations of volcanic imports, in the Geneva area to the west, and around Basel to the north (Anderson *et al.* 2003: 67, fig. 75). In this sense the Gallo-Romans of the Swiss Plateau also remained "faithful" to their local or regional shell-rich sandstone.

12.4.1.3. The western granite rotary quern sector

The distribution map of granite rotary querns in southern Spain is more nebulous and difficult to grasp (*cf.* Fig. 12.8). This is, in part, because I have not visited as many local museums in the granite fields as they are farther away from my base in Granada. Hence, for the spread of granite querns, I have had to rely for the most part on secondary written sources.

The largest assemblage of granite rotary querns I have been able to study is at the *Museo Nacional de Arte Romano* in Mérida (Badajoz, Extremadura). The precise number of these querns is not known because it was impossible, due to storage problems, to access most of the stones. Other smaller granite assemblages are known from texts at Quintana de Serena, Seville (León & Carmona 2006: 47), Marmorejo, Jaén (archaeological inventory: http://www.iaph.es/patrimonio-inmueble-andalucia/resumen.do?id=i5700) and in the museums of Gerena, Seville and Villanueva de Córdoba (oral information). Haba Quirós notes that around Medellín, to the east of Mérida, granite rotary querns were found at the sites of Las Galapagueras, Los Novilleros, El Casquero, Los Turuñuelos and Vega de Ortiga, Peña Lobada (among others) and that these querns are common on other *villae* (Haba Quirós 1998: 306, 321, 323, 412). Finally, a granite *meta* is mentioned at the entrance of the museum of *Italica,* Seville, in a letter written in the 1940s to the then director of the National Museum Fernández de Avilés (Berrocal Rángel 2007: 282).

The spread of these granite querns (*cf.* Fig. 12.8) stretches over a vast triangular area between Mérida (Badajoz), Gerena (Seville) and Marmorejo (Jaén). Granite distribution continues even further north-west, beyond the study area, with a large number at *Conimbriga* (Portugal) (Borges 1978).

This spread is totally logical. Granite, in the form of extensive surface outcrops and naturally sculpted surface blocks, is ubiquitous throughout vast areas of south-western Spain north of the Guadalquivir Basin in the Central Iberian, Ossa Morena and the South Portuguese geological zones.

In spite of the presence of extensive granite outcrops, no Roman rotary quern quarry has been identified. From the many granite querns in the depository of the Museum in Mérida, production must have taken place close by. There were certainly workings in the vast Pedroches district around Villanueva de Córdoba, Gerena (Seville) and Los Molares (Huelva) based in part on examples in local museums and on certified millstone production in later times.

Numerous granite quern quarries elsewhere in south-western Spain have left little, or no, evidence. Future research will probably confirm granite to have been one of the dominant regional rocks, following a centuries-old tradition of granite exploitation for grinding tools, as seen by the many saddle querns on display at the Early Iron Age settlement of Cancho Roano (Badajoz). Volcanic rock incursions into this granite district, as seen by a few examples at Mérida (Badajoz) and Villanueva del Duque (Córdoba), are probably incidental and did not have a pernicious effect on the local and regional granite workings.

12.4.1.4. The distribution of rotary querns of other sedimentary rocks

Production of rotary querns of other sedimentary rocks, concentrated for the most part in the central and eastern sectors of the study area, was never totally eclipsed by the volcanic imports (*cf.* Fig. 12.7). It is difficult to gauge, however, the extent of distribution of these secondary productions.

Roman limestone tufa (travertine) workings of the area of Almedinilla (CO-15), of Late Iron Age tradition, persisted after the arrival of the volcanic models as seen by three querns in the Almedinilla museum. The more notable example is a *catillus* with receptacles and radial cuttings (Fig. 12.10), typologically identical to Roman volcanic models. The scale of these local Roman workings was probably modest and limestone tufa rotary querns probably did not travel far from their source as no significant quern assemblages of limestone tufa have been identified in Roman contexts elsewhere in the study area.

Other Roman sedimentary rock productions seem more tenuous and most likely reflect local workings of surface blocks or small, modest quarry workings. Isolated sandstone and conglomerate rotary querns are found particularly throughout the eastern sector (Fig. 12.11) of the study area. Yet no such extraction sites can be confirmed in the field. The only quarry that could potentially be Roman is the site of Ibi (A-2) in Alicante.

Distribution from Ibi, a true extractive limestone quarry, is hard to gauge. No rotary quern survey, for example, in regional museums has been undertaken and, in spite of having produced several hundred querns, none can be traced to a settlement. At this stage, it must be assumed that the Ibi distribution probably did not surpass the local sphere. Its geographical position, about 100 kilometres from the volcanic quarries of Murcia and over 500 kilometres from those of Olot-Garrotxa in Catalonia, suggests that if Ibi had been active in Roman times, it would have provided, like the limestone tufa production of Almedinilla, no more than an economical local alternative to imported volcanic querns.

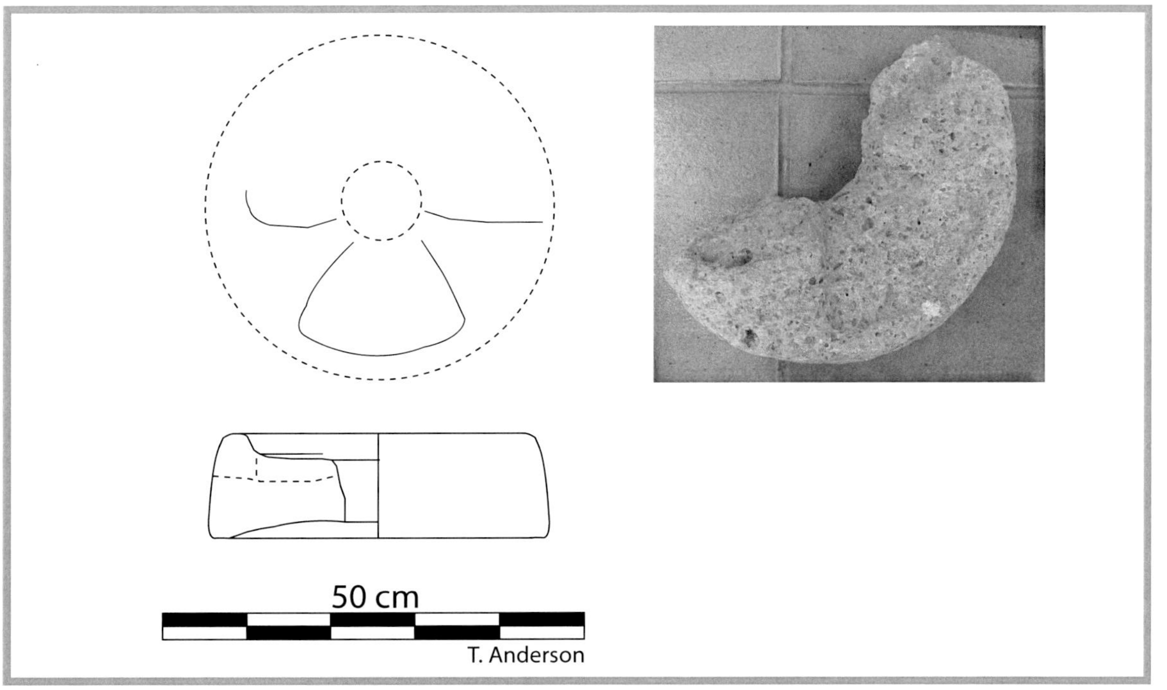

Fig.12.10: Example of a Roman rotary quern catillus *scored from porous limestone tufa in the Museum of Almedinilla, Córdoba. It is equipped with fittings (receptacles, radial handle and rynd cuttings) identical to those of volcanic models. This type of rock is known to have been exploited around Almedinilla since the Late Iron Age. There is no evidence, however, that querns of this rock were widely distributed.*

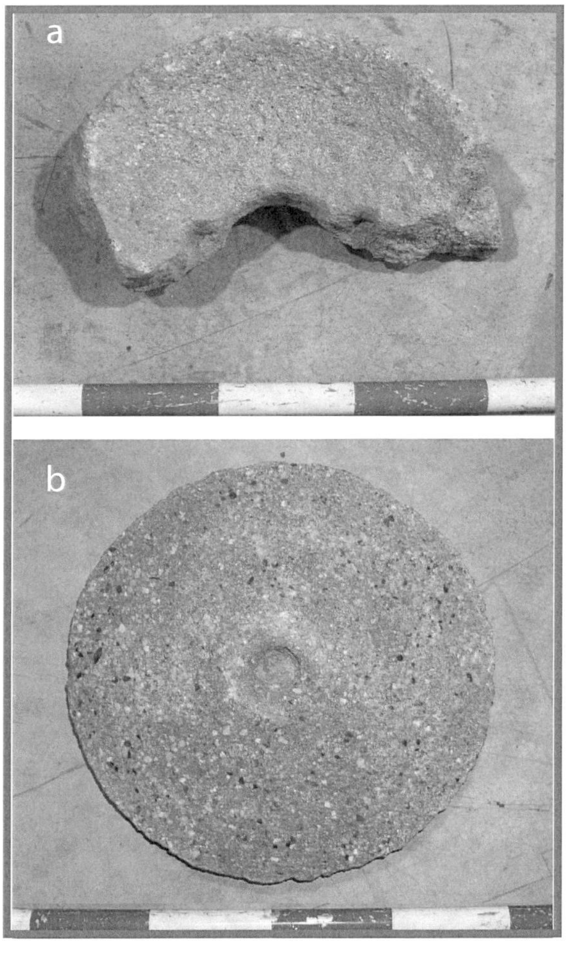

Fig. 12.11: Examples of Roman rotary querns in the Museum of Murcia scored from conglomerates: a) catillus *with opposite rynd cuttings on the upper surface. b)* meta *with a lip around the eye resulting from wear. These types of stones suggest local productions.*

12.4.1.5. Quantifying Roman rotary quern distribution

Throughout this work I have freely employed the terms "local", "regional" and "long-distance" in reference to commercialisation and distribution. In general, this tripartate concept is intangible and difficult to quantify. Therefore, so as to correlate these three terms with specific kilometric values, I have measured the distance of a number of Roman volcanic rotary quern "assemblages" (often, unfortunately, comprising only a single element) to their potential sources in the three different volcanic Spanish quarry districts. The approximate values are listed in Table 12.1.

For each of the volcanic production districts, I have arbitrarily defined a central reference point that serves as the basis for the measurement. The S-E Spanish volcanic district is divided into two sectors: 1) the small Cabo de la Gata volcanic querns (notably dacites and rhyolites) are measured from a point between the quarries of Cerro Limones (AL-1) and Hoya del Paraíso (AL-2); 2) the larger Murcia production, comprising a number of different volcanic outcrops (mostly lamproites), is measured from a central point chosen at the city of Mula (about 180 kilometres north of the Cabo de Gata). The reference point for the vast 3) Campo de Calatrava district is established in its centre, between the millstone workings identified near the current cities of Puertollano and Ciudad Real. These kilometric measurements are thus arbitrary values recorded in straight lines, "as the crow flies".

As an example, the single volcanic rotary quern from Almuñecar (Granada) (*cf.* Fig. 12.9d), the first on Table 12.1, is 140 kilometres from the point representing Cabo de Gata volcanics, 260 kilometres from the point representing the Murcia volcanics and 235 kilometres from the point representing the Campo de Calatrava volcanics (Fig. 12.12). It must not be forgotten that these values do not, in fact, reflect the reality of millstone transport. For example, the shortest straight-line distance from Almuñecar to the Murcia volcanics is 260 kilometres. But this theoretical inland transport does not take into account the obstacle of the Sierra Nevada Mountains which would obviously lengthen the journey. The distance from Murcia by sea, also possible to envisage, would be much longer. The second potential source, the Campo de Calatrava, is also north-south overland route at least 235 kilometres long. The most reasonable source, the third (highlighted in grey in Table 12.1), is the Cabo de Gata,140 kilometres away by sea.

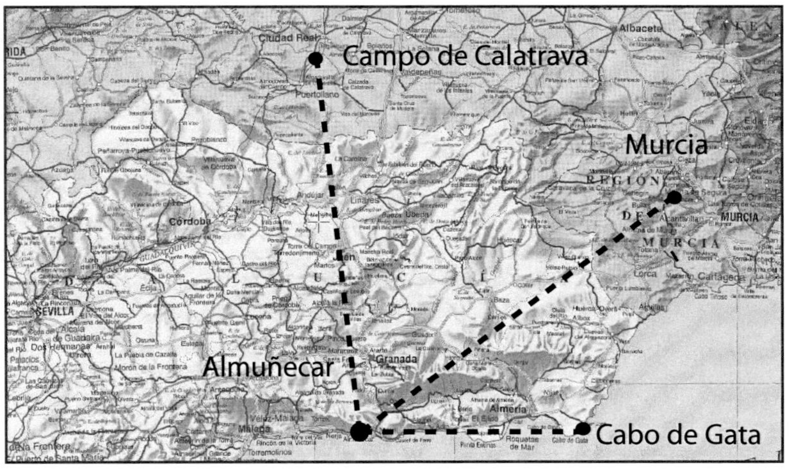

Fig. 12.12: Example of the theoretical "as the crow flies" distances between the volcanic rotary quern discovered at Almuñecar (Granada) and central points representing the volcanic quern production districts of Campo de Calatrava (235 km), Murcia (260 km) and Cabo de Gata (140 km).

Table 12.1: List of the Roman volcanic rotary quern individuals or assemblages and their estimated values of distances in kilometres between their find spot and a central point representing the three volcanic production centres. The distances are measured "as the crow flies". Real travel distances over land (avoiding natural obstacles) or by sea were obviously superior to the values in the table. The most likely volcanic source is highlighted in grey.

Volcanic querns from Roman sites	No. of querns	Cabo de Gata volcanics	Murcia volcanics	Calatrava volcanics
Almuñecar, GR	1	140	260	235
Galera, GR	1	115	110	180
Baza, GR	10	100	150	175
Almería area, AL	4	35	185	260
Los Cazadores,AL	5	12	-----	-----
Vélez Rubio, AL	4	105	90	220
Carmona, SE	2	320	385	205
Baelo Claudia, CA	1	330	445	340
Mérida, BA	2	440	430	200
Córdoba, CO	1	265	300	130
Priego de Córdoba, CO	2	200	260	155
Villanueva de Córdoba, CO	1	285	290	75
Villanueva del Duque, CO	2	310	325	100
Almedinilla, CO	1	185	250	160
Cabra, CO	1	220	285	155
Úbeda, J.	7	175	180	110
Jaén, J.	1	180	215	125
Castulo, Linares, J	3	195	205	90
Porcuna, J	1	220	250	100
Hellín, AB	~4	-----	20	----
Bolaños, CR	1	-----	----	5
Terrinches, CR		210	135	105
Oretum, CR	7	----	----	2
Mazarrón, MU	5	----	4-	----
Murcia area, MU	21	----	c. 20	----
Cartagena, MU	3	----	c. 20	----
Aljorra, MU	2	----	c. 20	----
Villar del Obispo, V.	2	350	165	290

Table 12.2 List of non-volcanic Roman rotary quern "assemblages" (sandstones, granites, limestone tufas and biocalcarenites) and the value of the estimated distance to the nearest production centre. These distances, on an average, are shorter than those of volcanic rotary querns.

Roman sites	No. of querns	Rock type	Source km
Baelo Claudia, CA	40	Biocalcarenite	25
Iulia Traducta, CA	11	Biocalcarenite	55
Carteia, CA	~5	Biocalcarenite	55
Mérida, BA	c. 30	Granite	20?
Medellín, BA (various sites)	c. 10	Granite	20?
Gerena, SE	c. 2	Granite	20?
Villanueva de Córdoba, CO	2	Granite	20?
Almedinilla, CO	5	Limestone tufa	5
Zujaira, GR	1	Sandstone	3

Similarly, a second table (Table 12.2) lists the theoretical distances of rotary quern assemblages of certain non-volcanic rocks (biocalcarenites, granites, sandstones and limestone tufas) to their sources.

The limestone tufa and sandstone rotary querns appear to be very near, less than 10 kilometres, from their reputed quarries or sources. The biocalcarenite quern assemblages, all near Gibraltar, come indubitably from outcrops along the 100 kilometre corridor between the cities of Tarifa and Chipiona (Cádiz) along the Atlantic coastline; the closest confirmed quern workings to the settlement of *Baelo Claudia* Trafalgar (CA-1) and Punta Camarinal/ Punta Paloma (CA-2a-b). The sites are, respectively, 25 and from one to five kilometres away. The two Roman cities by Gibraltar (*Iulio Traducta* and *Carteia*) are farther away from the potential quarries, at a distance of about 55 kilometres.

The value of the distances of the granite assemblages is the most difficult to quantify because potential granite outcrops cover thousands of kilometres throughout the Central Iberian, Ossa Morena and South Portuguese geological zones (representing about 20% of the Iberian Peninsula!). Granite, contrary to volcanic rocks, does not leave clear signatures of provenance that can be interpreted by means of geochemical analyses. The four quern assemblages, nonetheless, are each found in municipalities with a long tradition of granite extraction and it is not likely that these querns would have been imported from long-distance quarries when they had access to this rock in their own "backyard". For this reason, I have assigned these assemblages an arbitrary value of 20 kilometres qualified with a question mark (20?).

From the data listed in Tables 12.1 and 12.2, a stacked histogram was drafted combining the number of quern assemblages with the distance to their sources indicated in estimated kilometre values (increments of 10, with a range from 20 to 330 km) (Fig. 12.13).

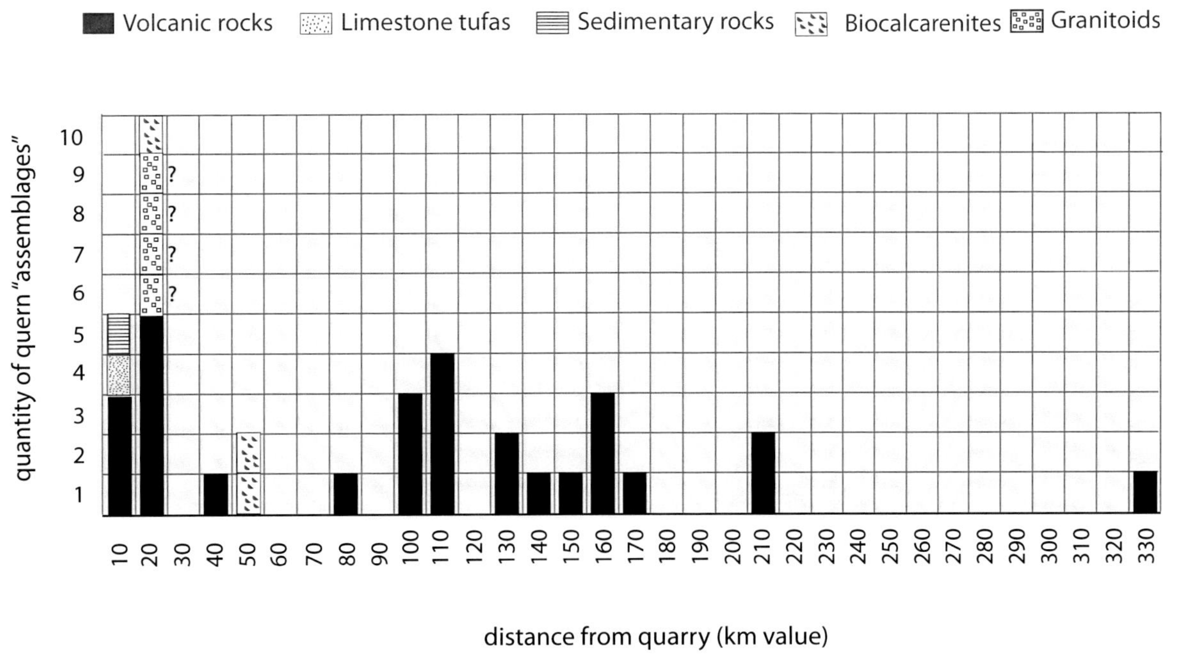

Fig. 12.13: Histogram illustrating the quantity of Roman rotary quern "assemblages" according to their estimated distance from their sources. The graph is based on the shaded values of Table 12.2. Five rock types are represented. The isolated example to the right of the chart is the volcanic quern at Baelo Claudia, *approximately 330 kilometres from the nearest volcanic source (Cabo de Gata). From this chart I interpret that in Roman times local quarries served a sphere of about 20 kilometres, whereas regional quarries supplied querns over a range of 40 to 50, possibly up to 80 kilometres. Long-distance quarries are represented by the values including and beyond the distance of 100 kilometres. The kilometre values are estimated "as the crow flies". Real overland or water transport distances were always greater.*

The histogram suggests the following tendencies for rotary querns in Roman times:

1) The first group concentrated between the sites with distances up to 20 kilometres, is tantamount to **local productions**. This group encompasses most of the sedimentary assemblages, a few volcanic assemblages (settlements near the volcanic outcrop) and the highly arbitrary estimated value of the granite assemblages.

2) The second group, encompassing the values of 40 and 50 kilometres (stretching possibly up to 80 km), corresponds to **regional productions**. The number of assemblages is modest and comprises part of the biocalcarenite and volcanic productions.

3) The third group, well-represented with values between 100 and 210 kilometres, corresponds to **long-distance productions** and comprises only volcanic quern assemblages. The single *Baelo Claudia* quern, with a value of 330 kilometres, is the case in the south of Spain that has travelled the longest distance from its source.

The results of this histogram suggesting a tripartate model of Roman rotary quern distribution could simply be considered a confirmation of the obvious. It is, nonetheless, the first time in southern Spain that quantitative values have been attached to rotary quern spheres of commercialisation. Yet these statistics are few and far between. The hope is that they serve as a reference to be confirmed or refuted by future rotary quern research.

The information gleaned from this "quantification" does not take into account petrographical analyses that are currently in an incipient phase (analyses of 10 volcanic samples from Roman assemblages in the Murcia Province are currently underway in collaboration with Jane H. Scarrow and Aitor Cambeses, of the Geology Department at the University of Granada). Preliminary results indicate, notwithstanding, that the Murcia querns and millstones are lamproites, a type of rock limited to a restricted number of outcrops in the northern half of the S-E Spanish volcanic province. This excludes the possibility that they were scored in the quarries identified in the Cabo de Gata, 100 kilometres to the south, placing the Murcia lamproite assemblages roughly in either local and regional spheres of distribution.

A comparison of these distances with recent studies on the circulation of Roman rotary querns in eastern France and Switzerland in Antiquity is interesting. Granite, schist and gneiss querns, most likely the result of local erratic block workings, did not travel far from their sources; the *grès de la Serre* in the Franche-Comté only circulated between 50 and 60 kilometres; the *grès coquillier* from the Broye district in western Switzerland (Cantons of Fribourg and Vaud) travelled between 60 and 80 kilometres; the *grès coquillier* of the Swiss Würenlos district in the Canton of Aargau circulated about 40 kilometres; most of the *Bundsandstein* of the Vosges travelled up to 40 to 80 kilometres (Jaccottey *et al.* 2007: 225-227). All of these local and regional productions were, nonetheless, complemented by imported volcanic querns arriving by fluvial transport from the Eifel in Germany and the Massif Central in France.

The distances of distribution of the local and regional Roman rotary quern productions in eastern France and Switzerland are therefore estimated to ranges not exceeding 40 to 80 kilometres - distances that are compatible with those of the study area in Spain. The long-distance volcanic imports arriving in eastern France and Switzerland hail from volcanic districts between 300 and 400 kilometres away, surpassing the span of 100 - 210 kilometres in the study area and comparable only to the distance of the single *Baelo Claudia* example. This suggests that the potential of long-distance distribution of rotary querns in Roman times can be far beyond the average observed in southern Spain.

To conclude this section, I return to the volcanic quern *catillus* from Almuñecar cited above (*cf.* Fig. 12.12). The question of its provenance is certainly more complex than reflected in the chart. The quern was discovered at the ancient coastal city of *Sexi*, a centre with strong Phoenician maritime trade links. On the surface, all would point to a source in the Cabo de Gata, 140 kilometres to the east. Although it does conform typologically (radial rynd-handle slots, receptacles) to the Cabo de Gata production, its dark, greyish vesicular surface does not resemble the reddish-violet dacites and rhyo-dacites productions identified in the Cabo de Gata at Cerro de Limones (AL-1) and Hoya del Paraíso (AL-2). Based on these macroscopic observations, there is either another unidentified quern quarry in the Cabo de Gata producing identical querns (see, for example, quarry AL-11 in the catalogue), or the quern was ferried from another volcanic source, for example in the Murcia Volcanic district. This is an example of the problem of provenance and distribution that can only be solved by geo-chemical analyses.

In any case, the study of distribution of Roman rotary querns in southern Spain is only part of a wider framework of a complex network of trade throughout Europe and the Mediterranean Basin of different types of volcanic Roman millstones, notably the larger, more sophisticated Pompeian and ring-mills, as well as water-powered models.

12.4.2. Roman Pompeian and ring-mill distribution

12.4.2.1. Roman Pompeian mill distribution

As noted previously (Chap. 3), it is not easy to assign large, bell-shaped lower stones (*metae*), a type common throughout the landscape of southern Spain, to either ring-mill upper stones or to Pompeian "hourglass" upper stones. It is more and more apparent that both Pompeian and ring-*catilli* share the same or similar types of lower stones.

The notion of a prevalence of Pompeian mills in southern Spain is still widespread in archaeological literature. However, I have recently defended an opposite hypothesis in the Colloquium of Lons-le-Saunier, France. That is, Pompeian "hourglass" mills were not widely distributed in Roman Spain and that flour grinding was taken on by ring-mills, as is the case in the Roman city of *Volubilis* in northern Morocco (Leduc 2008; 2011).

This notion is based on the fact that there are relatively few confirmed volcanic Pompeian *catilli* (about a dozen) spread over the whole of the Iberian Peninsula and there is no evidence of Pompeian production in the volcanic districts of Spain. The "hourglass" upper stones identified in Spain are in fact, for the most part, long-distance imports from volcanic districts elsewhere

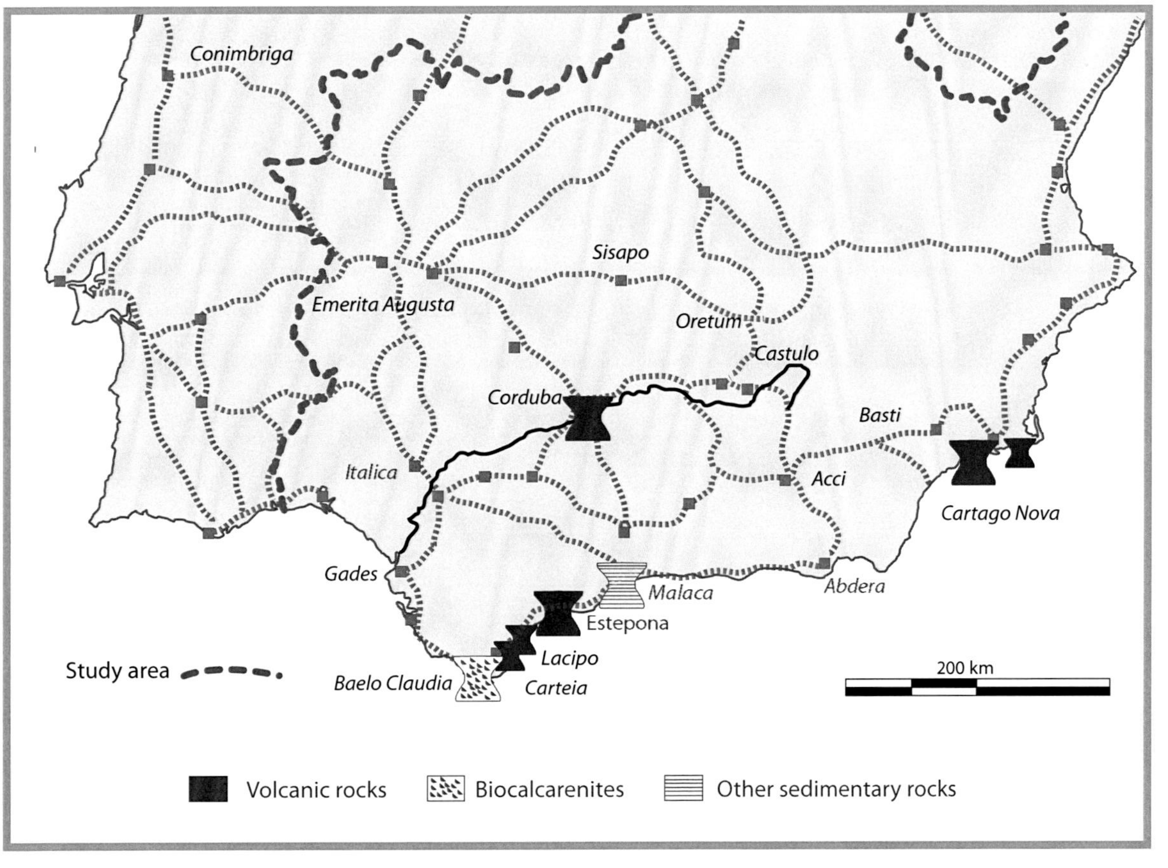

Fig. 12.14: Spread of certified Roman Pompeian mills in southern Spain (based on the presence of hourglass-shaped catilli*). In southern Spain most are volcanic and found along coastal settlements suggesting long-distance maritime imports. There is currently no evidence suggesting that volcanic Pompeians were produced either in the Campo de Calatrava or the S-E Spanish volcanic district. The non-volcanic examples of* Baelo Claudia *and* Malaca *are certainly local copies of imported volcanic models.*

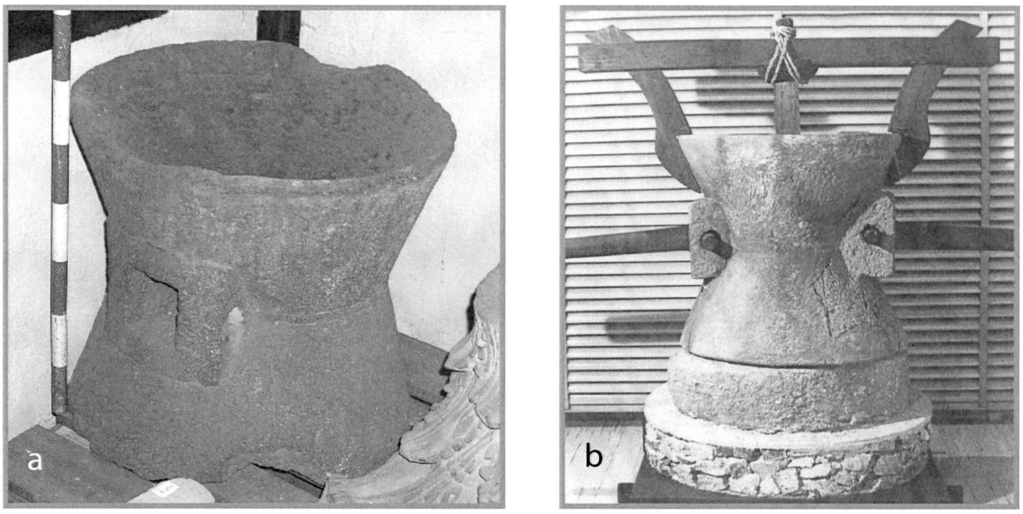

Fig. 12.15: Volcanic Pompeian catilli *from the Museo de Bellas Artes de Córdoba and from the Museo Arqueológico de Murcia. The typology and petrography of these stones point to a distant origin in the Orvieto volcanic district near Rome (right: photograph by the Museo de Murcia).*

in the Mediterranean, such as Mulargia, Sardinia or Orvieto near Rome. Only the cases of *Baelo Claudia* and *Malaca*, the first a roughout of local biocalcarenite (*ostionera*) and the second a local limestone (José Suárez, pers. comm.), can be considered to have been produced in *Hispania*. Furthermore, there is evidence that ring-mills, contrary to their "hourglass" counterparts, were fashioned in at least one of the Iberian volcanic districts.

In southern Spain, only eight hourglass-shaped upper stones have been identified (Fig. 12.14). A "classical" Pompeian (Fig. 12.15a), currently in the Museo de Bellas Artes of Córdoba (inv. DJ1069A) (Berrocal Rángel 2007: 281, fig. 4.4), is the only "inland" example. The director of this museum (José María Palencia) informed me that its provenance is not certain because the antiquarian Julio Romero de Torres, the largest donor to the museum, could have acquired it anywhere.

The other examples, including a small volcanic model from the Museum of *Carteia* (San Roque, Cádiz) (Irene Aranega, pers. comm.), as well as the exceptional volcanic *catillus* from *Lacipo* (described previously in this chapter), a precursor of the Pompeian type, are all distributed along the coastline, from the Straits of Gibraltar to Cartagena in Murcia.

Among these coastal finds is another "classical" Pompeian *catillus* (Fig. 12.15b), from Mazarrón near Cartagena, stored in the Museum of Murcia (inv. 957) (*cf.* Chap. 3, Fig. 3.21). Based on typological and geological criteria, notably the shape of the handle cuttings and the "band" around the girth, and the presence of leucite crystals (as noted by Jane H. Scarrow and Aitor Cambeses of the University of Granada), this upper stone is most likely a long-distance import from the outcrops around Orvieto, near Rome, a district that accounts for most of the mills at the bakeries of Pompeii and Ostia (Peacock 1989) and those spread throughout Narbonese Gaul in France (Jaccottey & Longepierre 2011).

Another example from the coastline, probably from a shipwreck, is a volcanic *catillus* stored in the ARQVA museum of Cartagena, Murcia (inv. 88021). This upper stone, like the *catillus* in *Carteia* museum, is minute (smaller than some hand querns) and equipped with simple lateral cuttings (no protruding lugs). To date these two examples have no typological parallel.

Fig. 12.16: Pompeian hourglass catilli *from the Museum of Melilla (photograph by Museo de Melilla) produced at Farkhana, a volcanic outcrop by Melilla in northern Africa. Farkhana is a potential source of millstones for Roman* Hispania *(photograph by Museo de Melilla).*

The provenance of another volcanic *catillus* in the Museum of Estepona, Málaga, is also unknown. Like the Cartagena example, it is probably from a shipwreck, suggesting a long-distance maritime import. Based on the similarity of its handle lugs with examples in the Museum of Melilla (Fig. 12.16), it could have been produced at the Farkhana volcanic outcrop by the north-African Spanish enclave (*cf.* Williams-Thorpe 1988: 275).

As noted in chapter 3, besides the long-distance imports, there are two locally produced Pompeian upper stones (Fig. 12.17). The unfinished biocalcarenite *catillus* at *Baelo Claudia* was probably recycled from a locally extracted construction feature (column drum). The second is an early find in the Málaga Museum (inv. A/CE4841) unearthed during excavation of the *Alcazaba*, a Moorish fortification covering the remnants of the old Roman city of *Malaca* (Berrocal Rángel 2007: 282, fig. 5a). This local limestone upper stone is interpreted, pending more precise analyses of its provenance, as a local copy of a Pompeian volcanic model (Anderson *et al.* in press).

All said, the portrait of the spread of Pompeian models across the Iberian Peninsula reflects a combination of a restricted number of long-distance imports seconded by a few local copies. This model is similar to the north of France where imported volcanic models from Orvieto in Italy are seconded by a few local copies hewn from Fosses-Belleu sandstone (Jaccottey & Longepierre 2011: 100-103, fig. 4).

Fig. 12.17: Pompeians catilli. *a) roughout of biocalcareninte (*ostionera*) from the Roman centre of* Baelo Claudia*; b) coarse limestone from the Roman centre of* Malaca *(Málaga). These are probably local copies of volcanic models.*

12.4.2.2. Roman ring-mill distribution

Apart from the Pompeian mill, the other large "industrial" Roman mill in southern Spain is the ring-mill (Fig. 12.18). This model, characterised by a cylindrical ring-*catillus* and a bell-shaped *meta* (Fig. 12.19), is at times referred to as the *Volubilis* type because of its high number at the Roman city of *Volubilis* in northern Morocco (Leduc 2008; 2011).

Michel Ponsich identified (although rarely illustrated) a large number of these mills in his different surveys of Roman settlements throughout southern Spain's Guadalquivir River Basin (Ponsich 1974, 1979; 1987; 1991). Peña Cervantes has more recently taken up the subject in her study of oil presses in *Hispania Baetica* and published a distribution map (Peña Cervantes 2010: 110, table 14; 155, fig. 19). Her map, devoid of indication of petrography, forms the backbone of the illustration presented in Figure 12.20 that is completed with examples I have observed in museums. In any case, it is evident that the ring-mill, as at *Volubilis*, was widespread throughout southern Roman Spain and outnumbered by far both the Pompeian model and the hydraulic mill, a type that is practically unknown (at least at the current stage of research).

*Fig. 12.18: Example of a volcanic ring-*catillus *from the Roman city of* Castulo *(Jaén), Ø: 64 cm (photograph Linares museum, inv. no. CE01430_R). The lower stone (of undetermined petrography) is not its original couple.*

Fig. 12.19: Roman lower stones (metae) *at the archaeological site of Oreto y Zuqueca, Ciudad Real probably of local or regional volcanic rock.*

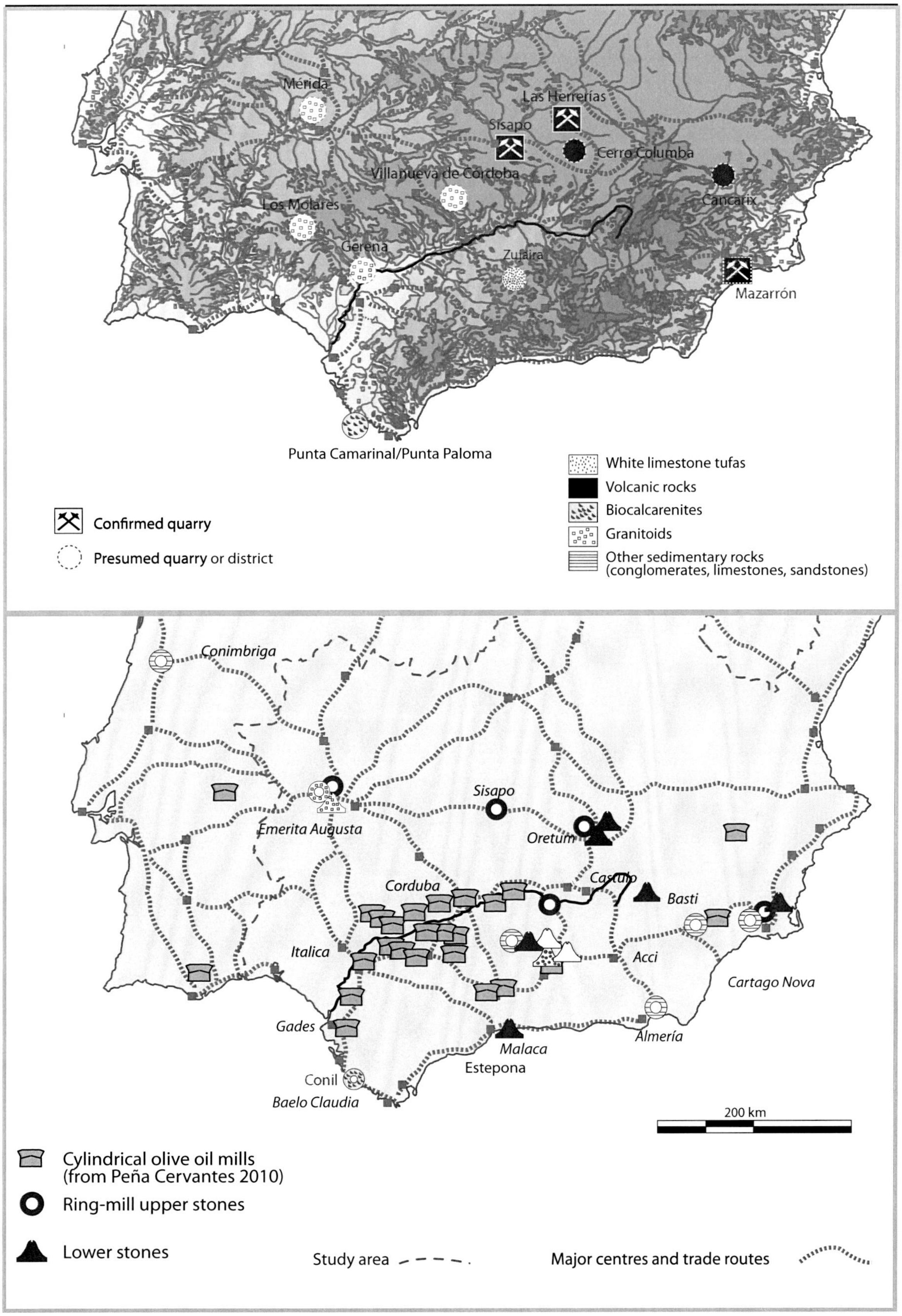

*Fig. 12.20: Top: Location of quarries and quarry districts producing ring-mills. Bottom: Spread of ring-mills. The grey symbols, concentrated for the most part along the Guadalquivir River Valley, represent ring-mills of unknown petrography found, reputedly, in the context of the olive oil industry (from Peña Cervantes 2010: 115, fig. 19). The other symbols are ring-*catilli *or* metae*, by rock type. Ring-mills are far more prevalent in southern Spain than Pompeian mills (see Fig. 12.14).*

The dominance of the ring-mill in southern Spain is perfectly reasonable since it is, in fact, the direct descendent, from the point of view of morphology, of the large rotary pushing mills unearthed in Iron Age settlements, often *in situ* on podiums.

Although the vision of the spread of the ring-mill is truncated by incomplete data (in particular petrographic determinations), it is possible to distinguish a concentration of volcanic models in the eastern sector. This probably reflects volcanic ring-mills originating in quarries like those of *Sisapo* (CR-1) and Las Herrerías (CR-2) in the Campo de Calatrava district. There are certainly many other still unidentified Calatrava outcrops that produced and exported this type of mill.

Ring-mills are also known to have been made in the lamproite exploitations in the northern sector of the S-E Spanish volcanic province (based on preliminary analyses by the geologists Jane H. Scarrow and Aitor Cambeses). The exact location of these quarries in Murcia and possibly Albacete has not been identified. In the Cabo de Gata to the south of Murcia, by contrast, there is no evidence of large ring-mill workings. The columnar joints of the Cabo de Gata outcrops were probably too small to permit scoring these larger mills.

In the museum depository of the Roman city of Mérida, as expected, there are a number of granite ring-mills and a few volcanic models, suggesting that this type of mill was supplied by local or regional granite workings and, less often, by long-distance volcanic imports (most likely from the Campo de Calatrava).

Elsewhere in southern Spain, ring-mills were at times hewn from sandstones and conglomerates. An example is the single Roman biocalcarenite model with lateral handle lugs in the museum of Conil in Cádiz. Another example is from Priego de Córdoba scored from a very hard, porous limestone called *carriola*, a rock that, on the surface, resembles travertine, but comes from a much older (tertiary), more compact, geological unit (Rafael Carmona, pers. comm.) (Fig. 12.21). Both of these mills are found in local contexts close to their rock sources.

It is noteworthy that Roman ring-mills are unknown in the northern half of the Iberian Peninsula. This rules out production of this type in the volcanic fields of Olot/Garrotxa (Girona). The reason of its absence from northern Spain is a problem still to be solved, especially considering the prevalence its Iron Age counterpart, the Iberian pushing mill, on many Iron Age settlements in Catalonia.

Fig. 12.21: Los Zurriones, Priego de Córdoba. A complete carniola *Roman ring-mill brought to light* in situ at a settlement. *The stone is local compact porous limestone.*

12.5. Millstone distribution in the Middle Ages

The corpus of millstones dating to the Middle Ages in museum depositories and archaeological reports is extremely modest and made up exclusively of rotary querns, ironically probably the least common type of mill for grinding flour in the Middle Ages. The fact that only these smaller models are published or in museums is not surprising, and this is not only related to the logistical problem of storage of the larger models. An example is the assemblage from the settlement of Cerro del Castillón outside Montefrío, Granada dating from the 9th to the 12th century (Motos Guirao 1987: 480). The fact that only small hand-driven models are published is, in a sense, logical, because these mills are domestic, found in houses of the settlement located on the top of a plateau about 1100 m high. The major mill of the settlement, most likely a hydraulic mill (windmills were still not known), would not have been on the top of the plateau near the living quarters but downhill, a few hundred metres away, on a stream along the valley bottom. Its millstones therefore would not have been recovered from the settlement. This site illustrates that little can be said from the archaeological record regarding the distribution of millstones that equipped the Visigothic and later Moorish watermills beyond the settlements along rivers, streams and irrigation ditches.

Historical documents for this period are almost nonexistent (or unavailable). The only known record is a contract dating to 1114, in Islamic times, related to the lease of a watermill along the River Guadalbullón near Jaén. The archive notes that four composite millstones (in 8 segments) would be brought from Arnedo after the restoration of the mill (Lagardère 1991: 109). The location of Arnedo is problematical. A quarry has been identified in northern Spain by Pilar Pascual and Pedro García beside a town called Arnedo in La Rioja (Pascual & García 2011: 286). But this quarry, 600 kilometres away, is not likely to be the source because Pascual and García describe it as a very modest site (*Atlas of European Millstones*; http://meuliere.ish-lyon.cnrs.fr/). Furthermore, in 1114, Arnedo in La Rioja was under Christian control, possibly complicating a potential exchange. A potential quarry is that of Villargordo (J-8), a few kilometres away from the mill. Yet this quarry is only certified to have been exploited three centuries later (Córdoba de la Llave 2003: 306, footnote 26) and cannot be linked with certainty to the source mentioned in the lease.

In any case, due to the widespread and deliberate abandonment of volcanic rocks in Late Antiquity or at the outset of the Middle Ages (Amouric 1997: 46-47; Brun & Borreani 1998: 301; Anderson *et al.* 2011a: 163; Longepierre 2011: 162-163), and the rise or return to sedimentary rocks (notably conglomerates with large clasts, coarse sandstones and limestones), it is difficult to envision the necessity in the Middle Ages for a vast commercial millstone trade network on the scale of that of volcanic rocks in Roman times. The ubiquity of sedimentary rock outcrops throughout the landscape of southern Spain, as seen through the sites of Rambla Honda (AL-3), Puerto de la Cadena (MU-1), Almadén de la Plata (SE-1) and Zagra (GR-5), should have sufficed to supply the mills and rendered obsolete the necessity of importing stones from long-distance workings.

There were, no doubt, certain quarries in the Middle Ages that possessed a reputation surpassing that of others. The millstones would have been coveted for the quality and the cost of their acquisition and transport would have been higher. This could be the case for the sites listed in the previous paragraph. Unfortunately, since there are no data on the end-locations of their products, it is not possible to paint a more precise portrait of the spheres of distribution of millstones from Middle Age quarries.

12.6. Millstone distribution in the Modern period

Written documents from the Modern period, roughly from 1492 to the French Revolution, at times record both source of millstones and their destination. The documents therefore provide a clearer vision of the commercial reach of certain millstone exploitations. Contrary to the Roman period, distribution maps cannot be drawn up based on petrographic criteria because the heavy and cumbersome millstones that equipped these watermills, windmills and *tahonas* are rarely stored in museums. Since the milling installations have disappeared or undergone transformations, most of these old millstones have vanished or now decorate public spaces.

The historical documents from 1491 to 1587 that I have been able to access, by means of molinological studies undertaken by other researchers, record the distance travelled by millstones from nine different production centres (Fig. 12.22). They are presented in ascending order of the estimated distances in kilometres ("as the crow flies") from the quarry to the mill.

1) In 1486, a historical document records a purchasing order of a segmented millstone for a mill in Córdoba from the quarry of Albaida (CO-7) (Córdoba de la Llave 1988: 843, footnote 23; 2003: 306, note 26). Albaida, known for its building stone, is just a few kilometres north of the city of Córdoba. Millstone production certainly benefited from the pre-established transportation network for the building blocks. Theoretically, a cart loaded with a millstone would have taken less than a day to cover the distance. Still to solve is the question of the volume of millstone production at Albaida and, if it were great, whether Córdoba served as a staging point to ferry this stone along the Guadalquivir River to other regions.

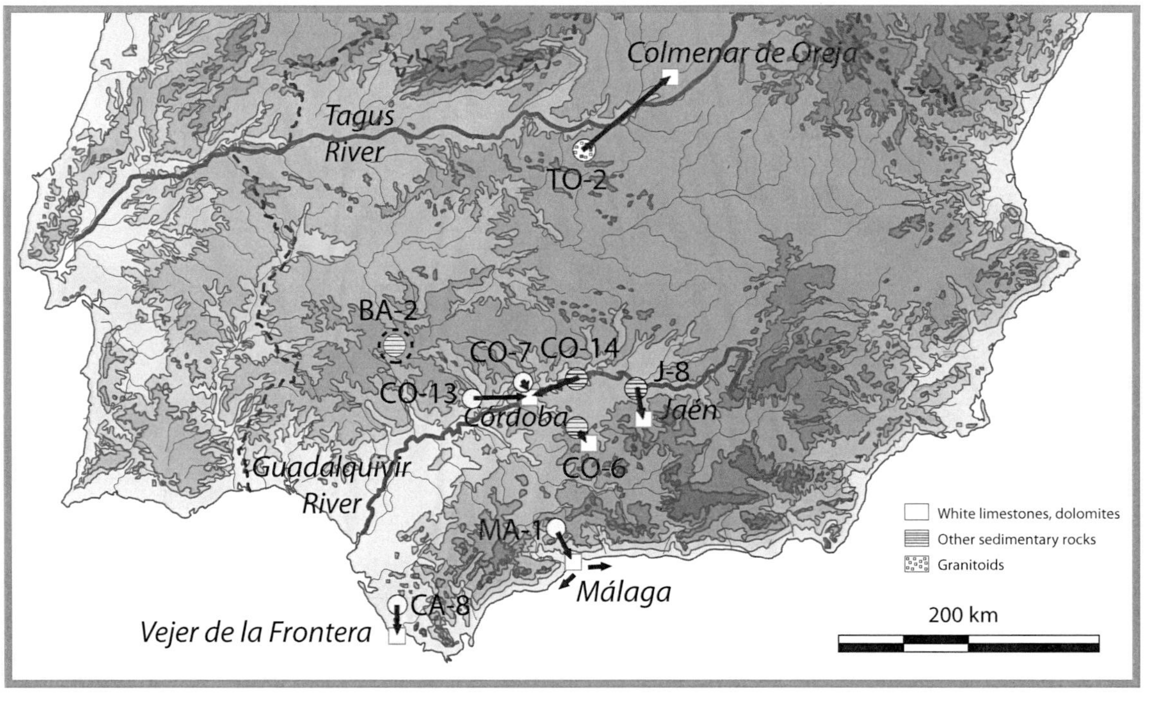

Fig. 12.22: Map indicating the ranges of distribution of nine millstone quarries from the Modern period (1481 to 1576). Most ranges do not surpass 40 kilometres. The sole long-distance example (approximately 100 km) is from the quarry of Ventas con Peña Aguilera (TO-2) to Colmenar de Oreja, Madrid. Most transport was by land. The option of fluvial transport along the Guadalquivir River was possible, although not documented, from Hornachuelos (CO-13) and Montoro (CO-14) to Córdoba. By contrast, transport by sea of millstones from Antequera (MA-1) via Málaga to nearby coastal towns is certified by written sources.

2) In 1557, a notarial protocol records the order of a millstone from the quarry of Izcar (CO-6), Córdoba for the Molino de la Piedra, a watermill nine kilometres away, to the south of the city of Baena (Córdoba de la Llave & Varelo 2011: 106).

3) In 1566, a Municipal Ordinance at Llerena, Badajoz, prohibits "foreigners" from extracting millstones unless they, like the locals, paid the official tariff (Chapter CXCVII of the ordinances; Maldonado Fernández website: http://manuelmaldonadofernandez3.blogspot.com.es/2010_04_01_archive.html). Although there is no specific indication of the distance of transport, the wording of this ordinance implied a radius of distribution of at least 10 kilometres, corresponding to the average distance I have calculated to the nearest towns in the neighbouring "foreign" municipalities.

4) In 1499, a notarial protocol records the order of a millstone for the Molino Nuevo (presumably in Jaén) from the *molar* (quarry) of Villargordo (J-8), about 20 kilometres away (Córdoba de la Llave 2003: 306, footnote 26).

5) In 1509, an archive records that the Berrueco quarry near Medina (CA-8) in the Province of Cádiz, supplied a millstone to the "Teja" watermill in Vejer de la Frontera, a town 20 kilometres away (http://www.patrimoniojandalitoral.es/f_vai_8.htm).

6) In 1500, a contention between the citizens of Antequera and Málaga was taken before the Catholic Monarchs in Granada. This feud indicates that millstones from the surroundings of Antequera *(i.e.* El Torcal, MA-1) were being transported at least 30 kilometres southwards, to the city of Málaga. It is possible that some of these millstones were then ferried from Málaga farther away by sea because a second notarial protocol (1508) records that El Torcal millstones were being traded by boat over short distances up and down the Málaga coast.

7) Two different protocols from 1481 and 1486 record that millstones from Hornachuelos (CO-13) were acquired for mills in the city of Córdoba, 40 kilometres away (Córdoba de la Llave 1988: 843, footnote 24; Córdoba de la Llave 2003: 306: footnote 26). Due to the proximity of the quarry to the Guadalquivir River, it is conceivable that these stones were ferried the distance from the quarry to Córdoba by boat.

8) In 1481, a *bermeja* (lower stone) from the quarry in Montoro (C-14), about 40 kilometres from Córdoba, was acquired for a watermill in the city of Córdoba (Córdoba de la Llave 1988, 843, footnote 22). The millstone was probably a *molinaza*, from the famous *Bundsandstein* outcrops of Montoro. Montoro's position along the Guadalquivir River suggests the stone could have been ferried down the river by boat.

9) In 1576, a text stresses that Ventas con Peña Aguilera (TO-2) manufactured the "best" granite millstones in all of Spain. The statement suggests that this production was widely distributed. The demand for Peña Aguilera millstones appears to be confirmed by a protocol in 1587, recording that the millstones for a restored watermill in Colmenar de Oreja (Madrid) "must be of *piedra berroqueña*" (granite stone) from Las Ventas con Peña Aguilera. The distance between the quarry and the mill is approximately 100 kilometres. It is of note that this route included crossing the Tagus River, adding to the cost of transport. Ironically two centuries later, quarries in the town of Colmenar de Oreja would be the producers of white limestone millstones known to have supplied animal-driven mills in the city of Madrid.

The image gathered from the histogram of the distances travelled by millstones from their sources in the Modern period is that of a series of quarries supplying local mills within a radius of about 20 kilometres (Fig. 12.23), a distance that could be attained by ox-driven cart in one or two days. Other more select workings supplied mills slightly farther away, on a regional scale, between 40 and 50 kilometres. Only Ventas con Peña Aguilera (TO-2), manufacturing the "best granite millstones in all of Spain" (Viñas & Paz 1951: 216), traded its stones a distance up to 100 kilometres. These were also transported by cart and possibly ferried, as could have been the case of those of Hornachuelos (CO-13) and Montoro (CO-14), along the Guadalquivir River or, as in the case of El Torcal (MA-1), by boat along the Mediterranean. This type of fluvial and maritime transport obviously facilitated long-distance distribution in Modern times.

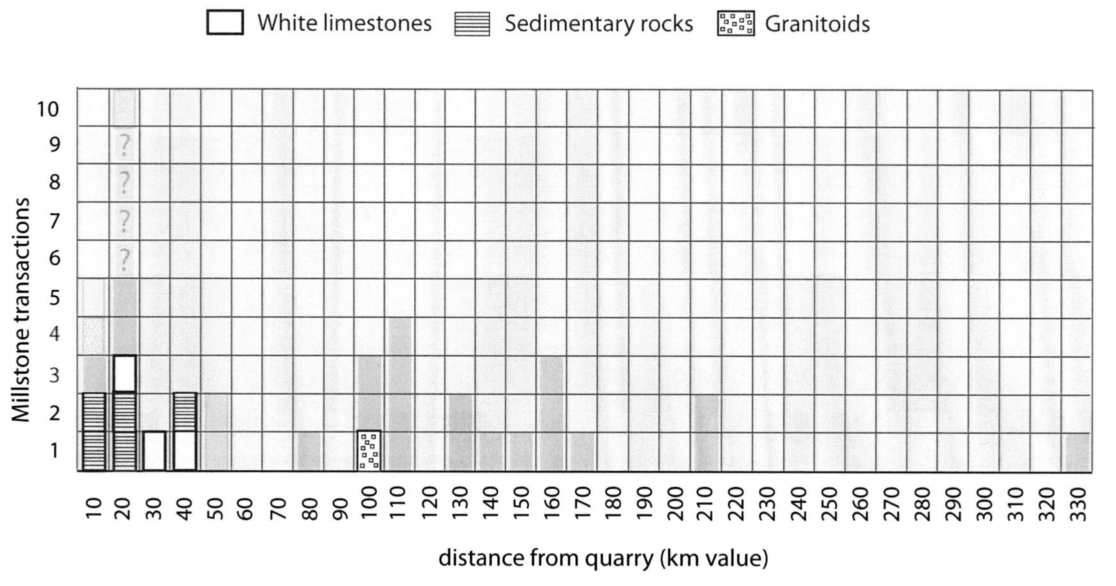

Fig. 12.23*: Histogram illustrating the quantity of millstone transactions in Modern times (1491 to 1587) based on written sources. The distances from the quarry to the mill are measured in increments of 10 kilometres. Compared to the graph illustrating the assemblages of Roman querns (in semi-transparency in the background, cf. Fig. 12.13), the number of cases (9) is extremely low and not statistically valid. It does, nonetheless, show differences in the choice of rock types from Roman to Medieval times (notably the absence of volcanic rocks) and sets apart one "long-distance" import of 100 kilometres (Ventas con Peña Aguilera, TO-2, to Colmenar de Oreja, Madrid) from the group with shorter distances of up to 40 kilometres.*

12.7. Millstone distribution in Contemporary times

Distribution of millstones from the end of the 18th century until the beginning of the 20th century is also based, for the most part, on written sources. The commercial reach of certain quarry sites devoid of written sources can also be placed in this category based on the estimation of their volume of production as seen in field surveys. For example, a quarry with hundreds of extractions obviously did not only serve a local market.

As in the case of the previous period, it is not possible to illustrate the Contemporary millstone distribution by mapping individual millstones and their petrography as millstones from this period are never stored in museums. Those that are potentially available for observation, for the most part, decorate public spaces and are mute as to their context of use. They are also of rock types, sedimentary for the most part, that are common and uncharacteristic and do not readily betray their sources.

It is noteworthy that the principal written sources for the Contemporary period are no longer municipal or notarial protocols but geographical dictionaries, in particular the works of Miñano and Madoz. Although many sites are identified in these massive dictionaries, the authors provide only meagre clues as to their commercial reach.

12.7.1. "Long-distance" millstone distribution (over ~85 km)

From the outset it must be clearly established that there is no single quarry or quarry district that dominated the long-distance millstone market of southern Spain in Contemporary times. No single site possessed the reputation and commercial strength of the Montjuïc quarry near Barcelona that traded its sandstone millstones along the Catalonian and Valencian coasts, even attaining the Balearic Islands, southern France and Italy. From written sources there is evidence, however, that a series of quarries in southern Spain did trade their products between 85 and about 210 kilometres (Table 12.3; Fig. 12.24). The values, although arbitrary, do indicate that these quarries had an excellent reputation, and that the high cost of transport of their products was merited. There are certainly other unidentified sites in southern Spain that exported their products similar distances. This is corroborated by a Royal Decree dating to 1829, forbidding the import of 42 millstones from Portugal into the Galician city of Pontevedra. The argument behind this prohibition was that there were sufficient millstone quarries on Spanish soil to supply the country, notably those in Álava in the Basque Country and in Andalusia, in particular in the Province of Granada (Ferrer 1830: 219).

1) TO-2: The quarry with the longest certified range of trade in Contemporary times is Las Ventas con Peña Aguiler, near Toledo. According to Miñano, this granite quarry supplied millstones 30 leagues and beyond (Miñano 1828, Vol. 9: 286), roughly between 120 and 210 kilometres (based on four to seven kilometres for one league). The larger 210 kilometre radius would include the city of Madrid. The commercial strength of Las Ventas con Peña Aguilera was already established in the late 16th century as it had the reputation of being Spain's "best" millstone quarry (Viñas & Paz 1951: 216).

2) CA-8: The white limestone quarry of Berrueco by Medina Sidonia, famous all over Cádiz for its millstones for watermills, windmills and *tahonas* (Madoz 1846: Vol. 4: 290), is reputed to have traded its products as far as Puebla de Guzman in the Province of Huelva (Garrido 2001: 167), a distance of about 170 kilometres. From different written sources dating to the 16th century, Berrueco was one of the most important quarries in southern Spain and is surely one of the "abundant" Andalusian quarries alluded to in the Royal Decree of 1829 prohibiting the entry into Spain of millstones from Portugal (Ferrer 1830: 219-220).

3) CA-3: The seaside biocalcarenite exploitation of Rota, Cádiz, exploited since Roman times for hand querns, traded its stones in the 19th century to an unidentified location in Portugal by two ships, the "*Felicidade*" and "*O Que Deus Quera*" (Ponce 1981). The nearest Portuguese port is 110 kilometres away. Today, evidence of extraction of large millstones does not give the impression of a former large-scale production. It must be noted, nevertheless, that much of the site could have been destroyed during construction of the new port.

4-5) BA-1 and J-1: The quarries of Alconera in Badajoz and Castillo de Locubín in Jaén are included in this group (in spite of the absence of any evidence of distribution inside Spain) because their millstones were exhibited at the Universal Exposition of Paris in 1867 (*Comisión Régia* 1867: 187, 190). This suggests that they were of excellent quality and their owners saw a potential to market them outside their region.

Table 12.3: List of Contemporary quarries that presumably commercialised their millstones over long distances (beyond the range of 85 kilometres).

Code	Millstone quarry	Rock type	Bibliography
TO-2	Ventas con Peña Aguilera	Granite	Miñano 1828, Vol. 9: 286
CA-8	El Berrueco	White limestone	Madoz 1846: Vol. 4: 290
CA-3	Rota	Biocalcarenite	Ponce 1981
BA-1	Alconera	White limestone	*Comisión Régia* 1867: 187
J-1	Castillo de Locubín	Sedimentary	*Comisión Régia* 1867: 190

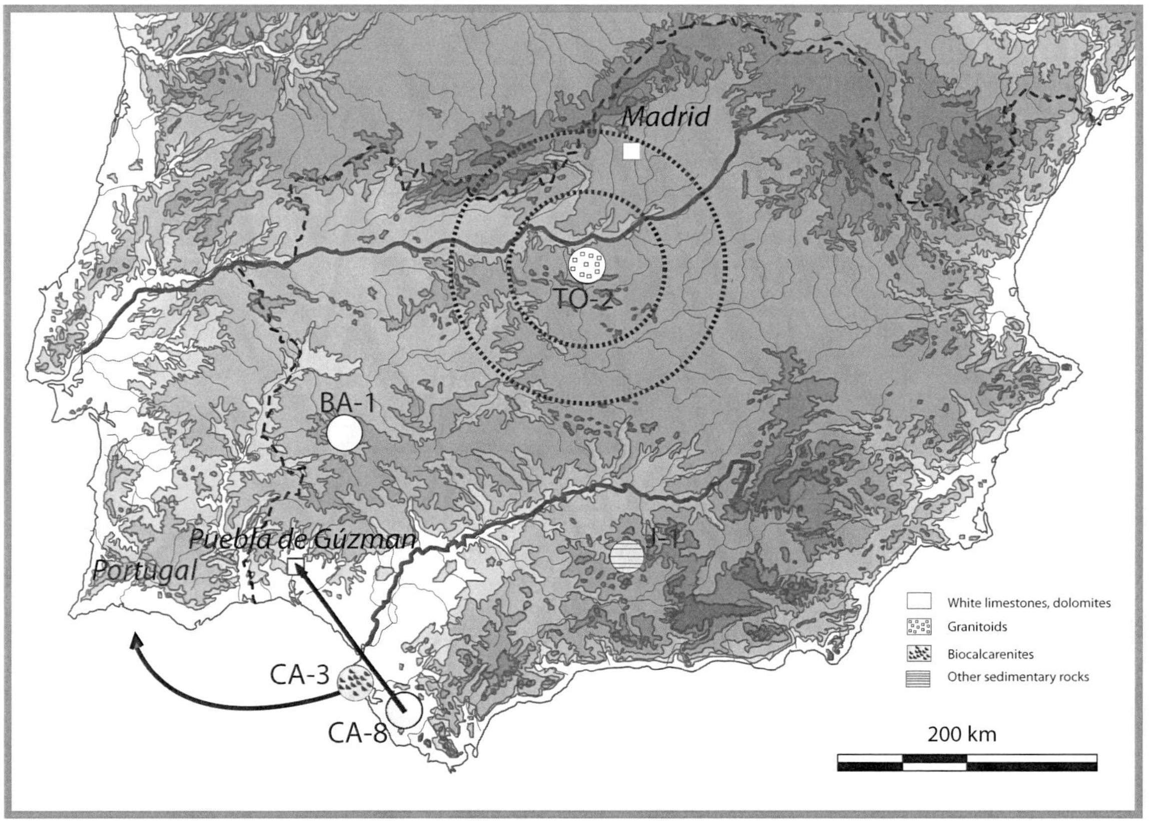

Fig. 12.24: Map of Contemporary millstone quarries with distribution of products certified by written sources surpassing the distance of 85 kilometres. Sites: Ventas con Peña Aguilera (TO-2), El Berrueco (CA-8) and Rota (CA- 3). Castillo de Locubín (J-1) and Alconera (BA-1) are included because their products were exhibited at the Universal Exposition of Paris in 1867.

12.7.2. "Longer-range" regional millstone distribution (~50 to ~85 km)

There are several Contemporary millstone workings with certified distribution ranging between 50 and 85 kilometres (Table 12.4; Fig. 12.25). These cannot be considered veritable "long-distance" exporters since the distance, and the cost of purchase and transport, would have been considerably less than that of acquiring millstones from sites such as Montjuïc in Barcelona or the burrstone workings in the Paris Basin in France. However, investment in the purchase and transport was nonetheless considerably larger than that of the "shorter-range" regional quarries.

Table 12.4: List of Contemporary quarries that presumably commercialised their millstones over distances of about 50 to 85 kilometres.

Code	Millstone quarry	Rock type	Bibliography
M-2	Colmenar de Oreja	White limestone	Madoz 1847, Vol. 6: 525
GR-1	Moclín	White limestone	Madoz 1847, Vol. 8: 480
Co-1	Los Frailes	*Rosso ammonitico*	Ezquerra del Bayo 1856: 385

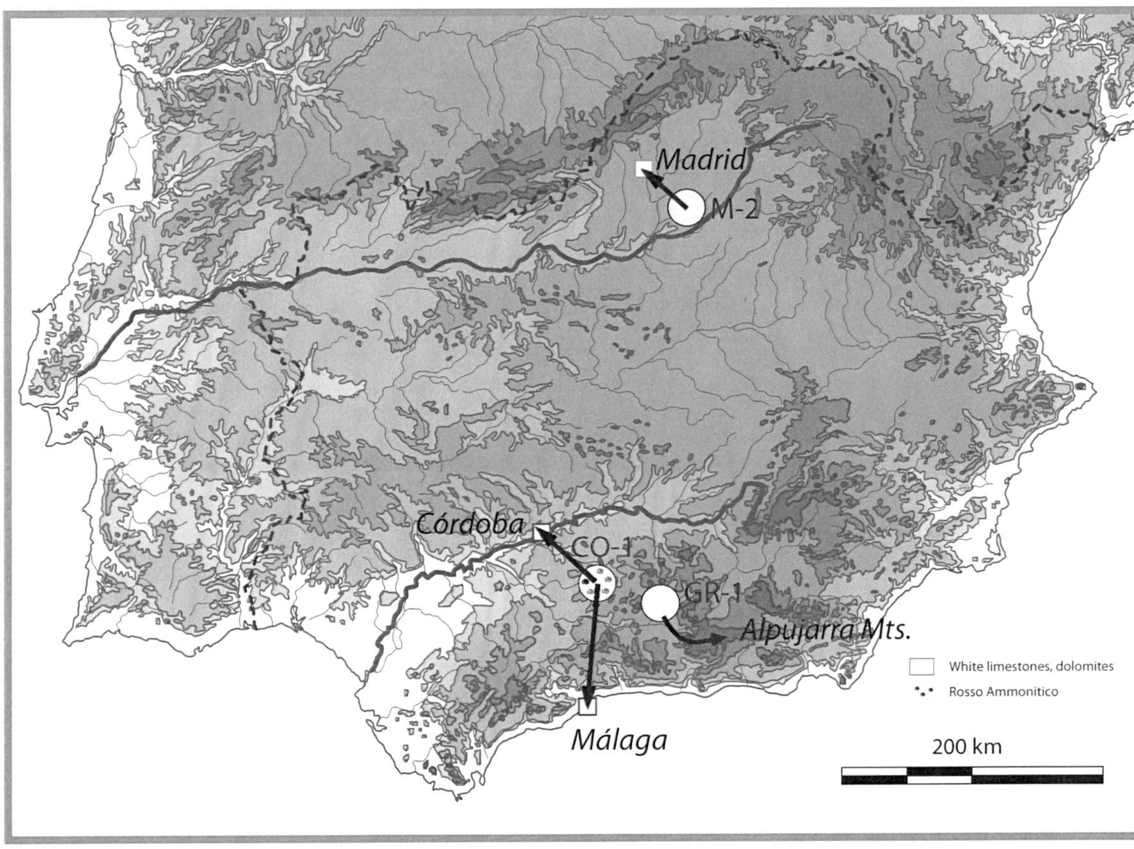

Fig. 12.25: Map of the Contemporary millstone quarries that exported their products over distances ranging from 50 to 85 kilometres. Sites: Los Frailes, Cablra (CO-1), Moclín (GR-1) and Colmenar de Oreja (M-2).

1) M-2: The white limestone quarry of Colmenar de Oreja (M-2), Madrid, is known to have supplied millstones to the *tahonas* of the city of Madrid about 50 kilometres to the north-west (Madoz 1847, Vol. 6: 525). The quarry would have benefited from the network of transport of construction material extracted for building sites at other Colmenar quarries, well-established since the 16th century.

2) GR-1: The white limestone quarry of Moclín in the Province of Granada supplied millstones at the provincial level (Madoz 1847, Vol. 8: 480), such as upper stones to watermills in the Alpujarra Mountains (Rodríguez Monteoliva 1989: 705), a distance of about 60 and 80 kilometres. The white upper stones were probably coupled with local conglomerate or sandstone lower stones, the dominant rocks in the Alpujarra Mountains. Moclin´s large size is also indicative of a wide distribution. It is possibly the site in Granada alluded to in a Royal Decree forbidding the import of millstones into Spain from Portugal (Ferrer 1830: 219-220). If this reference were proven to be true in future research, then Moclín would require "upgrading" to the status of a long-distance exporter.

3) CO-1: The pink *rosso ammonitico* limestone exploitation of Los Frailes, Cabra in southern Córdoba exported its millstones to watermills in the city of Málaga, 85 kilometres to the south (Ezquerra del Bayo 1856: 385) and to the city of Córdoba, 60 kilometres to the north-west (Montero 2008). In addition, the size of the quarry with its many tall vertical tubular hollows is perfectly compatible with a large volume of exports and a wide demand. As in the case of Moclín, future research might result in the upgrading of this site to that of a long-distance exporter.

12.7.3. "Shorter-range" regional distribution (~20 to ~50 km)

The mere appearance of a written reference to a millstone quarry in publications of the status of the dictionaries of Miñano and Madoz is indicative that their products were noteworthy, surpassing that of other modest, anonymous quarries. The fine line separating the concepts of local and regional production is, nonetheless, difficult to quantify, especially when the written sources provide little information as to the final destination of the products. Rarely do 19th-century texts go beyond the simple identification of the quarry, and qualify, although vaguely, their commercial reach (Fig. 12.26; Table 12.5). Based on the study of the distances between the Pinilla de Jadraque quarry (GU-2) and the neighbouring contemporary watermills (gleaned from the Madoz dictionary), the limit between "local" and "regional" quarries can be roughly set at 20 kilometres, a distance that could be attained in a day by an ox-driven cart.

1) GU-2: Pinilla de Jadraque is described by Miñano as follows: "excellent millstones are extracted from its precious quarry for other towns" (Miñano 1827, Vol. 7: 24). I have attempted to quantify this statement by recording the number of flour mills in the hamlets and towns in a radius of 20 kilometres. To accomplish this I have consulted the entries of 21 towns and hamlets in the Madoz dictionary (1845-1850) (Miñano's texts are not complete enough). In the 21 cases, there are 14 flour mills (see annex 4). Supposing that each mill required one millstone replacement every year (which was not necessarily the case), then the yearly production of 14 is a number that appears to be very low for a site cited in a geographical dictionary. Even doubling that number would imply only a few months of labour for only a small team of workers (based on one extraction every four to seven days, *cf.* Maestro 2011: 43-44). I therefore infer that in his use of the words "for other towns" Miñano alluded to a distance beyond 20 kilometres, placing the Pinilla de Jadraque millstone distribution outside of the strictly "local" sphere.

2) GR-6: The situation of the white limestone quarry of Alhama de Granada, Fuente de los Morales, is similar to that of Pinilla de Jadraque. Madoz states that its millstones were exported all over the area, even as far as some towns in the Province of Málaga (Madoz 1847, Vol. 8: 216). The towns referred to must be to the south-west (Alfarnate, Perirana, Viñuela, Canillas de Aceituno), over the Málaga mountains, at an average distance of about 25 kilometres.

3) CO-10: The conglomerate quarry along the Albardado stream near Belmez produced millstones, according to Madoz, both for towns of the vicinity and for towns "even farther away" (Madoz 1846, Vol 4: 131). In this case, the closest towns are Belmez, Peñaroja-Pueblonuevo and Villanueva del Rey, less than 10 kilometres away, whereas a handful of more distant towns (Fuente Obejuna, Villanueva del Duque, Pozoblanco, Espiel) range from 18 to 30 kilometres.

Table 12.5: List of Contemporary quarries that presumably exported their millstones over distances between about 20 to 50 kilometres.

Code	Millstone quarry	Rock type	Bibliography
GU-2	Pinilla de Jadraque	Limestone tufa	Miñano 1827, Vol. 7: 24
GR-6	Fuente de los Morales	White limestone	Madoz 1847, Vol. 8: 216
CO-10	Albardado	Conglomerate	Madoz 1846, Vol 4: 131
BA-2	Llerena	Sedimentary	Caso Amador 2008: 133
CC-2	Villars de Plasencia	Granite	Miñano 1828, Vol. 9: 433
GR-10	Vélez de Benaudalla	White limestone	Madoz 1847, Vol. 8: 480
GR-2/3	Loja	White limestone	Madoz 1847, Vol. 8: 480
GR-11	Rambla de las Canteras	Conglomerate	Madoz,1846, Vol. 5: 461

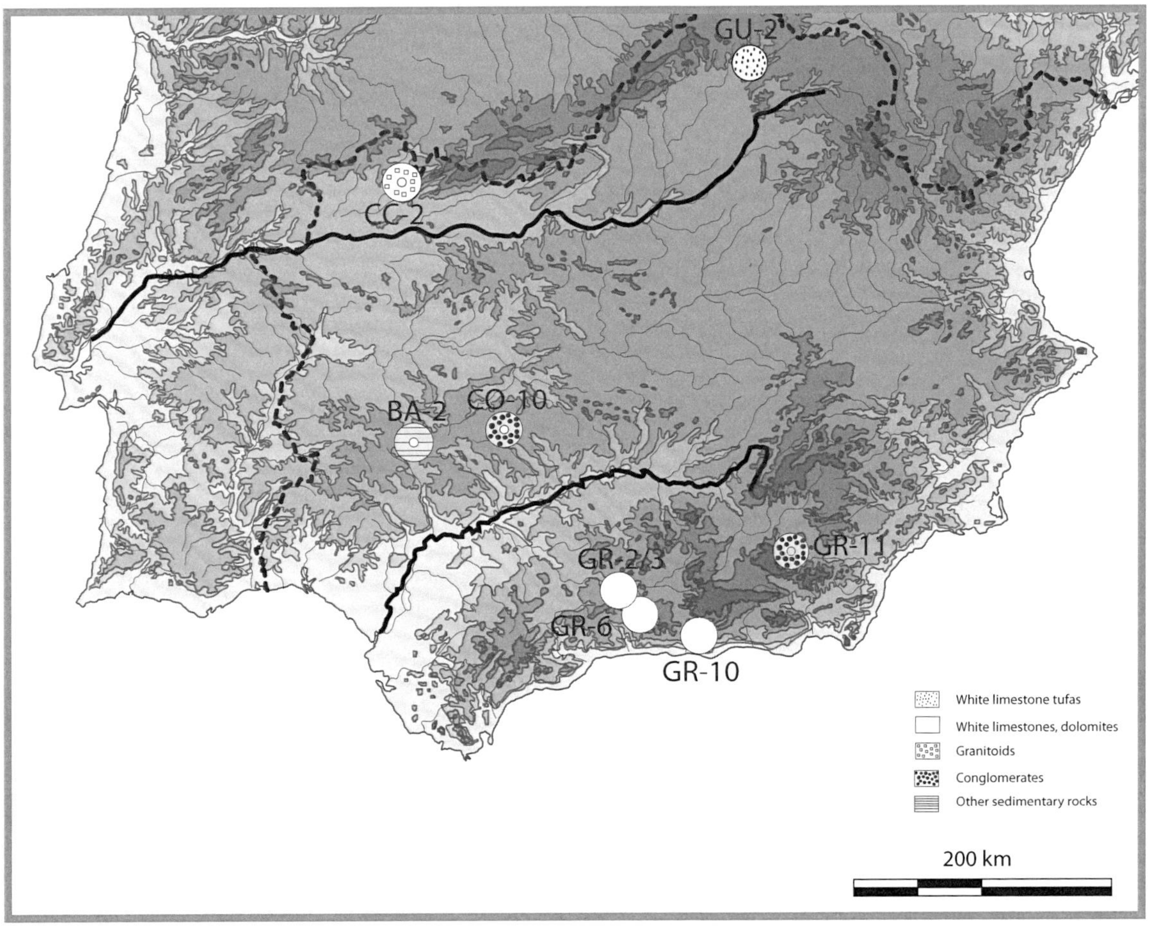

Fig. 12.26: Map of Contemporary millstone quarries that appear, according to 19th-century texts, to stand out above local millstone quarries with a sphere of distribution of between approximately 20 and 50 kilometres.

4) BA-2: The sedimentary rock (limestone, limestone tufa or arkose) quarry of Llerena, Badajoz, according to the census of 1791 (Caso Amador 2008: 133), supplied millstones to "many towns in the area". The few immediate towns are in a radius of 10 kilometres. To attain a higher number of towns the radius must be expanded to between 15 and 30 kilometres.

5) CC-2: Miñano states that the granite quarry of Villar de Plasencia, where millstones were hewn from surface rocks "... and transported up to four and six leagues away" (Miñano 1828, Vol. 9: 433). The town is on the *Via de la Plata* (Silver road), an ancient and important north-south thoroughfare dating from Roman times. The statement by Miñano indicates a distance ranging from 16 to 42 kilometres. Due to access to the important commercial road, it is logical to assume that the quarry supplied millstones beyond the local sphere.

6-7) GR-1, GR-10, GR-2/3: Madoz, when describing the main rock quarries of the Province of Granada, mentions the millstone workings of Moclín (GR-1), Vélez de Benaudalla (GR-10) and Loja (GR-2/3) (Madoz 1847, Vol. 8: 480). Since Madoz specifically cites these three productions, all exploiting white limestone or dolomite, then the three stand out over the other quarries in the province and must have had a wider range of distribution.

8) GR-11: The pebble conglomerate quarry of Rambla de las Canteras, outside of Caniles, Granada, was known for millstones that yielded dark flour (Madoz 1846, Vol. 5: 461). I cannot conceive that the distribution of the millstones of this easily accessible quarry, with its long, parallel trenches and huge production, could have been limited to the local sphere.

12.7.4. Local millstone distribution

Local distribution of millstones in Contemporary times is defined roughly as the spread of millstones from their source inside a sphere measuring a radius of about 20 kilometres. This distance is equivalent to that attained by an ox-driven cart in one or two days. These types of quarries are characterised by a modest volume of production.

A number of sites of this type have been identified in the study area. The quarries of Guájar Faragüit (GR-8) and Carchuna (GR-9) in the Province of Granada are examples with only about a dozen visible extractions. The workings of Cambil (J-4) and Molino de la Piedra (CO-5) are even smaller. None of these, without a doubt, can be considered anything but local workings and would not have merited inclusion in the pages of the 19th-century geographical dictionaries.

Yet there are 32 certified sites from Contemporary times by means of old texts, for the most part Madoz, that do not show any evidence that their products circulated beyond the local sphere (Fig. 12.27; Table 12.6). Hence, at this stage of research, I cannot assume that these workings served anything but local needs.

Table 12.6: List of Contemporary millstone quarries or quarry districts cited by written sources from the middle of the 19th century. There is no evidence of distribution from these sites beyond a local sphere. The sheer mention of the sites in the volumes of Madoz, however, sets them above other anonymous productions.

Code	Millstone quarry	Rock type	Bibliography
MA-1	El Torcal	*Rosso ammonitico*	García de la Leña 1789: 106
MA-7	El Jabonero	Coarse limestone	García de la Leña 1789: 106
M-4	Chapinería	Granite	Miñano 1826, Vol. 2.: 83
GU-1	Brihuega	Limestone tufa	Vallejo 1833: 387
AL-8	Adra, Cerro el Chispa	Conglomerate	Madoz 1845, Vol. 1: 88
AL-9	Vera	Sedimentary	Madoz 1850, Vol. 15: 670
J-3	Huelma	White limestone	Madoz 1847, Vol. 9 : 260
J-5	Andújar, Los Morales	Granite	Madoz 1845, Vol. 2: 305
J-6	Andújar, El Pedroso	Granite	Madoz 1845, Vol. 2: 305
J-7	Linares district	Granite	Madoz 1847, Vol. 10: 290
CO-9	Los Arenales	Sedimentary	Madoz 1849, Vol. 15: 136
CO-16	Minas de Espiel district	Sandstone	Madoz 1849, Vol. 14: 382
CO-17	Fuente Obejuna	Undetermined	Madoz 1847, Vol. 8: 230
MA-2	Teba	White limestone	Madoz 1849, Vol. 14: 752
MA-3	Alhaurín el Grande	Sedimentary	Madoz 1845, Vol. 1: 604
MA-5	Guaro	Sandstone	Madoz 1847, Vol. 9: 56
MA-6	Alozaina	Conglomerate	Madoz 1845, Vol. 2: 186
CA-10	Peña Harpada	White limestone	Madoz 1845, Vol. 1: 376
SE-3	Lora de Estepa	Coarse limestone	Madoz 1847, Vol. 7: 690
SE-5	Alanís	Undetermined	Madoz 1845, Vol. 1: 190
HU-6	Zalamea la Real	Granite	Madoz 1850, Vol. 16: 450
CR-3	Chillón	Granite	Madoz 1847, Vol. 7: 327
GU-6	Corduente	Granite	Madoz 1847, Vol. 7: 9
GU-7	Castilnuevo	Limestone tufa	Madoz 1847, Vol. 6: 172
AB-1	Fuentealbilla	White limestone	Madoz 1845, Vol. 1: 51, 256
BA-4	Salvaleón	White limestone	Madoz 1849, Vol. 13: 711
BA-5	Jerez de los Caballeros	Granite	Madoz 1847, Vol. 9: 627
CC-4	Bohonal de Ibor	Granite	Madoz 1846, Vol. 4: 376
CC-6	Logrosan	Granite	Madoz 1847, Vol.10: 355
M-3	Comenar Viejo	Granite	Madoz 1847, Vol. 6: 530
SE-7	Gerena	Granite	Madoz 1847, Vol. 8: 348
GU-5	Tobes	Limestone tufa	Castel 1881: 158

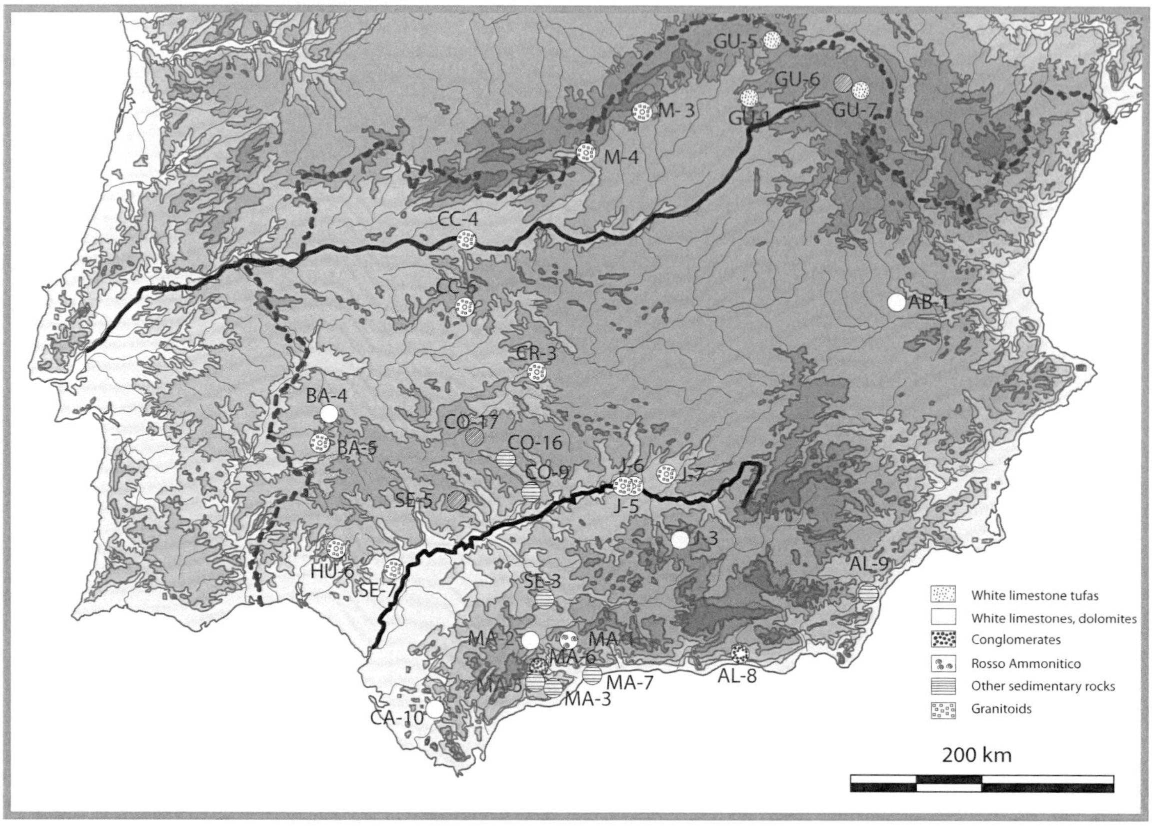

Fig. 12.27: Map of Contemporary "local" millstone quarries with no evidence of distribution over a range of about 20 kilometres.

12.7.5. Interpreting the distribution of Contemporary millstone quarries

The map illustrating the spread in Contemporary times of millstone quarries by their category of distribution potential and their rock type in the southern half of Spain can be divided into three main zones based on density of the sites and their petrography (Fig. 12.28).

Zone 1

The first zone, south of the Guadalquivir River Basin, covers much of Andalusia, from the Bay of Cádiz to the Sierra Nevada in Granada, and tapers off in the Province of Almería. This corresponds to a large portion of the Bétic geological range. Here, there is an important density of millstone quarries exploiting sedimentary stones.

The dominant stones in this zone in Contemporary times are hard, fine white and rose limestones that reputedly yielded a white flour. These outcrops are found from Cádiz to Granada with important productions at Berrueco (CA-8) in Cádiz, Los Frailes, Cabra (CO-1), in Córdoba and Moclín (GR-1) in Granada. These sites were seconded by similar but apparently smaller productions at Alhama de Granada (GR-6), Loja (GR-2/3) and Vélez de Benaudalla (GR-10).

The nature of the site of Castillo de Locubín (J-1) that exhibited its millstones at the Universal Exhibition of Paris in 1867 is difficult to explain. Its coarse yellowish texture would not seem to be as efficient in milling as the white limestone quarry from nearby Moclín (GR-1). The field survey, however, indicates a large production.

The density of quarries decreases to the east. Caniles (GR-11), like other older sites in the area, exploited a sort of puddingstone that yielded a "brown bread" (Madoz 1846: Vol. 5, 461). This region appears to be, from a geological point of view, devoid of suitable white limestone outcrops.

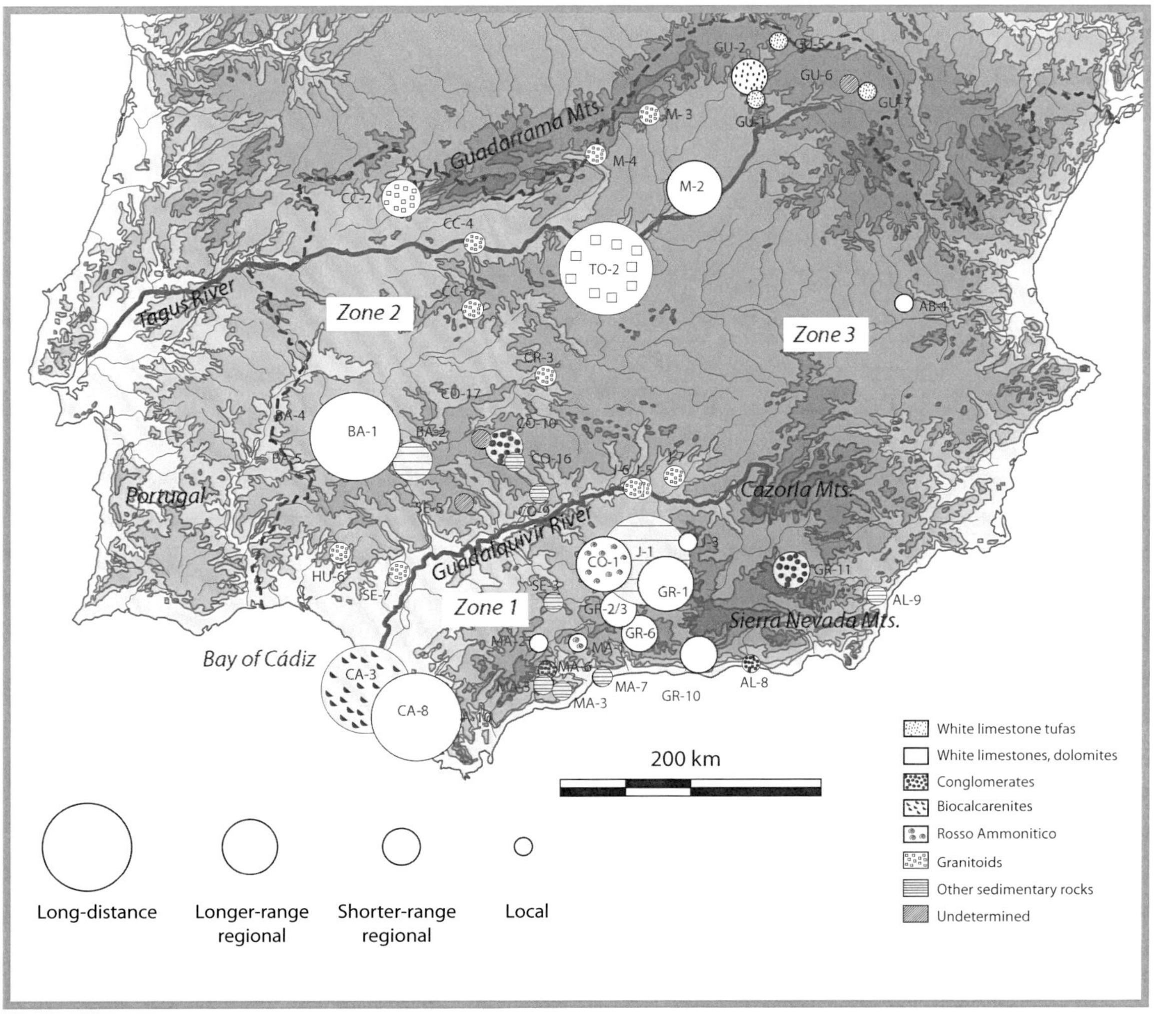

Fig. 12.28: Map combining the spread of the four classes of millstone quarries from Contemporary times (late 18th to early 20th centuries) according to their spheres of distribution and rock types.

The long-range distribution of production of the *ostionera* stone in the south is also difficult to explain and could be anomalous. The only site of this type dated to Contemporary times is Rota (CA-3) where millstones were extracted for transport to Portugal. This stone may have been destined for grinding a different type of product because for cereal grinding, the nearby Berrueco quarry (CA-8), with its white limestone, certainly yielded a much finer flour.

Zone 2

The second zone is marked to the south by the Guadalquivir River Valley and stretches from the Cazorla Mountains of Jaén, to the northern Bay of Cádiz, to the Portuguese border, to the Sierra de Guadarrama in the north from Cáceres Province to Guadalajara, passing through Madrid. This encompasses much of the western Iberian Massif, as well as the Ossa Morena and South Portuguese geological zones.

This zone is dominated by granitoids in the form of vast extensive outcrops and rounded surface boulders. The site of Ventas de Peña Aguilera (TO-2), a certified millstone producer and exporter since the 16th century, is its major exponent (Viñas & Paz 1951: 216; Baltanas 1998: 36-37).There is also a spread of smaller granite workings from the Guadalquivir Valley to the north of the Province of Cáceres. There were certainly more unidentified workings of this rock in Contemporary times.

Inside these granite-dominated zones, there is a "pocket" formed by four quarries that exploited sedimentary rocks, three of which were presumably important. These are Alconera (BA-1) to the south in the Province of Badajoz that produced millstones that yielded a white flour for white bread (Caso Amador 2008: 132-133). The fact that it, like the Castillo de Locubín (J-1), sent millstones to the Universal Exhibition of Paris in 1867 (Comisión Régia 1867: 182) suggests its products had a good reputation and distribution was spread over a wide area.

A second quarry within this pocket surrounded by granitoids is Colmenar de Oreja (M-2). This site supplied white limestones to the *tahonas* of the city of Madrid. It is interesting that in spite of the availability of granite from scores of closer quarries, animal-driven flour mills in Madrid chose the Colmenar de Oreja limestone, probably because it was reputed to yield a white flour for white bread (Vallejo 1833: 387).

The third pocket is made up of a series of sites exploiting porous limestone tufa. Brihuega (GU-1) in the Province of Guadalajara is noted for millstones that yielded a very white flour, as white as that of Colmenar de Oreja (Vallejo 1833: 387). There are also within this pocket other more modest limestone tufa sites in Guadalajara on the northern fringe of the area such as Pinilla de Jadraque (GU-2) and Tobes (GU-5) by Sienes.

A fourth pocket is in the south of the zone, in the Sierra Morena. It is composed of a series of sedimentary stone exploitations, with its major exponents being firstly the quarry along the Albardado stream (CO-10) near Belmez, Cordoba, a conglomerate outcrop with large rounded clasts (a sort of puddingstone) and, secondly, Llerena (BA-2) an unidentified sedimentary stone. It is doubtful that the quality of these millstones was superior to that of the dominant granite productions. Both stone types are reputed to yield dark bread. In any case, production of this group was not widespread and probably not relevant outside the local sphere.

Zone 3

The third zone is a vast millstone quarry "no man's land" covering the eastern half of the Iberian Massif. It stretches from the north of the Province of Jaén, through the Autonomies of Murcia and Valencia, and large parts of the Provinces of Albacete, Cuenca, Toledo and Ciudad Real. The sole millstone quarry documented in this area is Fuentealbilla (AB-1) in the Province of Albacete. The site is known since the middle of the 18th century by means of the Census of the Marqués de Ensenada. Its workings are certified later in the 19th century by Madoz (1845, Vol. 1: 256).

The existence of this huge area devoid of millstone workings is astonishing. Since the identification of these sites is based principally on 19th-century geographical studies that applied a systematic means of collection of information throughout Spain, the large gap is surely not due to a lack of rigour on behalf of the geographers Madoz or Miñano and their collaborators.

The gap, in fact, could be related in part to geology. Soft Quaternary and Tertiary material, usually unsuitable for millstones, abound in this zone. Another possible explanation, totally hypothetical, is that the mills in this region relied on small, anonymous workings that did not merit the attention of the 19th-century geographers. Another explanation for this gap is that the more important watermills, windmills and *tahonas* were supplied by millstones imported from the Montjuïc quarry of Barcelona. It is recorded that Montjuïc exported its products long distances around the western Mediterranean (France, possibly Italy) and along the Catalonian coast as far south as Murcia (Español 2009: 967; Barberá Miralles 2003: 194). The Montjuïc workings therefore certainly supplied millstones to the coastal cities of Valencia, Castellón and Alicante. For the inland mills in this "millstone quarry free region" the situation is unclear. A clue comes, however, from the description of Fuentealbilla (AB-1). Madoz, comparing Fuentealbilla's millstones to those of the Montjuïc, asserts the following:

"... y en le Molar, térm. de Fuente-albilla, del mismo part., la de piedras de molínos, de grano poco compacto, por lo que se aprecia menos que la barcelonesa" (Madoz 1845, Vol. 1: 256).

This translates roughly as: "... and at the placename *Molar*, in the area of Fuente-albilla, there is a millstone quarry with a stone that is not very compact, that is less valued than that of Barcelona".

We can therefore infer from this comparison that the imported "Barcelona" (Montjuïc) stone was present in this area in Contemporary times. The presence of this famous type of millstone, along with the geological arguments, may explain the scarcity of local millstone workings.

The long-distance exports up and down the Mediterranean coast from the Montjuïc of Barcelona, the Spanish exploitation with the widest certified commercial reach, was in fact a precursor of a phenomenon, the arrival of French millstone imports, that was soon to shake the foundations of the Spanish millstone industry in Contemporary times.

12.7.6. The French millstone "invasion"

The French millstone "invasion" introduced siliceous, "intelligent" *meulas francesas*, for the most part from the Paris Basin. These stones were responsible for yielding a fine, white flour free of grit for a bread called "the best of breads" (Belmont 2006a, Vol. 2: 148-153). This industry, in constant expansion for several centuries, was progressively gaining a stronghold throughout Europe since the end of the 18th century (Belmont 2006a, Vol. 2: 132-134). Its spread by fluvial, maritime and land transport has no equal in the history of millstone distribution, even surpassing by far the distances recorded in Roman times for Pompeian mills. French burrstones, in fact, crisscrossed oceans reaching the Americas and even New Zealand.

It must be noted that no qualitative study of these French millstones has been carried out in Spain. Yet they are mentioned in practically every molinological Spanish article I have consulted regarding Contemporary watermills published in the last decade. This in itself validates their omnipresence.

13. FROM STONE TO BREAD

The procedures of grinding and sieving flour before the baking process were complex and entailed many variables that evolved over time, depending on the type of grain (wheat, barley, rye, millet, etc.), the type of stone, the milling technology and the type of bread (García Sánchez 1983; Rodríguez Monteoliva 1989: 684-685). This brief chapter will not examine all of these processes, a task that is well beyond the scope of this work, but focuses on two aspects related directly to quarries and millstones. The first is the problem of the grit released from the stones into the flour that had an adverse effect on teeth (hence the health). The second, linked to the first, is the aspiration, appearing presumably in the Late Middle Ages, to obtain a fine white bread, free of "impurities".

13.1. Grit in the bread

Grit in the bread resulted, first of all, from milling cereals that were not properly cleaned. In the early 18th century, Agustí, in his chapter on the secrets of milling wheat and making bread in a treatise on agriculture, underscored the importance of cleaning the grains before grinding, because unwashed grains released pernicious dust and other impurities into the flour that ended up in the bread (Agustí 1722: 187).

Grit coming from the stones, however, was the major culprit of tainting bread. Gómez Ruíz, in his study of watermills along the Odiel River in Huelva, cites a Municipal Archive of Aroche dating to 1574 that refers to this problem. The document records that the *Cabildo* (church authorities in this case) ordered "*granno*" millstones to be coupled with "*concha*" millstones because the flour from a pair of "*concha*" stones was "harmful" (Gómez Ruíz 2003: 31). Gómez Ruíz interprets the *granno* as granite upper stones, *granno* referring to the dark "specks" of biotite in the white rock. *Concha* (shell) millstones, on the other hand, were less compact, darker-coloured, sedimentary lower stones with marine shell inclusions. The choice of granite for the upper stones is not surprising since Aroche is in the heart of a vast granite district. A millstone quarry is, in fact, recorded at Fuente de Aliseda (HU-7) (de Dios Ayuda 1794: 348) near Aroche. In any case, the historical record not only reveals concern by the authorities for the health of its citizens, but establishes a solution to remedy the problem. The record, however, does not shed light on the type of ailment the bread produced. It is reasonable, nonetheless, to suppose that it was related to the wear of the teeth.

Agustí also stressed that millstones must be *"fuerte"* (strong, meaning hard) and not *"blandos"* (soft), because the softer stones release *"polvo"* (dust) into the flour (Agustí 1722: 186). The author also noted that stone grit found its way into the flour especially during the first round of milling following a millstone's *"picada"* (dressing of the working surface) (Agustí 1722: 187). In the Alpujarra mountains of Almería, a technique to rid the stones of the impurities resulting from dressing was to run sawdust through the stones. This practice at times led to conflict with the farmers, who mistakenly believed that the grains they entrusted to the miller were being sacrificed to clean the miller's stones (Cara Barrionuevo *et al.* 1999: 153).

Grit in the bread certainly had a harmful affect on the population of southern Spain, as elsewhere. This pernicious effect was heightened by the necessity of certain millstones to be dressed frequently, in certain cases on a daily basis, thus increasing the amount of dust finding its way into the bread. There are records, for example, that indicate that the stones from Colmenar de Oreja (M-2) required dressing every six hours (Vallejo 1833: 387). Millstones in the Alpujarra Mountains of Granada and in Jaén also had to be dressed daily (Ordoñez 1993: 3; López & Cabrera 2004: 204). It remains unknown if the grit of certain stones was more harmful than that of others. In any case, the problem practically disappeared in the 19th century with the introduction of the hard siliceous French burrstones, very hard rocks that only required a minimum of dressing.

13.2. White and dark rocks, millstones and bread

In southern Spain there are a series of terms or expressions associated with quarries, millstones, flour and bread that have multiple interrelated meanings that, at times, are difficult to interpret. *Piedra blanca*, for example, in its general sense, denotes a white-coloured stone exploited in a white-stone quarry. But it also designates the upper stone (runner) of a millstone pair. *Bazo or baza,* meaning dark, and *baxo,* meaning low, denote dark-coloured stones, as well as the lower stone of a millstone pair. The adjectives *blanco* and *bazo*, in milling terminology, also designate the colour and type of flour, hence the colour and type of bread. *Pan blanco*, therefore, is white bread, whereas *pan bazo (or baxo)* denotes dark bread. The connotations of these colours therefore go beyond simple definitions of stones, quarries, flour and bread and have social and economic implications.

13.2.1. White and dark millstone rocks and quarries

The Berruecos Mountain (CA-8) in Cádiz was described as a "... mine of white, grainy stone exploited for mills and animal-driven flour mills" (Martínez y Delgado 1875: 129). Similarly, Salvaleon (BA-4) in Badajoz was described as a "white" exploitation (Madoz 1949, Vol. 13: 711). The rock of El Hacho in Seville (SE-3) was "rough and white and apt for flour mills" (Madoz 1847, Vol, 7: 609). Around Loja (GR-2/3), in the Province of Granada, there were also "white" stone quarries (Madoz 1847, Vol. 10: 360). The petrography of all of the preceding exploitations, as seen by field surveys and from the study of geological maps, varies from limestones to dolomites, rocks that are whitish in colour.

On the other side of the colour spectrum are the quarries exploiting coarse, dark-coloured rocks. Tomás López, in the late 18th century, mentions the *"piedras bazas"* from Otívar (GR-13), in the southern mountains of Granada (Reyes Mesa 2000b). García de la Leña also uses the term "*bazo*" to describe the rock at a quarry in the Jabonero Valley (MA-7) (García de la Leña 1789:106). These sites have not been identified in the field, but from geological maps, Jabonero is in an area dominated by coarse limestone, whereas the area of Otívar is dominated by puddingstone.

Jerez de los Caballeros (BA-5) in Badajoz is unique in that it is the only site reported to have both white and dark millstone quarries ("*... hay tambien canteras de piedras morena y blanca para molinos harineros ...*", Madoz 1847, Vol. 9: 627). It is unclear, however, if this is an allusion to the exploitation of different types of rocks in this predominantly granite area (granites and limestones, for example) or if it refers to a different qualities of granite.

In any case, it appears that white stones were not as compact or hard as dark stones, and that even among the similar rock types there could be great differences of quality. A miller of the Alpujarras noted that white stones could grind about 12 *fanegas* (bushes) of flour (*circa* 600 kg) before requiring dressing, whereas dark millstones could grind ten times the quantity (120 *fanegas*) (Cara Barrionuevo *et al.* 1999: 155). If these proportions can be extended more generally to other white rocks, such as fine limestones and dolomites, and dark rocks such as coarse limestones, conglomerates and sandstones, then one might wonder why millstone makers bothered to exploit white rocks, if they did not possess other properties, such as yielding white flour.

13.2.2. "White" upper stones and "*baza*" lower stones

A second definition of the terms "*piedra blanca*" (white stone) and "*piedra baza*" (coarse stone) designate respectively the upper stone (runner) and the lower stone. These terms appear under these designations in a number of historical archives throughout Andalusia from the 15th through the 19th centuries (Córdoba de la Llave 1988: 842-843, 853-854, 871, footnotes 24, 62, 68, 137; López & Martín-Caro 1989: 1022; Gómez Ruíz 2003: 31; Córdoba de la Llave & Varela 2011: 148, 183; Rodríguez Monteoliva 1989: 705).

The origin of these designations is not defined in molinological literature. It is plausible that they come from the using finer "white" stones (limestones) for runners, and coarser, darker or "reddish" stones (sandstones and conglomerates) for lower stones. A document in the Municipal Archives of Aroche (Huelva) dating to 1574, alluded to previously, corroborates the use of two different rock types for millstone pairs, such as a granite runner (white biotite) coupled with a coarse limestone lower stone (Gómez Ruíz 2003: 31).

A link between these definitions and rock types could be evidenced by the fact that certain quarries, presumably of finer material, manufactured only upper stones, whereas others, with more coarse properties, made only lower stones. The quarry of Hornachuelos (CO-13), for example, only supplied upper stones, while Albaida (CO-7) specialised in delivering lower stones (Córdoba de la Llave 2003: 305). The problem with these two examples in the Province of Córdoba is that their geological definitions are identical (limestone/dolomite) so each should theoretically have produced "*piedra blancas*". In spite of their geological similarities, it is highly likely that the rocks at these two sites did in fact possess different grinding properties, as is the case of the twin sites of El Berrueco (CA-8) and La Pila de Casares (CA-12) near Medina Sidonia (Cádiz) that although exploited what appears to be identical rocks, according to an old text yielded flours of different quality and colour (Martínez y Delgado 1875: 129).

13.2.3. Dark and white bread millstone quarries

Since the Middle Ages, a white bread has been associated with the "power and refinement" of the upper classes (Belmont *et al.* 2011: 221-223). In the study area, as a consequence of centuries of Islamic rule, I cannot help but evoke the symbolism at the beginning of the 16th century of a whiter bread linked to the Christian Reconquest. It is tempting to speculate that the desire for an ever whiter and finer bread ran parallel with the desire of the Catholic rulers to maintain a pure and orthodox faith, symbolised by the pure white eucharistic body of Christ, untainted by impure elements of *conversos*, converts to Christianity from Islam and Judaism.

The role of millstone quarries in attaining a fine, white flour for a soft, white bread is a subject that Alain Belmont has been investigating in the last years in France through historical texts and petrographical analyses in the framework of archaeological excavations (Belmont 2006b: 86-89; Belmont 2011: 14-16).

In southern Spain, as would be expected, white bread was reserved for the upper classes and dark bread was for the poor. For instance, in a sermon dating to 1558, the Dominican Friar Domingo Baltanás of Seville described how the well-born "*gran señores*" shared their food with their underlings. The older members in the service of nobles, who had been at their side in times of war and peace, ate from the same plate as the nobles. The poor, in turn, due to their condition, also received food, but not "*pan blanco*" (white bread) from the noble's table, but "*pa baço y duro*", that is, hard, dark bread (Fig. 13.1).

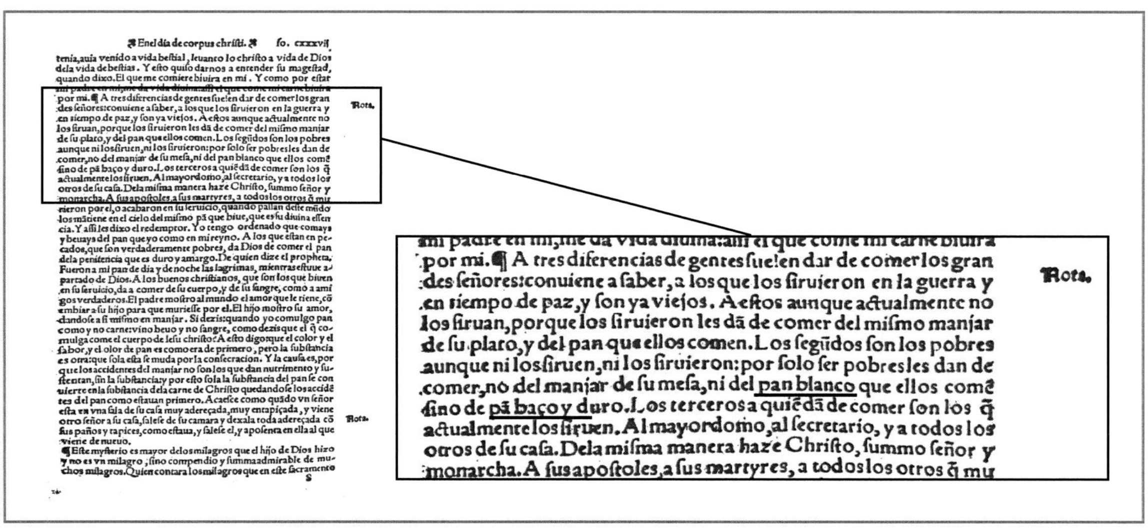
por mi. ¶ A tres diferencias de gentes suelen dar de comer los gran des señores: conuiene a saber, a los que los siruieron en la guerra y en tiempo de paz, y son ya viejos. A estos aunque actualmente no los siruan, porque los siruieron les dã de comer del mismo manjar de su plato, y del pan que ellos comen. Los segũdos son los pobres aunque ni los siruen, ni los siruieron: por solo ser pobres les dan de comer, no del manjar de su mesa, ni del pan blanco que ellos comẽ sino de pã baço y duro. Los terceros a quiẽ dã de comer son los q̃ actualmente los siruen. Al mayordomo, al secretario, y a todos los otros de su casa. Dela misma manera haze Christo, summo señor y monarcha. A sus apostoles, a sus martyres, a todos los otros q̃ mu

Rota.

Fig. 13.1: Extract of a sermon of the Dominican Friar Domingo Baltanás from Seville. The text alludes to both "pan blanco" and "pan bazo" (written "pa baço") (from the Exposicion de los Evangelios con sermones desde primero domingo del Aduiento hasta el domingo XXV despues de la trinidad, sacada de los sanctos doctores la yglesia by Domingo Baltanás, folio cxxxvi, 1558; from Google Books). Interpreted transcription: "A tres diferencias de gentes suelen dar de comer los gran señores: conviene a saber, a los que los sirvieron en la guerra y en tiempo de paz, y son ya viejos. A estos aunque actualmente no los sirvan, porque los sivieron les da de comer del mismo plato que ellos comen. Los segundos son los pobres aunque ni los sirven, ni los sirvieron: por solo ser pobres les dan de comer, no de manjar de su mesa, ni del pan blanco que ellos comen sino de pan baço y duro. Los terceros a quien dan de comer son lo que actualmente los sirven. Al mayordomo, al secretario, y a todos los otros de su casa".

Dictionaries and religious writings throughout the 17th century are indicative of the multiplication of the use of the terms "*pan blanco*" and "*pan bazo*", suggesting a growing difference of bread products.

White bread was also associated with the ecclesiastical class as seen by an anecdote recorded by Gerald Brenan in the 1920s in Yegen, an isolated village in the Alpujarra Mountains of Granada. It tells the story of the village priest and his housekeeper named *Pan Blanco* (White Bread):

"[The priest Don Indalecio] ... brought with him his housekeeper, who was familiarly known as Pan Blanco *or White Bread. The story that lay behind this nickname was that many years before, when* Pan Blanco *was a girl in a village close by, she had been engaged to be married. A few days before the ceremony she had gone to Don Indalecio, who was then young too, for confession. He fell in love with her and set his heart on getting her at any price. Going to his mother, therefore, he told her that he would kill himself if he could not have her, so she agreed to speak with the girl. 'María,' she said , I have a question to ask you. 'Would you rather eat black bread all your life or white bread?' 'White bread,' said the girl. 'I'm glad to see you so sensible ,' the priest's mother replied. 'All you have to do now is give up this marriage, which is quite beneath you anyway, and take service as housekeeper with my son.' She did so".*

(*South from Granada: A Sojourn in Southern Spain* by Gerald Brenan, 1998, p. 70).

This white/dark dichotomy was also reflected in the colour of the bread obtained, or reputedly obtained, from certain millstones. An early reference (1776) to the role of the millstones in making white flour comes from Zubieta, Navarra in a notarial protocol that records that the stones of a watermill were worn and were yielding a "dark and rough" flour, causing the mill, owned by the authorities, to lose clientele and revenue. The document therefore advises that the stones be replaced with a new pair from Macaia, France (*i.e.* Labort region), stones known to yield a whiter and softer bread (Alegría Suescun 2007: 8, footnote 15). The petrography and the exact source of the Macaia stones (siliceous stones from Dordogne?) is unknown. In any case, the archive demonstrates that 18th-century mill owners were willing to go beyond their local quarries and import superior millstones that yielded a finer white flour.

It is obviously not the stone *per se* that produced a fine, white flour. An array of factors played a role in the quality and colour of the daily bread. The choice of cereal, to begin with, was fundamental. Of the available cereals, bread wheat, with its starchy endosperm, was the more commonly accessible white bread grain (Lagardère 1991: 64-65). Other grains such as rye and barley, by contrast, served for making darker bread.

A fundamental factor in the obtention of white bread was the miller's technical ability. The miller determined the level of humidity of the grains (by sprinkling the grains with water) previous to grinding to facilitate the separation of the bran from the white endosperm. The Catalonian writer Agustí noted, for example, that adding salt to the water to humidify the grain before milling resulted in a whiter flour (Agustí 1722: 187).

The miller also controlled the tentering that regulated in the amount of "light" between the upper and lower millstones so as to ensure a coarser or finer grinding. He also decided when to accelerate or slow down the rotation of the stones, and when to halt grinding to remove and re-sharpen the stones. Finally, the miller determined if the flour needed a second round of grinding, and the choice of the sieves to attain the desired calibre of flour.

In spite of all his skill and experience, the miller could not attain white flour without the proper millstones. One of the advantages of the bright white stones from the limestone quarry of Claix, near Angoulême in south-western France, was that the powder it released into the flour was invisible to the eye and did not affect the flour's colour; dark grit from conglomerates, granites or sandstone, by contrast, did tint the bread (Belmont 2011: 15). In sum, the advantage of grinding with certain white rocks was not the absence of stone residue in the flour, but the perception of its absence.

In his research on the watermills of the Alpujarra Mountains south of Granada from the 15th-18th centuries, Rodríguez Monteoliva directly associates *"piedra blanca"* with *"pan blanco"* and *"piedra baza"* with *"pan bazo"*. Accordingly, the stone was not called *"bazo"* because of its brownish or yellowish colour, but due to the colour of the flour that it yielded. This colour, according to the writer, was due to the particles of bran that could not be separated by even the finest sieve (Rodríguez Monteoliva 1989: 701).

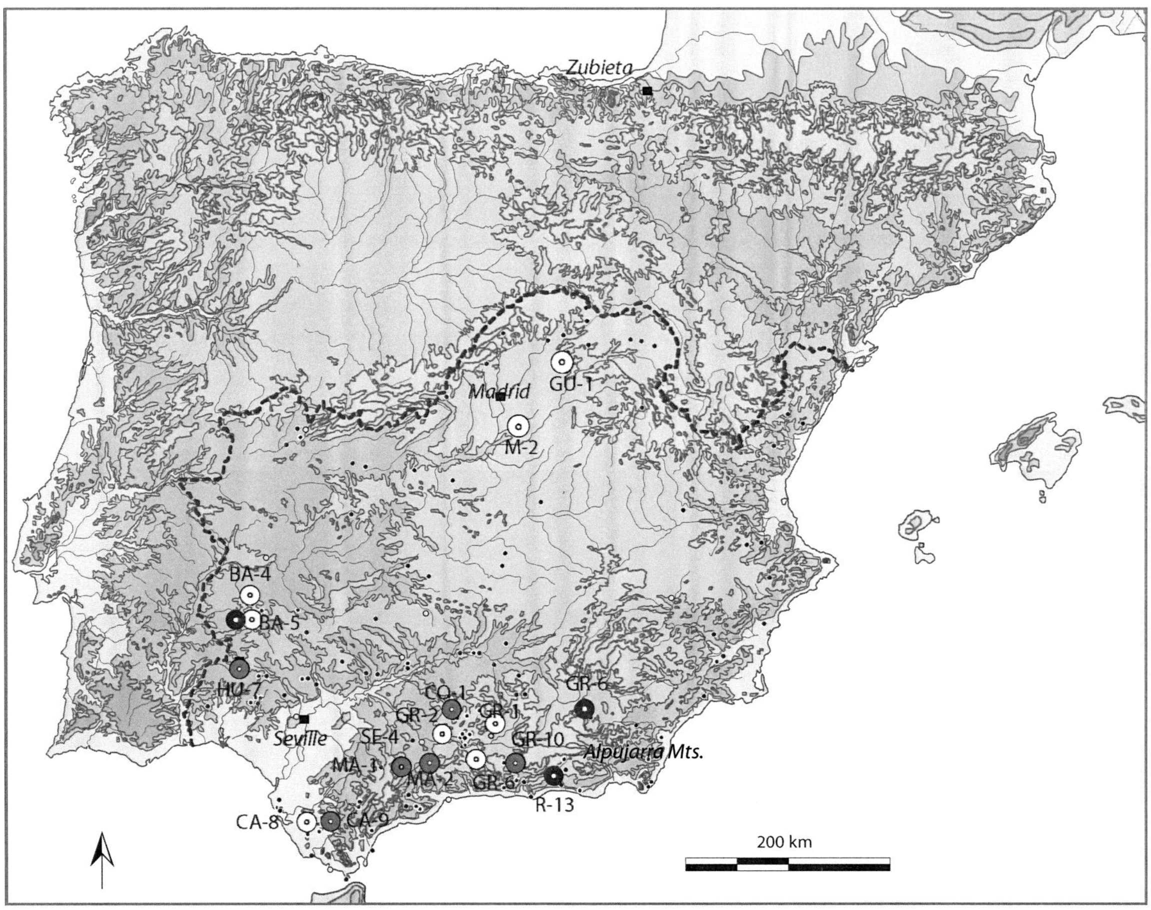

Fig. 13.2: Map of the sites and regions mentioned in this chapter. Sites designated as "white" quarries in old texts (in white): Colmenar de Oreja (M-2); Berruecos (CA-8); Salvaleon (BA-4); El Hacho (SE-3); Loja (GR-2); Moclín (GR-1); Alconera (BA-1); Alhama de Granada (GR-6); Brihuega (GU-1); El Berrueco (CA-8). Sites designated as "dark" quarries in old texts (in black): Otívar (GR-13); Jabonero (MA-7); Caniles (GR-6). Jerez de los Caballeros (BA-5), according to an old text, has both white and dark quarries in its surroundings. The sites in grey are, according to their petrography, presumably "white" quarries: Fuente de Aliseda (HU-7); La Pila de Casares (CA-12); Teba (MA-2); El Torcal (MA-1); Los Frailes, Cabra (CO-2); and Vélez de Benaudalla (GR-10).

Unfortunately, in the case of the Alpujarra Mountains, I cannot associate the types of bread with specific quarries. The Alpujarra sites appear to exploit coarse stones, most likely puddingstones with large clasts. It is probably for this reason that "white stones" had to be imported from Moclín (GR-1), at least 50 km away (Rodríguez Monteoliva 1989: 705).

Elsewhere in the study area, several millstone quarries are directly or indirectly linked to white bread by old written texts (Fig. 13.2). The Survey of 1791 of Extremadura, for example, records that the "*concha*" (shell) rock exploited at Alconera (BA-1) in the Province of Badajoz was destined specifically for flour mills because it yielded a "*pan blanco*" (white bread) (Caso Amador 2008: 133). From what can be gathered from geological sources, the Alconera rock was a sort of coarse limestone (Geological map 854, *IGME*) and must have had outstanding grinding properties because its millstones were exhibited in the Universal Exposition of Paris in 1867 (Comision Régia de España, *Catálogo Exposición Universal* 1867: 182, entry 13-57).

The Fuente de los Morales quarry by Alhama de Granada (GR-6), with its specific layer of hard white limestone, provided, according to Madoz, "... *las mejores muelas de molino para pan blanco*" ("... the best millstones for white bread"). These were exported both throughout the region of Granada and to the northern mountains of the Province of Málaga (Madoz 1847, Vol. 8: 216), a distance of approximately 20 to 30 kilometres.

Two other quarries associated with white bread are Brihuega in Guadalajara (GU-1) and Colmenar de Oreja in Madrid (M-2). A 19th-century treatise on hydraulic power records that millstones from Brihuega produced "... *una harina tan blanca como la de Colmerar*" (a flour as white as that of Colmenar). This reference likens Brihuega's white "pumice" with the white limestone of Colmenar de Oreja, a quarry known to have supplied stones to the animal-driven flour mills of Madrid (Vallejo 1833: 387).

The comparison of the grinding qualities of the limestone quarries of Berrueco (CA-8) and La Pila de Casares (CA-12) by the priest of Medina Sidonia (Cádiz) toward the middle of the 18th century reveals that there could be differences in the colour of bread, in spite of the similarity of their rocks. The rock at the Pila de Casares, "... *no es tan blanca como la de Berrueco, pero es más cortante y muele mejor... y porque aunque saca más harina de igual porción de trigo, sale prieta y queda el pan algo moreno* ..." [The Pila de Casares rock] "... is not as white as that of Berrueco, but cuts and grinds better ... and although more flour is yielded from the same amount of wheat, it comes out darker and the bread results somewhat darker ...") (Martínez y Delgado 1875: 129). From the geological maps, the rocks of these two sites are identical. But the text clearly indicates that the Berrueco (CA-8), probably the largest and longest production site in the Province of Cádiz, delivered millstones that yielded a flour superior to those scored at La Pila de Casares (CA-12). In any case, this example not only clearly confirms the role that rock plays in the colour of the flour, but illustrates that there were subtle differences of properties between what would appear to be identical rocks.

There are other sites not confirmed in texts that could be considered, based on field observations, as potentially "white quarries". Teba (MA-2) and Vélez de Benaudalla (GR-10) share bright white, homogeneous limestones. There are also rose-coloured stones near Loja (Madoz 1847, Vol, 7: 609), as well as *rosso ammonitico* limestones at El Torcal (MA-1) and Los Frailes near Cabra (CO- 2) that surely would not have gained their high reputation or large production if they did not share the same grinding properties as their "white" quarry counterparts.

Opposed to the "white quarries" are those directly linked to dark bread. Tomás López, in his 18th-century *Diccionario Geográfico*, identifies Otívar (GR-13) in the Province of Granada as a *"piedra basa"* (*basa* stone) quarry (Reyes Mesa 2000b: 2-3). Reyes Mesa, a specialist of watermills, notes that historians have interpreted *"baxo"* (or *"baso"*) bread in historical records, dating as far back as the 16th and 17th centuries, as a sort of low, unleavened flat bread when, in fact, it is bread made with *"harina baxa"* (*baxa* flour) yielded from dark, coarse *"piedras baxas"* (*baxa* stones). The author explains that in this case the bran was not separated by the *"baxa"* millstones, therefore yielding in a rough bread with a darkish tint, similar to the colour of the stones. The name given to the bread, therefore, proceeds from the name of the stone (Reyes Mesa 2000b: 2-3). From what I have been able to gather, it is likely that a large part of these "dark bread" quarries were in fact conglomerates (puddingstones) containing rounded pebbles.

This is certainly the case at the Rambla de las Canteras puddingstone quarry outside Caniles (GR-6), a site I have been able to survey in the field. Madoz is brief but succinct in his description: *"... hay una cantera de piedras de molino para pan moreno"* ("... there is a millstone quarry for dark bread") (Madoz 1846, Vol. 5: 461).

From the different labels given to quarries, there is direct connection between the colour of the rock and the colour of the bread. This is, however, highly simplistic. It is also reasonable to suppose that coarse limestones, sandstones and conglomerates resulted in a dark bread, not so much from the grit they released into the flour, but from a lesser capacity to separate the bran from the endosperm. Hard white limestones, by contrast, appear to be more apt to separate the bran from the endosperm and therefore yield a whiter flour. In addition, the fine white grit released into the flour was probably invisible to the naked eye.

Yet, as seen from the sites of Berrueco (CA-8) and Pila de Casares (CA-9) in Cádiz, there were certainly slight differences of grinding properties within the same lithography. An identical rock, at least from the macroscopic point of view, can therefore yield two different types of bread. These issues, still not adequately addressed, would certainly benefit from future research incorporating petrographical analyses, as well as experimentation by millers and bakers. Another factor to take into account on this subject is that the conception of white bread could have changed through time. A "white" 16th-century bread might not be considered a white bread in the 19th century.

In any case a revolution in bread making took place in southern Spain with the widespread adoption of French siliceous millstones in the second half of the 19th century. These new "intelligent" stones provoked a sharp decline in the demand for white limestone millstones and progressively brought about the end of the Spanish millstone quarry industry. It may not have been so much the French siliceous stone's ability to produce white flour, but the fact that French millstones did not require the long hours of dressing, often on a daily basis, that the local and regional white limestone millstones demanded. French millstones, dressed only every few months, permitted longer, uninterrupted hours of milling, enhancing dramatically the quantity of a high quality white flour and raising the affluence and status of their mills and their owners.

14. SUMMARY, CONCLUSIONS AND PERSPECTIVES

14.1. Introduction

The opening heading in the introductory chapter that millstone quarries are "everywhere" is, of course, an overstatement to underscore the notion that these types of workings are very common throughout the landscape of southern Spain (and Europe). However, when consulting the considerable molinological research from southern Spain published in the last two decades, focused principally on the features and mechanics of watermills and rarely on the question of millstones themselves, there is little indication of the omnipresence of these workings. Yet, as noted throughout this work, these workings are widespread, except in regions with unfavourable geological formations, such as large river basins with alluvial deposits and large extensions of soft or brittle rocks with properties that are unsuitable for millstones. To wit, the quantity of millstone quarries identified in southern Spain since the outset of my research in 2007 has increased from a handful to about 150. And there are probably still many that remain anonymous.

In this study I have also attempted to underscore that millstone quarries and their products played an essential role in the operational sequence of cereal processing, the base of the nutrition of the people, on a par with cultivating, harvesting and storing grain for human consumption.

I have also drawn attention to the problem of identifying these sites. It is not always simple, for example, to differentiate millstone workings from quarries producing cylinders for columns, sharpening stones or olive oil edge rollers. To bring this point home, I have included in the catalogue the Roman workings of Cerro Bellido, Seville (SE-4) and Punta Camarinal/Punta Paloma, Cádiz (CA-2a-b), sites that produced column "drums" similar in size to millstones. The inclusion of these sites in the catalogue is appropriate because there is evidence that some of these Roman drums were recycled in the Middle Ages into millstones. The problem of identifying the nature of millstone quarries, especially those that do not benefit from written texts, is particularly relevant in the south of the Iberian Peninsula due to the many quarries producing edge rollers and later cones for the vast olive oil industry.

14. 2. Sources, resources and fieldwork

Of all of the different sources available to identify millstone quarries, the most valuable are geographical dictionaries and other writings spanning the end of the 18th century to about the middle of the 19th century. The geographical dictionaries of Miñano (1826-1829) and Madoz (1845-1850) account for the identification of the largest number of sites (slightly less than half). It must be noted that the group pointed out by Madoz (about 40) is much larger than that of Miñano (5). I have nonetheless insisted on Miñano's role in identifying millstone workings, in spite of their modest number in the south of Spain, to advise future researchers on the many he identified in the north of Spain, outside the border of the study area.

Other works of regional character, such as the histories of García de la Leña (1789) for the Province of Málaga, and that of Martínez y Delgado (1875) for the city of Medina Sidonia, Cádiz, as well as a few early geological works such as that of Ezquerra del Bayo (1856), also offer relevant data.

All in all, these sources account for almost half of the sites that I have been able to identify. Nonetheless, in spite of their importance, these old texts rarely provide precise information as to their location. In this sense they most often serve simply as starting blocks from which to initiate research.

As elsewhere in Europe, notably in France, historical sources such as royal decrees, municipal ordinances or notarial protocols also identify millstone productions, and at times provide data, although usually very limited, as to the distances of commercialisation, means of transport and diameter of millstones. A large portion of these documents, dating to the end of the Middle Ages through Modern times, are especially relevant to establishing chronology because they predate the advent of the geographical and geological treatises mentioned above.

It must be noted that in the course of my research, these types of historical archives were barely exploited as primary sources, a task beyond my means that would have required months in local archives sifting through old documents and deciphering the old Castilian calligraphy. Hence the primary sources that I have quoted, except for a few that were available in digital format, are borrowed from molinological research undertaken by other historians. Noteworthy is the painstaking investigations of the medievalist Córdoba de la Llave on the watermills in the city and province of Córdoba.

Having said that, a large number of private, municipal, provincial and national historical archives are still waiting to be exploited from the perspective of millstone production. The vast private records of the aristocratic family of the Duke de Medina Sidonia in San Lúcar de Barrameda, Cádiz, for example, will certainly shed light on the vast millstone production of El Berrueco (CA-8), a quarry that the family counted among its possessions for at least several centuries. The Medina Sidonia archives will also, without a doubt, elucidate the relationship between the Berrueco workings and a number of watermills owned by the noble family, such as those in the town of Vejer de la Frontera. A writer in 1813, in fact, recorded that the *Duque* was remunerated with 18 millstones a year for a concession at the quarry of El Berrueco (CA-8) (Cruz y Bahamonde 1813: 91, note 1). It is conceivable that remuneration with millstones was more profitable to the *Duque* than currency because they served to replace the worn out millstones at the watermills he owned.

Since the beginning of this study, access to written sources via the Internet, from old geographical dictionaries to molinological articles, has soared. In the initial stage of my research, for example, only a few of the volumes of the Madoz dictionary were available in a format that allowed searches by keywords. Therefore, at that time I was required to read several volumes (700 and 1000 pages each) word for word, searching for even the most remote clue or allusion to millstone workings. Yet, this initial "low-tech" perusal of the texts, an extremely long and cumbersome task, was far from fruitless. It furnished valuable insight both into the way in which millstone quarries were recorded by this author and in the identification of keywords in 19th-century Spanish for subsequent digital searches in the remaining volumes.

This research in historical texts required that I develop certain skills that are characteristic of historians. As an archaeologist, I was obviously more comfortable digging a quarry or combing through dust-covered querns and millstones in museum depositories. In any case, these different skills went hand in hand because on the whole, the texts provided insight into Modern and Contemporary productions, whereas the study of querns and millstones in museum depositories provided data for the older production centres, from Protohistory to Antiquity.

The information gleaned from studying millstones stored in museum collections not only served to establish an initial general classification of querns and millstones (the products of the quarries), but also to identify a number of production centres and districts. The 40 *ostionera* querns at the Roman city of *Baelo Claudia*, as opposed to the single volcanic import in the collection, clearly suggested the exploitation of local and regional *ostionera* outcrops known along the coastline of the Bay of Cádiz. Likewise, the high proportion of volcanic stones among the large collection at the *Museo Arqueológico de la Provincia de Murcia* suggested the presence of unidentified exploitations of local and regional volcanic outcrops. Similarly, the great number of granite querns and millstones at the *Museo Nacional de Arte Romano* in Mérida pointed to the exploitation of local and regional granite outcrops.

The study of millstones in museums, at times, can go even further and point to specific millstone production centres. The unfinished and broken lamproite querns in Murcia, collected from the Municipality of Mazarrón, can be traced back to a lamproite outcrop at the Cabezo de la Oliva (MU-2), a short distance from the Roman villa where they were discovered. Likewise, a second set of roughouts in the Museum of Almería also have their origins in the quarries of the Cabo de Gata, sites that have been identified in the field (AL-1 and AL-2).

Although written sources and museum depositories are a wealthy source of information, they are limited almost exclusively to the chronological frameworks ranging from Prehistory to Antiquity and from the end of the Middle Ages to the 19th century. They provide little data regarding the millennium including the Gothic and Islamic domination. This shortcoming is felt throughout this study.

A third source of information about millstone quarries, labelled "unconventional sources", has proven to be extremely useful. As noted in chapter 8 about quarry infrastructure, millstone workings were often served by tracks, paths or roads, some of which have endured over time and are now integrated into modern hiking trails. Hence, a number of millstone quarries have been identified as a result of consulting a growing number of hiking or cycling itineraries posted on the Internet. These websites are often accompanied with descriptions of the landscapes and photographs of the different "curiosities" (*i.e.* millstone quarries) along their trails. These types of itineraries resulted in the identification of the workings at El Lanchar (J-2), El Berrueco (CA-8), Peña Harpada (CA-10) and Soneja (CS-1).

Other items on the Internet, such as photographs and videos, also serve as tools to identify millstone workings. The workings of Almadén de la Plata (SE-1) dating to the Middle Ages would never have come to my attention without a *You Tube* posting of an interview with the local archaeologist, Miguel de Vargas. The Internet has also provided amateur historians with a platform to communicate detailed information about their municipalities, information that otherwise would never have seen the light. Little would be known of the early 17th-century stonemason Juan de Bargas and the millstone quarry of Albardado (C-10) if it were not for the research in local archives and published on the Internet by a team of secondary school students from the town of Pozoblanco in Córdoba led by their teacher (González Peralbo 2008).

In short, these conventional and unconventional sources for the most part can account for the identification of a quarry. But rarely can they be considered preliminary research. They therefore require confirmation and description that can only be carried out by fieldwork. Of all the sites identified by these types of studies, field surveys were carried out in about a third.

In many cases, due to the scant information regarding a site's location, and to avoid loss of time, I relied on local contacts (amateurs and local historians) for guidance to the site. In other cases, no visit was attempted because the site was either too far from my base in Granada or because the information of its whereabouts was too vague, probably requiring days of fieldwork with no guarantee of success. In summary, it is obvious that if I had been in a position to survey more sites in the field, my database would have been larger and the conclusions gleaned from them more complete.

Finally, field surveys permitted me to gather precious observations not available from sources, such as extraction techniques (never described in the sources), estimates of volume of production and the presence or absence of different chronological phases. The physical survey of sites also permitted the collection of rock samples for several petrographical and geochemical analyses. On the whole, it must be noted, my visits were, alas, too short, and undertaken without the means of recording precise measurements so as to draft precise scaled drawings. This aspect was particularly frustrating due to my past as a field archaeologist, where it was a priority to draft accurate records of archaeological features.

14.3. Millstone quarry products and milling installations

Chapter 3 marks a slight "departure" from the principal subject of quarries to briefly describe the different quarry products as well as their mechanisms or installations, from small hand-driven saddle querns to large watermills with cylindrical millstones powered by vertical or horizontal waterwheels and a complex gearing system. The descriptions of these mills are based, for the most part, on observations made in museum depositories, and on descriptions gleaned from molinological literature. It is obvious that the understanding of the product, its dating and its evolution, goes hand in glove with the understanding of the chronological development of quern and millstone quarries. Since the different types of millstones, fittings and means of traction were described in some detail in chapter 3, I will avoid repetition here, and focus only on placing the different millstone types into the context of their quarries based on the classification presented in chapter 6.

As to the question of the introduction of rotary millstone technology, what is noteworthy for the south of Spain, is that the transition of the "to-and-fro" motion of the saddle quern to the full circular motion of the rotary quern (and larger millstones) cannot be singled out in time (due to lack of research) as it has been in Catalonia (middle of the 5th century BC) (*e.g.* Alonso 1999). It is apparent, however, that both small rotary querns and larger rotary pushing mills were introduced simultaneously some time in the Iron Age. This is a major difference with the arrival of the rotative power in Central Europe where only small rotary querns are known in Iron Age contexts. It is also logical to assume, although the evidence is scant, that the introduction of water-powered mills in southern Spain took place, as elsewhere, under Roman domination and that wind power came much later, in the Middle Ages.

It must be noted that the descriptions of the different mill types presented in chapter 3 represent general notions and cannot be considered a formal typological classification. Conditions are still not present, at this stage, to establish an exhaustive, all-encompassing, typo-chronology of millstones in the south of Spain. Although positive steps have been taken and presented in this study for the Roman period, and for rotary querns throughout time, the corpus of stones from all the other periods in southern Spain is too poorly represented in the archaeological record to garner empirical statistics.

Saddle querns

These earliest cereal grinders are typical finds on settlements from the Neolithic to the middle of the Iron Age. They are common, for example, at Late Prehistoric sites such as Los Millares in Almería and at Early Iron Age sites like Cancho Roano in Badajoz. They served to grind a number of foodstuffs as well as other products such as pigments, as seen from traces of reddish ochre (*almagra)* at Los Pensadores, Granada (Anderson 2010, unpublished).

A great proportion of schist, granite or sandstone saddle querns were collected from surface sites (MQ-1a) such as riverbeds, ravines and taluses, in all likelihood located near the settlements. There is some evidence, although poorly documented, of some block extractive quarries (MQ-2b) during these early times. Two of these early extractive sites (Los Pensadores, GR-12; Cabezo de Oliva, MU-2) were "stumbled" upon because they coincided (presumably) with later Roman rotary quern production. The third, El Barronal (AL-10) in the Cabo de Gata volcanic district, presumably made querns from angular blocks pried out from the summit of columnar jointing (Carrión , *et al.* 1993; Haro Navarro *et al.* 2006). For more in-depth, specialised studies of quern procurement in these earlier periods in southern Spain, see Carrión *et al.* (1993), Risch (1995), Haro Navarro *et al.* (2006) and Delgado Raack (2008).

An important aspect to retain related to "to-and-fro" mills is that I have never observed Olynthian hopper rubber mills in museum depositories in the south of Spain. This is astonishing due to the many ports with very early trade links with the eastern Mediterranean world. This appears to corroborate that the Olynthian mill type was concentrated in the eastern Mediterranean. The closest examples to the south of the Iberian Peninsula are at the shipwreck of El Sec off the coast of Majorca (Arribas 1987), Lattes, in southern France (Py 1992), and Rhodes (today Rosas, Girona) in Catalonia (Genís 1986: 113).

Iron Age rotary mills

Millstone makers in the Iron Age manufactured both rotary hand querns and larger pushing mills. In fact, the two types shared, basically, the same morphology, fittings and rock types. The main difference was the means of traction. The smaller "domestic" models, on an average about 35 cm in diameter, were driven, as seen from ethnological studies, at floor level in a seated position. The larger models, about twice the diameter and often found *in situ* on podiums, were pushed by means of a lever from a standing position. These larger models obviously yielded a larger volume of flour and, in all likelihood, served more than just a family unit. The "inverted keyhole" or "L-shaped" lateral cuttings, an important typo-chronological feature, were present on both small and large models.

In Catalonia and Valencia, both models appear to go as far back as the middle of the 5th century BC (Alonso 1999), and represent, to date, the earliest manifestations of the rotary movement in milling. In the south of Spain, these mills are unknown in such early contexts. This absence is enigmatic, especially considering the high degree of stonework (architecture, statuary) in southern Spain's Iberian Culture. Therefore, it is still not certain if this absence is due to the state of research, notably the lack of modern excavations of Middle and Late Iron Age settlements, or if the Iberians of the south were simply reluctant to adopt this new technology.

At present there is no concrete evidence that either of these early type rotary models were produced in extractive quarries. Granite and conglomerate examples probably reflect surface workings. This appears to be the case, for example, at the site of Numantia (Soria) where most of the mills, both large and small, are identical to the surface blocks or outcrops found a short distance (a few kilometres) from the settlement (Checa *et al.* 1999).

In the early stages of rotary mill development, in the Sub Bétic range of Andalusia, there is a clear preference toward limestone tufa, a highly porous rock found in fluvial, carbonate-rich contexts. It was exploited around the town of Almedinilla (Córdoba) and furnished millstones to the nearby Late Iron Age settlement at Cerro de la Cruz (Quesada *et al.* 2010, Quesada *et al.* 2014). It is possible that Iron Age millstone makers profited from surface blocks detached from large tufa formations (MQ-1a). However, due to the softness of the rock, easily cut with a pick, it cannot be excluded that cylinders were hewn directly from bedrock in true extractive fashion (MQ-2a). In fact, a Roman true extractive limestone tufa exploitation for building material has been recently found near Almedinilla (Muñiz 2012). This find could be indicative of older millstone workings yet to be identified.

It is noteworthy that no true extractive Iron Age millstone quarry has been identified in Europe. Direct extractive work leaving circular hollows in the bedrock (MQ-2a), from the current evidence, was only introduced in Roman times. This idea was originally suggested by Runnels (1981: 72) and Luc Jaccottey, based on findings in France and Switzerland, has seconded the notion in his recent work (Luc Jaccottey, pers. comm.). It would not be surprising, nonetheless, that future work in the south of Spain might bring to light true extractive Iron Age millstone workings, borrowing from the pre-Roman tradition known throughout the Mediterranean Basin of scoring building blocks directly from bedrock.

It is also noteworthy that there is no evidence in southern Spain, at least to date, of volcanic rock workings dating to the Iron Age. This contrasts with the situation in Catalonia, where these hard, vesicular rocks originally from the Olot-Garrotxa volcanic district in the Province of Girona, were carved into rotary querns as early as the middle of the 5th century BC, as seen by the example at the settlement of Ca n'Olivé (Alonso 1999: 262, fig. 171).

Roman mills

The Roman conquest of the peninsula was accompanied by a diversification of mills driven by humans, animals and (presumably) water. Rotary querns and the larger ring-mills appear to have evolved directly from the indigenous Iron Age tradition, whereas Pompeian mills and watermills, poorly represented in southern Spain, were mostly imported. Rock type also changed with the large-scale introduction of volcanic material. A great advance in production, lending itself to the standardisation of querns and millstones, was the introduction of the true extractive scoring method (MQ-2a).

Roman rotary querns

Roman rotary querns, well-represented in the museum collections of southern Spain, were hewn from a variety of rocks, such as biocalcarenites, granites and sandstones, as well as different volcanic materials (dacites, rhyolites, lamproites).

Production took the form of collection of surface material (MQ-1a) and block extractive quarries (MQ-2b). Both of these techniques are known at the Cerro de Limones (Al-1) and Hoya del Paraíso (AL-2) quarries in the S-E Spanish volcanic district. Rotary querns were also produced on a large-scale in the true extractive quarries (MQ-2a), such as the biocalcarenite models scored directly from bedrock leaving the characteristic circular extraction hollows at the coastline sites of Trafalgar (CA-1) and Rota (CA-3). True extraction was not limited to sedimentary stones as querns were hewn directly from bedrock from the lava flows of the volcanic domes outside the Roman city of *Sisaspo* (CR-a).

Production techniques of granite querns are less documented. Some were certainly carved from surface blocks (MQ-1a). It cannot be excluded that extensive granite outcrops were exploited in the form of either large blocks (MQ-1b) or true extractive bedrock quarries (MQ-2a), following the examples of quarried rectangular blocks for architectural features. These granite millstone workings, however, are yet to be identified in the field.

Roman rotary ring-mills

Roman rotary mills with their ring-*catilli* and bell-shaped *metae,* best known by the examples at the Roman city of *Volubilis* in northern Morocco (Leduc 2008), are turning out to be more and more commonplace in the south of Spain. Their role in the history of milling is still not clearly defined. They are often associated with the olive oil industry, as seen through a number of examples unearthed in excavations in or near features associated with oil presses (Peña Cervantes 2010: 66). Yet they are also clearly associated with the bakeries (*pistrinae*) at the Roman city of *Volubilis* in Morocco (Leduc 2008; 2011).

The recent research by Peña Cervantes (2010), like Frankel (1999) before her, associates these mills with the *mola olearia* of the ancient writer Columella, the mill that ground olives into a pulp before being pressed between mats to extract the oil. Peña Cervantes points to the findings of *Volubilis* (Akerraz & Lenoir 2002), the surveys of Ponsich (1974, 1979, 1987, 1991) in the Guadalquivir Valley and recent archaeological excavations to corroborate this view (Peña Cervantes 2010: 36). Furthermore, the tentering gear of this type of mill (contrary to the vertical edge roller) would have offered the option of regulating the space between the stones so as not to crush the olive pit, an action, according to Columella, that reduced the flavour and the quality of the oil (Peña Cervantes 2010: 39, note 50). Columella, presumably born and raised in ancient Cádiz, would have been perfectly aware of the different milling mechanisms of southern Spain.

Yet placing the ring-mill exclusively in the sphere of the oil industry is not doing it justice. Akerraz and Lenoir at *Volubilis* interpreted a dual function of this type of mill according to its rock type. For these authors, the *grès coquillier* (shell-rich sandstone) millstones dressed with elongated S-shaped furrows served for crushing olives, whereas the basalt models, devoid of furrows but highly porous, were for grinding grains (Akerraz & Lenoir 2002: 203). Brun seconds the "polyvalent" nature of these mills (Brun 1997b: 71-72) while Leduc, in turn, unequivocally places these types of mills in the context of the bakeries of *Volubilis* (Leduc 2008).

There is still much work to be done on distinguishing the types of millstones destined for different types of industries (and other foodstuffs, metals, pigments, etc.). At the present state of research, in my opinion, it is not possible, as was done for the mills of *Volubilis*, to specify the function of ring-mills based exclusively on petrographical and grinding surface dressing criteria.

From the present state of the research, this study defends the "multifunctional" aspect of these mills based on the following arguments:

1) Ring-mills are, in fact, larger versions of the rotary hand querns and pushing mills used for grinding grains and other products since the Iron Age. Their adoption in the oil industry on a large-scale only began after the Roman conquest. It is therefore unreasonable to suppose that their principal initial use (for grains), prior to their adoption by the olive oil industry, would have been completely abandoned, particularly in regions of southern Spain rich in cereal production.

2) Due to the scarcity of Pompeian and hydraulic mills, the ring-mill was the only other model adaptable to the "industrial" needs of the bakeries of Roman *villae*, towns and cities. This is seen, for example, at the numerous *pistrinae* of *Volubilis* where ring-*catilli* outnumber Pompeian models 39 to 3 (Leduc 2008: 483). It is difficult to imagine that this industrial role could have been played out exclusively by rotary hand querns, a type usually attributed to domestic grinding.

3) The ring-*catillus* of the ring-mill is very similar from the typological point of view to the Roman Haltern-Rheingönheim mill, a model known in Saalburg (volcanic rock) in Germany (Baatz 1995: 12, fig. 12) with a spread as far south as Basel (*Bundsandstein*) and Avenches (yellow sandstone) in Switzerland (Hurbin 1982; Castella & Anderson 2004: 129-130). This central European grain mill obviously was not destined to grind olives, a fruit that did not grow in the region. Besides, the examples from Basel were brought to light in a Roman bakery (Hurbin 1982).

4) If ring-mills had been dedicated exclusively to the oil industry, they would have remained inactive about half of the year, outside of the span of the olive harvest that lasted, if the same as today, from autumn to the end of winter. Contrary to cereal grains, which can be stored for extended periods of time, olives had to be processed in the briefest time possible after their harvest (Pablo Casado and Nanén López, pers. comm.). It is therefore plausible to imagine that some ring-mills were at certain times of the year grinding cereals and at other times grinding olives into pulp. This would require a thorough cleaning of the stone between uses and adjusting the space (called light) between the stones to the requirements of each type of milling.

These arguments are obviously highly speculative and highlight the importance of future research, in particular on plant macro-remains on millstone surfaces. All said, the production of cylindrical ring-mills was widespread throughout the study area. Volcanic models are known to have been made in the Campo de Calatrava at both *Sisapo* (CR-1) and Las Herrerías near Bolaños (CR-2). At *Sisapo*, as seen by circular hollows, they were hewn directly from bedrock (MQ-2a). At Las Herrerías, the extraction technique is not known because the Roman quarry faces have been destroyed by modern workings. Other sedimentary rock examples could have been carved from either surface material (MQ-1a or MQ-1b) or scored or detached from bedrock (MQ-2a or MQ-2b).

Pompeian mills

The situation of Pompeian mills in Roman Spain has been the subject of a paper presented in the colloquium of Lons-le-Saunier (Anderson *et al.* in press). In fact, from the current evidence, these mills are poorly represented in Spain. They are for the most part volcanic imports from districts elsewhere in the Mediterranean Basin. In the years since the Lons-le-Saunier meeting, several new Pompeian *catilli* have surfaced in the collections of Carteia (Cádiz), Empúries (Barcelona), Huesca (Aragón), as well as two additional examples at Pallaruelo de Monegros

(Huesca) (currently under study in collaboration with Luc Jaccottey). Most appear to be leucite rocks imported from the Vulsini volcanic district near Orvieto, Italy. Yet these finds (and others that will certainly follow in the future) do not alter the hypothesis that most Pompeian mills on Iberian soil are long distance imports. The only certified case of a Pompeian *catillus* produced on Iberian soil is an unfinished biocalcarenite *catillus*, at *Baelo Claudia* (Cádiz). This example, probably a local copy of a volcanic model, and probably hewn from a recycled construction column drum, would fall into the category of exploiting large, detached surface blocks (MQ-1b).

Roman watermills

The subject of Roman watermills in *Hispania* merits a certain amount of ink. It is perfectly reasonable to assume the presence of watermills throughout Roman *Hispania*, especially taking into account the large number of hydraulic works such as the dam of Proserpina near Mérida and the aqueducts that crisscrossed the landscape with their best representations in Segovia and Mérida. These different features are fine examples of the rapid transmission of technology throughout the Roman world.

Closer to the hydraulic grain mill technology were the numerous *norias*, large wooden vertical waterwheels discovered a century ago at Rio Tinto in Huelva that elevated water at the Roman mines (Delgado & Regalado 2012). Nonetheless, there are no traces of watermill installations themselves in southern Spain. And there are hardly any characteristic millstones in Spanish museum depositories, contrary to the situation as seen through recent work in museum depositories in France and Switzerland (Anderson *et al.* 2004).

The only certified Roman watermill in the Iberian Peninsula is at *Conimbriga,* Portugal, where Jean-Pierre Brun recognised architectural features related to a vertical waterwheel (Brun 1997a). Furthermore, at *Conimbriga* itself, among the 55 millstones in a catalogue compiled by Borges two decades before Brun's article, there are only two *metae*, both of siliceous sandstone (Borges 1978: 127, nos. 54-55), that share certain typological features with hydraulic millstones, notably totally perforated eyes, a slight rim around the edge of the *metae* and grinding surface angles. A setback to the watermill interpretation of these millstones is that their diameter (48.4 and 47.3 cm) is smaller than even the smallest known Roman water millstones at 55 cm in diameter (*cf.* Castella 1994; Anderson *et al.* 2004: 6). Furthermore, they also do not bear the characteristic cramp-iron (*anille crampons* in French) perforations or channels characteristic of their driving mechanism.

A second potential watermill was pointed out by Philippe Leveau at the Roman villa of El Munts, Altafulla, in Tarragona (Jean-Pierre Brun, pers. comm.). With his experience at the hydraulic works of Barbegal, near Arles, in southern France, Leveau is very well-placed to identify this type of installation. Inquiries among archaeologists based at Munts, however, have not confirmed this identification.

Elsewhere, recent archaeological research in southern Spain has identified a few Roman water millstones. Three are pointed out by García Romero (2002: 623-624, figs 132-135) in his study of Roman mining in southern Spain. All three, unfortunately, hail from uncertain contexts, and although their diameters are compatible with Roman models, their discoidal morphology points to a dating from the Middle Ages. Two other examples appear in a recent article. One is actually a rotary quern *catillus* 34 cm in diameter with radial cuttings (Palomo & Fernández 2007: 517, fig. 8). The nature of the second is less evident. From the photographs (Palomo & Fernández 2007: 514-515, figs. 6-7) and from oral information from Silverio Gutiérrez, director of the Villanueva de Córdoba Museum, the second appears to be a conglomerate measuring about 50 cm in diameter. Its upper stone is not typical. It has inwardly curved edges and squarish handle cuttings at the four cardinal points, similar to those of cylindrical ring-mills. A fifth dovetail cutting is probably a repair. Its *meta*, not visible in the photograph, has an eye about 3 cm in diameter that is perforated from top to bottom. This is the only feature (although very small) that this mill shares with certified hydraulic millstones. Gutiérrez nonetheless reports that the mill was discovered *in situ* in a settlement with no indication of a source of water. Hence, at this stage, there is not enough evidence to certify that this mill was powered by water.

The sole example identified on Iberian soil that shares the typological traits of the stones of Barbegal (France) and Avenches (Switzerland) is a pierced *metae* measuring 65 cm in diameter (see drawing in chap. 3) decorating the entrance of the archaeological site of Oreto y Zuqueca in Ciudad Real (Anderson *et al.* 2011a: 161). Its typology and find-spot, on the outskirts of the ancient centre of *Oretum*, near the Jabalón River, recalls that of millstones of En Chaplix outside the city of Roman *Aventicum* along a canal, near Lake Murten (Castella 1994). The volcanic rock of the Oreto y Zuqueca millstone points to a local or regional production from one of the many neighbouring volcanic outcrops in the Campo de Calatrava.

As mentioned previously, it is very surprising that no unquestionable watermill installation has been brought to light in an archaeological excavation in Spain, especially in the Roman centres of *Corduba* or *Emerita Augusta,* or among the many *villae* excavated in recent years during rescue operations. It must be acknowledged that finds of this type, as Jean-Pierre Brun has noted (Brun, in press), are usually extremely poorly conserved (often of wood) and difficult to identify due to the absence (except for the millstones) of characteristic features. It may be just a question of time before these mills begin to appear in the archaeological record of southern Spain.

The absence of characteristic Roman hydraulic millstones in museum depositories also remains a mystery. I have already evoked the problem that large millstones create for storage and that excavators at times leave them behind in the field. There is also the problem of recognition of their nature: it is not always easy to recognize a hydraulic millstone in a fragmented state.

A final explanation of the "absence" of Roman millstones in *Hispania* is that they could have been of a different morphology, unrelated to those identified north of the Pyrenees. In this case the examples presented by García Romero (2002) and Palomo and Fernández (2007) take on a new interest. All said, there is still much research to be undertaken on this subject.

Medieval mills

Medieval millers throughout Europe inherited a wide array of rotary querns, animal-driven, and water-driven mills from the Roman period. The novelty in this chronological period is the introduction of the windmill. In southern Spain this took place sometime during Islamic rule.

Rotary hand querns continued in use throughout the Middle Ages. They differed from the earlier models in that their grinding surfaces became flatter, resulting in a general discoidal shape. They were driven by means of a vertical handle set in a blind socket cut into the upper surface. Rynd cuttings moved from the top of the upper stone to the bottom. The few quarries identified producing these models such as the Arroyo de las Calzadillas (SE-1a) in Seville, Zagra (GR-5) in Granada, Rambla Honda (AL-3) in Almería and Puerto de la Cadena (MU-1) in Murcia are true extractive sites (MQ-2a). The first two (GR-5 and SE-1a) yielded standardised querns measuring about 50 cm in diameter.

Querns in the Middle Ages, however, do not always follow a standard diameter. The largest published assemblage is that of El Castillejo, Montefrío (Granada), dating from the 9th to the 12th century (Motos Guirao 1987). Here querns vary between 20 cm and 50 cm in diameter. From their uneven, rough aspect they reflect the exploitation of surface material (MQ-1a) instead of true extractive quarries (MQ-2a). One must therefore envisage the coexistence of both surface workings (MQ-1a) and true extractive bedrock extractions (MQ-2) for querns of the Middle Ages, as was the case in Roman times. However, it is not known if the proportion of medieval surface workings (MQ-1a) was feeble, as in Roman times, or if it represented a large proportion of the production.

At several quarries of the Middle Ages, quern extractions are side by side with larger extractions for animal or water-driven mills (AL-3, MU-1). The larger animal-driven *tahona* mills were the natural heir to the Roman Pompeian mill (also called "donkey"mill) and were installed where access to running water was difficult. The major difference in the Middle Ages, compared to Roman times, is the lack of evidence of these installations. The medievalist Cordóba de la Llave (1988: 839-840) reports that although present in the Province of Córdoba, none were in the capital city of Córdoba. Most *tahonas* took on a subordinate role to that of watermills and no specific quarry site of the Middle Ages can be attributed to them. Nonetheless, it is likely that part of the production of some true extractive quarries, such as that of Rambla Honda (AL-3) and Puerto de la Cadena (MU-1), both in very dry regions, could have ended up in *tahonas*.

Our knowledge of watermills in the Middle Ages is founded on a large amount of molinological research, notably that conducted by Cordoba de la Llave (1988; 2003; 2011) in Córdoba, Reyes Mesa (2000a-b, 2006) in Granada and Cara Barrionuevo (1999) in the Alpujarra Mountains of Almería. A very early written reference to a watermill is in an article citing an old Islamic narrative authored by Ajbar Maymu'a (Palomo & Fernández 2007: 499). This narrative records that the ruler Abul-Jattar, during internal conflicts in 743 AD, took refuge in the Kulayb mill along the Guadalquivir River (Palomo & Fernández 2007: 499). Little is known, however, of the millstones that equipped these early watermills or where they were procured. They are neither in museum depositories nor illustrated in archaeological publications. Their absence can probably be attributed (following the Roman model proposed by Brun) to the fact that excavations of settlements focus on the urban sectors and not on the settlement periphery, where the watermills were installed.

From research undertaken elsewhere, notably in Switzerland and France, the tendency is for these larger millstones to evolve from a characteristic bell-shape with steep grinding surfaces to cylinders with flat grinding surfaces (Castella & Anderson 2004: 134-135). As to fittings, the eyes of both upper and lower stones are pierced (to accommodate the spindle) and rynd cuttings are restricted to the base of upper stones.

The quantity of millstones from secure medieval contexts is nonetheless too low to serve as a base to establish a chrono-typology. The few certified examples also fall into diameter ranges coinciding with more recent Modern and Contemporary millstones, annulling any attempt at dating based solely on diameter.

With the growing number of mills in the Middle Ages, especially water-powered mechanisms, there was certainly a larger demand both for new stones and stones to replace those worn out. It is therefore hard to imagine millstone workings not exploiting bedrock either in the form of true extractive (MQ-2a) or block extractive (MQ-2b) quarries. However, since these stones share sizes with those of later Modern and Contemporary workings, it is very difficult to identify them in the field based simply on their morphometrics.

Larger millstone extractions that shared the same outcrops with smaller quern extractions at sites such as the Rambla Honda (AL-3) and Puerto de la Cadena (MU-1) can probably be dated to the Islamic domination. The absence of Latin-based place names and the abandonment of volcanic rocks in favour of conglomerates, are also indicators that support a date from the Middle Ages. It is also highly likely that more recent workings at quarries from Modern and Contemporary times, such as Moclín (GR-1a) and El Berrueco (CA-8), erased traces of earlier extraction phases.

Most windmills in Spain were raised in areas poorly supplied by water. In the southern half of Spain there are two main windmill concentrations. The first is in Don Quixote's La Mancha where they were generally equipped with rectangular vanes of wooden slats. The second concentration is in the south-east in the Murcia and Almería Provinces, one of the driest regions in Spain. The windmills of this second district differ from the first in that their vanes are canvas sails of triangular shape (Rojas & Amezcua-Ogáyar 2005). In any case, there is no evidence that extraction techniques of millstones destined to windmills were any different from their contemporary watermills. Wind-powered mills were therefore supplied essentially, like their hydraulic counterparts, by bedrock quarries (MQ-2).

Modern and Contemporary mills

Modern and Contemporary mills are practically identical, from the standpoint of petrography, to those of the Late Middle Ages. The Frenchman Piot, in his treatise on milling from the late 19th century, records that the constant wars and "civil troubles" hindered the progress of Spanish milling (Piot 1860: 73). The author goes on to state that in many regions rocks other than granites and sandstones were still commonly exploited and, although French siliceous stones were imported from La Ferté-sous-Jouarre, social conditions slowed the introduction of French personnel (millers and millstone dressers), presumably to uphold the quality of maintenance of both the mills and the millstones (Piot 1860: 12, 73).

This is the heyday of millstone quarries in southern Spain because they had to provide thousands of stones to mills spread throughout the landscape. As noted previously, Madoz, in his geographical dictionary dating to the middle of the 19th century, recorded 22,492 flour watermills, 676 windmills and 1476 animal-driven *tahonas* throughout Spain (Lizarralde 2010). It is obvious that watermills, in particular *rodeznos* with their horizontal waterwheels, were by far the dominant type. Lizarralde's quantifications are seconded by the work of Reyes Mesa who registers in the Province of Granada only two windmills compared to 540 watermills (Reyes Mesa 2006: 213). This is, of course, a radical departure from the impression offered by the tale of Don Quixote.

Rotary hand querns, still produced in these times, were certainly most often relegated to grinding animal fodder. These models, slightly bigger than their earlier counterparts, and always dressed with complex furrows, are included in this study because they were known at times to have served to grind foodstuffs for human consumption, in particular during and after the Civil War.

Bedrock quarries (MQ-2) in the later periods account for most of the millstone quarry sites. The choice depended on the nature of the outcrop. In some of the larger exploitations such as Moclín (GR-1a) and Cabra, Cantera de los Frailes (CO-1), the millstone makers, due to the homogeneous compact rock, could score directly from the rock mass with picks (MQ-2a), whereas at other sites, such as Alhama de Granada (GR-6) and Carcabuey, Cudillas (CO- 3), millstones were hewn from angular slabs previously detached by levers (MQ-2b).

In these periods, in addition to the large millstone production centres, there were other more discrete sites producing a very limited number of millstones. This is the case, for example, of the large surface blocks that served as quarries (MQ-1b) at Arbunuel (J-4) in Jaén and at the Molino de la Piedra (CO-5) near Baena. The extractions from these two sites were certainly destined to local nearby watermills.

Finally, no quarry of the animal fodder rotary querns has ever been, to date, identified in the field. This suggests that these querns were exploited in angular block workings (MQ-2b) that leave little or no extraction marks in the field. Since these Contemporary querns were for the most part hewn from hard limestone, a rock exploited massively for building material, it is possible that they were manufactured at sites that also produced building blocks.

14.4. Millstone quarry geology

Millstone makers throughout southern Spain, as elsewhere, sought out hard and abrasive rocks that were either porous or contained crystals or inclusions such as shell fragments that increased the rock's "bite". Some were worked in specific time frames, while others were exploited over long stretches of time.

Most of the petrographic identifications for this study are macroscopic or based on geological maps. The only geochemical analyses were undertaken at the Norwegian Geological Survey (NGU) in Trondheim, Norway, on samples from the Cabo de Gata and Calatrava volcanic districts (Anderson *et al.* 2011a: 157). A dozen other volcanic samples are currently being studied by Jane H. Scarrow and Aitor Cambeses of the Geological Department of the University of Granada taken from Roman querns and ring-mills in the Museum of Murcia.

The following are the nine major types of millstone rocks identified in the study area:

Sedimentary rocks

Sedimentary rock types exploited for millstones can be broadly divided into coarse and fine rocks. Among the coarse rocks are yellowish, porous, shell-rich (*ostionera*) biocalcarenites, exploited since Prehistory until the 20th century in quarries along the Bay of Cádiz on the Atlantic coast. Since Roman times, this stone was also highly prized for construction material. Brownish pebble conglomerates, or puddingstones, were also exploited on a large-scale for millstones. Most of these outcrops were found in the Bétic Range, south of the Guadalquivir River Basin. The size of the pebbles could vary from 2 to 10 cm. This rock was less valued in Romans times but became widespread from the Middle Ages through Contemporary times.

Highly porous, cream-coloured limestone tufas known as "travertines" were sought for querns and millstones in the Iron Age and, to a lesser extent, in Roman times. In the Province of Guadalajara, to the north, there are a number of exploitations of this rock dating to Contemporary times. In fact, the name of the hamlet of Tobes (GUA-5), a town with millstone workings, derives from the name of the rock (*toba*).

Among the finer compact sedimentary rocks are white limestone, off-white or greyish dolomite and pink *rosso ammonitico*. These bedrock outcrops, located for the most part to the south of the Guadalquivir River Basin, were ostensibly exploited from the end of the Middle Ages to the 19th century. The most celebrated are the workings of El Berrrueco (CA-8), Moclín (GR-1), El Torcal (MA-1) and Cabra (CO-1). The drawback is that their products required constant dressing, sometimes on a daily basis. Yet they were presumably appreciated because they yielded white flour.

Metamorphic rocks

The only metamorphic rock identified in the study is schist. Mica schists with garnet crystals projecting from the grinding surfaces enhanced the rock's "bite" and were highly appreciated for saddle querns in Prehistoric and Protohistoric contexts in south-eastern Spain. These were most likely gathered from simple surface workings (MQ-1a). The only Contemporary schist quarry producing large cylindrical millstones is El Campillo in the Province of Huelva (HU-1). The modest use of schists in southern Spain in recent time contrasts with massive Medieval garnet schist workings at Hyllestad and the Modern staurolite schist workings at Selbu in Norway (Grenne *et al.* 2008).

Igneous rocks

The first igneous rock group comprises a variety of very hard, granular, crystalline rocks that fall globally under the term granitoids. In Spain, these outcrops are generally referred to as "*rocas berroqueñas*". Variations of the term "*berruequeño*" are often found as toponyms. These outcrops are spread throughout a vast area north of the Guadalquiver River Basin from the Province of Extremadura to the Province of Jaén. In certain areas, nature has sculpted these extensive surface outcrops into rounded surface boulders that lend themselves to millstone workings. Granitoids were extensively exploited since Prehistory until the 20th century but their quarries remain elusive in the field.

The second igneous group comprises a range of black, grey, or reddish-purple vesicular volcanic rocks. In the south of the Iberian Peninsula there are two volcanic districts where millstone workings have been pinpointed: 1) South-eastern Spanish volcanic district covering the Provinces of Almería, Murcia and part of Albacete and 2) the Campo de Calatrava district in the Province of Ciudad Real. Although they can be roughly identified by the trained eye, drawing their precise portrait (rhyolites, dacites, basalts, lamproites, melelinites) requires specialised geo-chemical analyses. Although exploited since the end of Prehistory, it was in Roman times that they saw their greatest use for a number of different types of millstones known to have been traded, at times, very long distances.

To conclude this section, it is noteworthy that no exploitation of porous, hard siliceous, rock comparable to French burrstones has been identified in the Iberian Peninsula. The closest Iberian rival to the French exports was the hard sandstone of Montjuïc in Barcelona, a quarry that exported its millstones up and down the Mediterranean coast and beyond.

14.5. Millstone terminology, topography, techniques

The terminology used to describe extractive techniques of millstone quarries is borrowed, in part, from that of construction quarries. The research of Jean-Claude Bessac (1982; 1996; 2003) on ancient quarry tools, "*gestes*" (actions) and extraction sites, is pervasive throughout this study. Some definitions such as "direct" and indirect" percussion", are rough (and awkward) translations of the techniques Bessac has defined as "*percussion posée*" or "*percussion lancée*".

For more specific millstone quarry terminology, I have borrowed from Crawford (1955), Runnels (1981), Tucker (1985), Peacock (1986; 1987) and Jobey (1986). The English terminology is also strongly influenced by the French publications of Alain Belmont (2006a; 2011; 2011 *et al.*) and the *Groupe Meule* (Buchsenschutz *et al.* 2011). For example, the term "tube" or "tubular", continuously used throughout this work in reference to multiple, superimposed, circular extraction hollows, is a direct translation from the French. For the Spanish terms I have relied on personal communications and publications of Pilar Pascual and Pedro García (Pascual & García 2011) based on their work in La Rioja and Joaquín Sánchez in Minorca (Sánchez 2011). The recent research of Maestro Hernández (2011) on the millstone workings in the mountains of Palencia has also provided useful Spanish vocabulary. Due to the growing international interest in querns, millstones and millstone production, a multilingual glossary, developed already in part by F. Jodry of the *Groupe Meule* (Jodry 2011a), is vital for the future research of a subject that reaches far beyond present political and linguistic boundaries.

The topography of millstone quarries

Part of the descriptive language used to define millstone quarries is related to their topographical features. These are defined as follows: a) slope quarry; b) cliff or scarp quarry; c) ravine bottom or riverbed quarry; d) valley or ravine edge quarry; e) plateau quarry; f) coastline quarry. These are features where either bedrock is exposed on the surface (or very near the surface) or where there are suitable accumulations of surface blocks, for the most part by natural processes. Quarry classification and morphology, developed in chapter 6, are dependant, in part, on these facets. I have, nonetheless, retained these descriptions in this chapter so as not to multiply the variables that go into establishing millstone quarry classification.

Production: tools and techniques (extraction and fashioning)

As to the description of the millstone maker's tools, I have been limited to a few tool marks visible at quarries, to old photographs of millstone workers in the mountains of Palencia and at the Montjuïc of Barcelona (notably the pick and the lever) and to quarry literature. It must be remembered that quarry tool marks, as seen today, are usually too poorly conserved to permit fine analysis. Serious tool mark study can only be undertaken in the framework of modern archaeological excavations where well-preserved examples are protected under working debris. Hence, the extractive techniques presented in this chapter rely heavily on the work of Bessac (1996) and are highly influenced by personal observations during archaeological excavations at the millstone workings of Châbles in Switzerland (Anderson *et al.* 2003), as well as Claix (Charente) and Mont Vouan (Haute Savoie) in France (under the direction of Alain Belmont).

Parting from the point of view that "a picture is worth a thousand words", I have attempted to visually reconstruct the different extraction techniques. Each technique depended greatly on whether the bedrock was naturally compact and homogeneous or if was subject to fractures. In general, picks served to cut through massive, homogeneous bedrock, whereas levers or crowbars were used to pry out angular blocks from fractured bedrock.

The "*chaîne opératoire*" of quern and millstone fashioning and the choice of tools, also described in this chapter, therefore depended on whether the rock source was 1) a simple surface block, 2) an angular block detached from a block quarry, or 3) a rough cylinder scored from bedrock in a true extractive quarry.

In this chapter, I also describe composite millstones, that is, millstones assembled from a number of different segments (halves, quarters) and bound with iron bands. Several quarries producing composite millstones are recorded in the study area by historical sources. Yet, the only site identified in the field producing these models is Fuente de los Morales (GR-6), near Alhama de Granada.

Based on experimental work undertaken in Switzerland (Anderson *et al.* 2003: 47), making the upper and lower stones of rotary quern, from extraction to fashioning, required between two to three days. By contrast, a larger single millstone took longer, between three to seven days, as indicated by Maestro Hernández (2011) based on work in the early 20th century in Palencia. Otherwise, from data gleaned from Madoz, two per week could be produced by a team of four to five workers at El Berrueco (CA-8) (Madoz 1846, Vol. 4: 290). Of course, variables such as the hardness of the rock and the skill of the millstone makers would have influenced these rough estimations.

14.6. Millstone quarry classification

The route leading to the system of classification of millstone quarries in the south of Spain, considering its simplicity, has been extremely circuitous. I first attempted to include a wide number of variables, that is, topography, extractive techniques and sites morphology, as well as technical observations of workings in France, Switzerland and Germany. After a series of tests, I came to the conclusion that a classification comprising all of these variables resulted in an incongruous system with almost as many categories as number of quarries.

I therefore opted for an extremely simple classification rooted in the nature of the rock. Hence, millstone quarries exploiting loose surface material are labelled with the number 1 (MQ-1), and those exploiting bedrock are labelled with the number 2 (MQ-2). Each of these classes is then subdivided into two parts as detailed below.

Small surface block collection (MQ-1a)

These quarries exploited small surface blocks from specific topographical features such as riverbeds, ravines, taluses, screes and moraines. The blocks or boulders could be *repeatedly* collected without difficulty and without requiring an extraction process. They were usually sculpted by natural processes into a shape close to that of a quern or millstone. The word "repeatedly" is stressed since isolated, random collection of surface material does not constitute a quarry. Although these sites were without a doubt numerous, they are almost impossible to identify in the field because they leave no traces.

Although also difficult to identify, querns from these workings can be inferred at times from their general shape and their surface patina. The rounded bases of saddle querns and the hemispherical shape of rotary querns suggest, for example, that they were fashioned from rolled surface boulders. This is confirmed when a difference of texture or patina is observed between the original rolled surface (cortex) resulting from natural processes and the unnatural surface chipped or pecked during production (Anderson & Castella 2007: 123-125, inv.10).

From the large quantity of saddle querns in museum collections from Late Prehistory, the Bronze Age and the first half of the Iron Age, it is reasonable to assume that these types of quarries were ubiquitous throughout the landscape. These sites were probably near settlements, so as to reduce the effort and time of transport. Their petrography is varied. In the eastern half of the study area, mica schists, sandstones and conglomerates, common surface materials, are dominant (Delgado Raack 2008: 138). In the volcanic districts in the Province of Almería, dacites and rhyolites were also available as surface material. In the western areas, the dominant surface rock type was granite, as evidenced by the large number of saddle querns of the Early Iron Age site of Cancho Roano (Badajoz) (Jiménez Ávila 1997: 149).

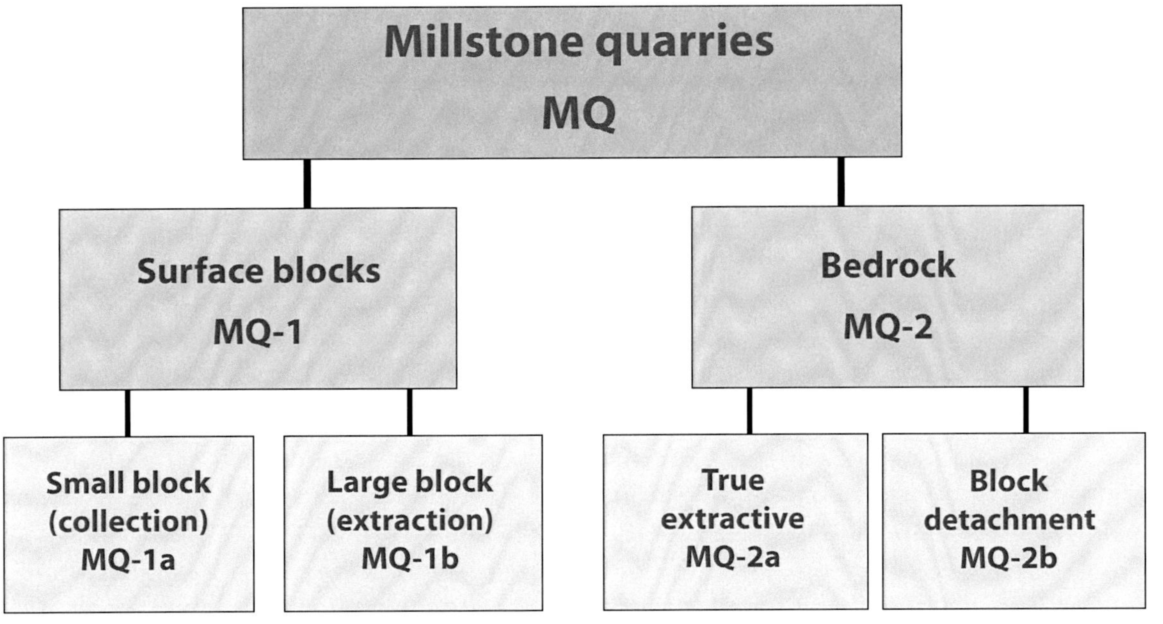

Organisational chart of the classification of millstone quarries identified in southern Spain.

For the more recent phases of the Iron Age and the Roman period, no exploitation of surface blocks has been documented in southern Spain. Although extractive quarries began progressively to dominate millstone production, especially in Roman times, surface material continued to be exploited as a local alternative to importing exogenous products. In La Rioja near a Roman Villa, there is evidence of surface boulder workings for rotary querns from nearby alluvial deposits (Pascual & García 2011: 287-288). At a number of sites in Roman Switzerland, granites, schists, gneisses and *verrucanos* were exploited as surface boulders and served as local alternatives to querns hewn from extractive workings (reddish breccia, *grès coquillier,* and *Bundsandstein*) (Anderson *et al.* 2003: 64-67). In Norway, in these early times, erratic block working seems to have been the norm and outnumbered that of extractive quarries (Dahlin & Anderson, 2014a: 43). The ubiquity of gneiss, granite and schist in the form of erratic blocks, may have even slowed the emergence of the mica schist true extractive quarry industry until the 8th century AD, as was the case at Hyllestad (Grenne *et al.* 2008: 48, 64).

All said, exploitations of surface blocks, although now invisible to the eye, were the type of quarry *par excellence* of Prehistory and Early Protohistory. Although they continued to exist in later periods, their production was proportionally low compared to that of extractive quarries. At the technical level, these workings required for the most part only knowledge of the fashioning stage. More importantly, these exploitations did not require the skills and tools of bedrock productions.

Extracting from large surface blocks (MQ-1b)

This second type of quarry differentiates itself from small boulder collection in that the detached blocks exploited were much larger and heavier. These blocks, in fact, played the role of true extractive quarries and required the skills and tools (in particular the pick) of true extractive workings. This type is divided into four groups (MQ-1b/1-4). The first two (MQ-1b/1-2) correspond to large blocks displaced by natural processes. The third (MQ-1b/3) corresponds to large surface blocks formed naturally *in situ*. The fourth, corresponds to the reuse of blocks originally moved by the millstone maker.

MQ-1b-1 is equivalent to topographical features such as taluses made up of multiple blocks from which a millstone can be extracted. Examples are the sites of La Perdrizas near Moclín, Granada (GR-1b), Castillo de Locubín, Jaén (J-1) and Vélez de Benaudalla, Granada (GR-10).

In the case of the second type, MQ-1b/2, the exploitation consists of only a single large block resulting from landslides, such is the case at Arbuniel, Jaén (J-4) and Molino de la Piedra, Córdoba (CO-5) or large glacial erratics, a type that is frequent elsewhere in Europe but has not been identified in southern Spain. Although the extraction techniques are identical, the number of extractions from one type of site to another is very different. Talus productions (MQ-1b/1) are potentially larger than single block productions (MQ-1b/2).

The third type, MQ-1b/3, is that of a large natural surface slab and pedestal boulder workings found in particular geological units. The difference with the previous type is that they were not transported by landslides, torrents or glaciers but engendered, *in situ*, after millions of years by agents of erosion. The best examples are the karstic slabs at El Torcal (MA-1), Teba (MA-2) and Fuensanta, Loja (GR-2). Another example of this type of exploitation is that of *bolos* or *piedras caballeras* (pedestal rocks), rounded surface boulders also formed *in situ* by agents of erosion in units of limestone and granite.

The fourth type (MQ-1b/4), probably the least common of the four, corresponds to blocks originally destined to architecture that were recycled into millstones. A specific example is the Roman Pompeian upper stone at the city of *Baelo Claudia* probably scored from a recycled biocalcarenite block or column drum. A similar example is the presumed use for millstones in the Middle Ages of Roman column segments quarried at Cerro Bellido (SE-4).

It is difficult to quantify the different productions. The MQ-1b/1 type is usually an alternative production related to a nearby extractive quarry, as is the case of the large talus blocks of Pedrizas, Moclín (GR-1b), less than a kilometre from the mother quarry of Moclín (GR-1a). Some examples of the MQ-1b/2 type appear to be very local productions near or beside watermills, and may have been a means to rapidly replace a broken millstone while waiting for a permanent replacement. These extractions, probably of lower quality, may also have been destined to more menial activity, such as grinding animal fodder. The MQ-1b/3 type, due to the limitless quantity of surface material, could potentially be vast productions, attaining a wide sphere of distribution as is presumably the case at the sites of El Torcal, Málaga (MA-1) and Loja, Granada (GR-2/3). Recycling blocks from architectural features (MQ-1b/4) is almost "anecdotal" and, apart from the Pompeian *catillus* at *Baelo Claudia* (CA-1b), is poorly represented in the study area. Elsewhere in Europe it also only represents a small proportion of millstone production.

Bedrock exploitations (MQ-2)

The second general division of millstone quarries corresponds to that of sites that exploit bedrock (MQ-2). These sites differ from surface block exploitations in that they require an extractive process that entails at times a high degree of skill, as well as a panoply of tools that demand constant maintenance and repair. In these cases, millstone makers can be considered "professionals". These sites represent the most economic manner to exploit rock and allowed massif outcrops to be worked intensively and over an extended period of time. They are subdivided into two large classes according to the extractive process. In the case of the first, true extractive or direct extractive quarries (MQ-2a), workers cut directly into the rock mass resulting in quarry faces with characteristic circular hollows. For the second, block detachment quarries (MQ-2b), by contrast, workers detached angular-shaped slabs by means of levers resulting in planar quarry faces that are much more challenging to identify in the field.

True extractive millstone quarries (MQ-2a)

True extractive millstone quarries, a term borrowed from Runnels (1981: 72), are sites where querns and millstones were scored by cutting them directly from massive and homogenous bedrock outcrops resulting in quarry faces with characteristic circular extraction hollows. These productions required a skilful knowledge of direct percussion with the pick. The typical site of this type, when conservation is ideal, presents tubular quarry faces covered by diagonal pick marks resulting from multiple, superimposed, horizontal extractions. The quarry floors, when conserved, bear the marks of cylinder splitting, whether these be wedge holes, wedge channels, chisel or pick marks.

Detached block millstone quarries (MQ-2b)

Block detachment quarries differ from true extractive quarries in that blocks or slabs are not hewn directly from the bedrock but are detached or pried out from the mass, profiting for the most part from natural fractures. Therefore the extraction hollow is not cylindrical but angular. The main extractive tool at these sites is not the pick but the lever or bar. In these cases the bedrock is not homogeneous, but characterised by natural fissures or factures. The planar quarry faces at these sites do not readily reveal anthropic cuttings. Block extractions can, in fact, be mistaken for natural mechanical processes such as gelifraction. Since they are difficult to identify in the field, it is usually the presence of unfinished or broken cylinders that betray their existence.

The morphology of extractive millstone quarries

The definitions of the morphology of extractive millstone quarries are based on terms borrowed from the vocabulary of construction quarries. Several terms such as bench, pit, trench and edge quarries are common types of open air workings. The term "pocket" quarry in this study is associated with multiple small exploitations on slopes. Subterranean quarries, although not uncommon in southern Spain for building blocks, are only represented by one site. The two remaining types, extensive contiguous shallow quarries and extensive dispersed quarries, terms inspired by the work of Jean-Claude Bessac (2003), are less common in the realm of construction but lend themselves to certain millstone workings. These terms are all directly related to the intrinsic features of the outcrop including its extension, accessibility, inclination and depth, direction of work progression, and whether the extractions were contiguous or dispersed.

The question of subterranean millstone quarries deserves a few extra words. The identification of only one subterranean millstone quarry in such a vast study area is surprising, especially considering the long tradition of underground mining, starting with flint shafts from the Neolithic at Vicálvaro in the outskirts of Madrid (Consuegra *et al.* 2004) to the Roman underground galleries of silver and gold in Murcia and Almería. There are also a number of Roman underground block quarries at Peñatejada outside Córdoba (Penco Valenzuela *et al.* 2004: 235-238) that supplied the Roman city of *Corduba*, as well as underground exploitations from the Middle Ages at Puerto de Santa María, Cádiz, that provided the blocks for the Cathedral of Seville (Jiménez Martín 2006: 179).

Elsewhere in Europe, volcanic millstones were extracted in underground galleries at Mayen, Eifel (Germany) (Hörter 1994; Harms & Mangartz 2002), and at Mont Vouan in Haute Savoie, France (archaeological excavations directed by Alain Belmont). Nearer to the study area, in southern Portugal, is the underground site of Alvito (Lidia Catarino, pers. comm.) and in Morocco, on the opposite side of the Straits of Gibraltar, is the celebrated *"Grotte d'Hercules"* (Tangiers), a quarry visited by Curwen in the 1950s. This pioneer of millstone studies actually witnessed quern makers at work in the cave by candlelight (Williams & Peacock 2011a: vii).

The sole underground exploitation identified in the study area is the site of La Merced, Loja (GR-4b). An explanation for the absence of other subterranean sites in southern Spain is puzzling. Future research on this question will certainly unveil sites that are now caved in and hidden from view. In any case, La Merced (GR-4b) is very small, only a few metres deep. Hence the millstone makers at this site did not have to cope with the problems linked to deep underground galleries, such as illumination, ventilation, rise of the water table and removal of the working debris and the products.

Comments on the chronology of millstone quarry types

In general, millstone quarry types adhere to a chronological pattern. Simple surface workings (MQ-1a), exploiting loose surface boulders, is the oldest type, covering the periods from Prehistory and Protohistory. There is also evidence of surface workings in both Roman and Medieval times as seen through certain hand querns, at times poorly made, especially of rock types that are not commonly exploited in quarries. But these Roman and Medieval workings were not the norm, and surface workings of these periods was outnumbered by far by true extractive (MQ-2a) and block or slab detachment (MQ-2b) quarries.

Block detachment quarries (MQ-2b) are also known since early times. The sites of Zujaira (GR-12) (Anderson 2010, unpublished) and El Barronal (AL-10) (Haro Navarro *et al.* 2006) presumably supplied slabs pried with levers from bedrock for prehistoric saddle querns. Once again, these early workings were probably the exception and secondary to simple surface workings (MQ-1a). It is in the Late Iron Age, with the production of rotary querns, that block detachment quarries reach a new, unprecedented level. Although no site of this type has been identified in southern Spain, Late Iron Age rotary quern block detachment sites are known throughout Europe, notably at Lodsworth in England (Peacock 1987), in the Lovosice district in the Czech Republic (Frölich & Waldhauser 1989), at Fossotes, La Salle, in the French Vosges (Farget & Fronteau 2011) and at Schweigmatt in the Black Forest of southern Germany (Joos 1975).

True extractive millstone quarries (MQ-2a), as Runnels suggested three decades back in his work on Greek millstones of the Argolid (Runnels 1981: 74), are an innovation dating to Roman times. There is in fact no evidence of the use of this technique for millstones in the Late Iron Age. This concept is seconded by the research of Luc Jaccottey (pers. comm.) based on recent field surveys of Iron Age quern production sites in France. Roman true extractive exploitations, in turn, are evidenced by several rotary quern quarries, notably the Serre workings in the Jura (Jaccottey 2011: 300) and three sites in Switzerland (Châbles FR; Chavannes-le-Chêne VD; and Würenlos AG) (Anderson 2006).

An excellent example of the transition from the Iron Age block extraction technique to the Roman true extractive technique is found in the Upper Rhine Valley along the borders of Germany, France and Switzerland. Block detachment for querns is exemplified by the Iron Age Schweigmatt site exploiting a reddish breccia in the German Black Forest (Joos 1975). From all the evidence, the Schweigmatt production was subsequently supplanted by Roman querns hewn directly from regional reddish sandstone *(Bundsandstein)* in true extractive quarries (Anderson *et al.* 2003: 64, 66). There are certainly several reasons for this change. The location of the Schweigmatt quarry, in the hills of the Black Forest, was less accessible and less extensive than the *Bundsandstein* quarries, potentially established to facilitate transport along or near the shores of the Rhine River. Finally, another reason the true extractive *Bundsandstein* workings probably eclipsed the Schweigmatt workings was due to the higher quantity and standardisation of their products (Anderson *et al.* 2003: 64, 66).

In southern Spain, examples of true extractive workings dating to the Roman times are evidenced at the volcanic workings of *Sisapo* (CR-1) in Ciudad Real and at the biocalcarenite exploitations along the Bay of Cádiz, such as Trafalgar (CA-1) and Rota (CA-3). Therefore, at this stage of research in the south of the Iberian peninsula, all the evidence reinforces the model that true extractive quarries (MQ-2a) only appeared after the Roman conquest. But it would not be surprising that future research, due to the long tradition of stone work for statuary and architecture in the Iberian Iron Age Culture of southern Spain, identify pre-Roman true extractive quern and millstone quarries, exploiting certain types of rock outcrops such as limestone tufa.

14.7. Millstone quarry toponymy

Although place name study in the framework of millstone quarry research is essential, the results in southern Spain have not been as conclusive as those undertaken in France and in the north of Spain. The French toponyms *Molière* and *Moulière* from the Latin "*mola*" (meaning mill) have resulted in the identification of many millstone quarries in the French Dauphiné, some dating as far back as the 13th and 14th centuries (Belmont 2006a, Vol. 1: 56-59). Likewise, the work of Pilar Pascual and Pedro García in the north of Spain, tracing the name *Molares* (millstone quarry) and its derivatives, has also resulted in the identification of a number of workings (Pascual & García 2011: 285-286).

The searches based on the name *Molares* in southern Spain, by contrast, have only identified sites at Almonaster la Real, Huelva (HU-8), Molares, Cuenca (CU-1) and Tobes, Guadalajara (GUA-05). Another derivative of *Molares* presumably linked to millstone working, although not confirmed in the field, is the site of *Piedras Moleras* (CO-11) meaning "millstone stones" by Villanueva de Córdoba. There are also a half dozen sites linked to *Moles*, *Mola, Molar* or *Muelas*. But these names may have nothing to do with millstone workings because they are also known to be associated with certain natural topographical features, notably buttes or flat topped hills (Sanz Elorza 2012).

Conversely, there are a number of millstone exploitations linked to the name *Cantera*, meaning simply "quarry": *Las Canteras*, Colmenar de Oreja (M-2), *Las Canteras*, Granátula de Calatrava (CR-5), *Cantera Honda*, Paterna (CO-8), *Las Canteras*, Moclín (GR-1a), *Rambla de las Canteras*, Caniles (GR-11), *Las Canteras*, Ugíjar (GR-14), *Cantera de las Pilas* (CA-9), *Las Canteras*, Castillo de Locubín (J-1), *Cantera de los Frailes*, Cabra (CO-1), *Cerro de la Cantera*, Huelma (J-3), *Las Canteras*, Guadalquitón (CA-13) and *Las Canteras*, Alhaurín el Grande (MA-3). It is interesting that most of these sites, despite their name, show no sign of extraction of anything but millstones.

The small proportion of *Molares* names, as opposed to *Canteras,* is intriguing. This difference with France or the north of Spain could possibly be a result of a new terminology arriving after the fall of Islamic rule and the influx of stonemasons (*canteros*) from other areas of the Peninsula to work at the numerous secular and religious architectural projects initiated after the Reconquest. It may have been these professionals that also ventured into the field of millstone production and applied the name *Canteras* to these sites.

14.8. Millstone quarry infrastructure

Research in southern Spain regarding millstone quarry infrastructure is nonexistent. As stated on several occasions, no quarry site in this area has benefited from an archaeological excavation that might shed light on the features linked to its operation. This obliges me to rely on field observations and on written sources. I also depend heavily on research undertaken in the north of Spain, as well as the results of excavations of sites elsewhere in Europe.

Tool repair and maintenance

The picks, chisels and wedges of millstone makers, tools that were used to carve and split the rock, required constant maintenance. Some tools simply suffered from wear at a rate that depended on the quality of their metal and the abrasiveness of the stone. Other tools broke during use. Hence, the repair and maintenance of these tools required the skills and infrastructure of a blacksmith.

Small millstone productions did not have access to a smithy directly at the quarry. In these cases they had to organise in advance the number of tools taken to the site so as not to interrupt their work as a result of an unwanted break. These productions either had to rely on a local blacksmith or had to possess the skills and tools to undertake repairs themselves, as was the case of some of the *moleros* in the Palencia mountains (Maestro Hernández 2011: 38, 43). Other sites, such as Châbles in Switzerland, a small Roman quern quarry, had access to a smithy a stone's throw away (Anderson *et al.* 2003: 59).

There is evidence, by contrast, that certain millstone quarries had smithies working specifically for them. This was the case of the intensive underground exploitations at Mont Vouan in Haute Savoie that merited possessing their own smithy. This is evidenced by an important layer of black coal-rich sediment containing smithy slag and hammer scales in the isolated mountain quarry of Grand'Gueule (Belmont & Anderson 2010: 104-108). It may also have been the case at Gardom's Edge, Baslow, Derbyshire in England (Radley 1964) and Vioménil in France (Vosges) (see Atlas of European Millstone quarries http://meuliere.ish-lyon.cnrs.fr/php/results2.php).

In southern Spain I have not observed any evidence of smithies directly at any millstone quarry. This is not surprising since smithy work does not necessarily leave behind evidence that can be readily observed on the surface as all the smithies at millstone workings elsewhere in Europe have only been discovered underground in the context of archaeological excavations.

The large production at Moclín (GR-1) would probably not have needed an independent smithy because the workers could have accessed a smithy in town, just a short walk away. By contrast, it is difficult to imagine that the vast workings at El Berrueco (CA-8), 10 km from the nearest town (Medina Sidonia) and manned at one point with up to 50 workers (Cruz y Bahamonde 1813: 91, note 1), did not employ a full time blacksmith.

Debris management

Working debris at certain millstone productions, such as simple surface workings, dispersed exploitations and extended discontiguous quarries, did not present a particular problem as it did not result in large quantities and did not hinder the progression of work. Debris from steep slope workings also were easily evacuated. Seaside workings simply dumped their debris in the water. At other sites, by contrast, in particular at bench, pocket, pit, trench and subterranean quarries, the volume of debris grew rapidly and became a logistical problem for the millstone makers. They were obliged to make a good choice of where to deposit it so as to avoid having to move it a second time.

Transporting debris entailed investing the least amount of effort possible. The most logical solution was to evacuate debris downhill, taking advantage of gravity to facilitate transport in baskets or in a wheelbarrow, and placing it in a heap. This was relatively simple in the case of bench and pocket quarries that by definition were situated on slopes.

In the case of trench quarries, debris was placed just outside the trench. Over time, the heaps joined together forming long cordons parallel to the trenches. For both pit and trench quarries, depending on their depth, paths or ramps had to be built to remove the debris. At times, the quarrymen raised drywalls backfilled with debris that served as ramps, both to enter and exit the workings.

In very deep quarries the millstone makers were obliged at times to install ropes (possibly with pulleys) to raise sacks of debris. In very recent times, some mechanisms like capstans, as seen at the quarries in the Eifel (Germany) and Selbu (Norway), were put to use to facilitate lifting debris and millstone blocks or roughouts.

In other cases the workers backfilled sectors of the quarry downhill when millstone extraction was completed. This is the case of several subterranean exploitations at Mont Vouan in France, recently the object of archaeological excavations (directed by Alain Belmont). Clearing these backfilled sectors in the framework of an archaeological dig not only brings to light well-conserved tool marks under the debris, but permits archaeologists, by means of vertical stratigraphy, to study the sequence of how the sector was backfilled.

All said, the onerous task of moving debris from the working areas was probably reserved for the newer members of the working crew. It is also possible to imagine that at times this might have been carried out by women and children.

Residences and shelters

Making a single large millstone required several days' work. If the workings were located far from town, then the millstone makers were obliged, at least during certain times of the year, to improvise provisional lodging to pass the night. If work was intensive and permanent, the workers had to find seasonal or permanent accommodations.

In fieldwork, I have observed that millstone makers a certain sites built simple huts or hovels that would have served as shelters for short periods of time. Drystone hovels, with space for just a few men, were raised at the sites of El Lachar (J-2) and El Torcal (MA-1). At El Torcal there even appears to be a corral that, hypothetically, could have enclosed the animals used for the transport of the millstones. At subterranean sites, the cave itself could have served for workers to pass the night. The small post-framed house beside the quern quarry at the Roman site of Châbles-Les Saux in Switzerland (Anderson *et al.* 2003: chap. 8), in all of its simplicity, would have nonetheless provided a comfortable setting for the quern maker.

The situation of worker lodgings at larger productions was different. The Moclin (GR-1) and Cabra (CO-1) sites, close to towns, did not need to house its workers. The cutters could either return to their houses at night, or in the case of itinerant workers, find lodging in a local inn.

The situation at the vast Berrueco quarry (CA-8) was very different. The narrative of Cruz y Bahamonde dating to the early 19th century records that this site, about 10 km from the nearest town, housed 50 workers (Cruz y Bahamonde 1813: 91, note 1). This is the only example in the study area where there is written evidence of lodgings provided for millstone workers. There is no information, however, regarding the living conditions or the rent the workers had to pay for the lodgings.

Millstone transport

Millstone transport comprises two basic steps. The first was removing the stone from its place of extraction to a staging point. It was at this second point, always in or near the quarry, where the first leg of a longer journey was initiated toward either a workshop where the millstone was finished or to its final destination at a mill. This journey, depending on its distance, could comprise more than one means of transport either by land, by water or, more recently, by rail.

Moving millstones from their place of extraction was not simple. This depended on their size, the morphology of the quarry and the surrounding terrain. A number of extraction sites were not necessarily accessible by animal-driven carts. In these cases, millstone makers built slipways and ramps with stone debris and rails or tracks with wood. When draught animals could reach the extractive zones, they used sledges of different forms to drag the stones to a waiting cart. During this process the millstones, in order not to be damaged, were sometimes wrapped in reeds (Maestro Hernández 2011: 46).

At times, the transport to the final destination was very short so the same sledge used to haul the millstone out of the quarry could be used to drag it to the mill. There are even cases where a few men, sometimes assisted by animals, reputedly used the "rolling stone" method, inserting a wooden beam through its eye and rolling it like a wheel. These cases, attested orally on a number of occasions and illustrated by Gómez Ruíz (2003: 85, fig. 7), were certainly exceptions and not the norm.

Most millstone transport was undertaken by flat-bed carts pulled by either mules or oxen. This could be organised by the millstone makers themselves, who as farmers could also be the owners of the draught animals. In other cases, they were transported by professionals (*carreteros*).

Fluvial and maritime routes served for long-distance transport since ancient times. The 4th century BC shipwreck at El Sec, off the Bay of Majorca, was loaded with a cargo of Olynthian mills and a few Morgantina models (Arribas 1987). Likewise, the 2nd century BC shipwreck at Illa Pedrosa, off the coast of Catalonia, was transporting a cargo of 130 rotary querns (Vivar 2004). In southern Spain, to my knowledge, no shipwreck with a cargo of millstones has been published. However, the presence of shell-rich biocalcarenite outcrops along the coast of the Bay of Cádiz at Trafalgar (CA-1) and Chipiona (CA-6) suggests Roman millstone transport by sea. There are also texts recording maritime transport from the Cádiz area both in the early 16th (Fernández López 1982: 221, note 10) and 19th centuries (Ponce 1981).

Transport of millstones by rail into the south of Spain was a very late development, dating to the second half of the 19th century after completion of the rail network. Although I have no textual evidence, producers of larger millstones in southern Spain probably profited from this new means of transport, as is known to have occurred for other rock quarries yielding construction material such as the quarries of Gerena (SE-7). Ironically, a consequence of this new rail transport was that it facilitated the arrival of imported French millstones (burrstones), which would eventually undermine the Spanish millstone industry.

14.9. The men behind the millstones

The profile of a millstone maker, his degree of specialisation and his place in society are concepts that are difficult to judge from the evidence at hand. The earliest saddle quern and rotary quern makers were first of all dedicated to farming and raising livestock. They certainly had an excellent knowledge of working rock, as seen through the long tradition of their stone tools, and building rock dwellings, defensive and funerary features. Since querns had long working lives, making them was far from an everyday activity and certainly not more venerated than other types of stone or metal work.

This situation changed during Roman times. Production of sophisticated, standardised rotary querns and millstones on a large-scale for a growing population, provided millstone makers with work at least on a seasonal, and at times full-time basis. Working at Roman quarry sites also introduced new skills, in particular, that of extracting cylinders directly from bedrock (MQ-2a). At this stage, millstone makers can be considered veritable specialists.

The trace of millstone makers in the south of Spain is nonetheless lost in the Middle Ages and during the Islamic domination. They certainly remained highly specialised workers due to the demand of large millstones for a growing number of watermills spread across the landscape along rivers, streams, and an ever developing network of irrigation ditches. During these periods there may have been millstone quarries providing permanent work. But the lack of use of one dominant type of rock (as was the case of volcanic rocks in Roman times) with far reaching spheres of trade, cut back the activity of specialised quarries and spread production to more common local or regional sedimentary outcrops. This "dispersal" of quarries probably obliged millstone makers to combine this activity with other types of stonework or farming.

The situation after the fall of Islamic rule does not seem to have changed. Juan de Bargas, in the Province of Córdoba in the early 17th century, according to written records, worked as both a millstone maker and a stonemason for construction projects (González Peralbo 2008). This duality was probably caused by the lack of regular millstone orders obliging him, and others like him, to engage in other rock-related professions.

The situation in Contemporary times probably continued as in former times with millstone makers working only to meet specific orders. There is evidence, however, of the rise of certain specific production centres that had the capacity to retain professional millstone makers either full-time or for at least large parts of the year. As noted, El Berrueco (CA-8) in Cádiz in the early 19th century employed 50 workers (Cruz y Bahamonde 1813: 91, note 1) and a half century later employed 23, split up in five different workshops (Madoz 1848, Vol. 11: 343).

Millstone making teams were divided into crews with a leader who saw to maintaining the rhythm and quality of work. These team leaders may even have acquired the right to exploit certain quarry sectors in the form of concessions and thus would have gained profit from the workings. This "wealth", difficult to quantify, combined with contact with the owners of the quarry, noblemen or clergymen, certainly raised their social status.

The intensity of work at millstone workings would have brought with it occupational hazards like the ever-present danger to the eyes, limbs and back. Accidents at quarry sites were certainly not uncommon. For long term workers, and depending on the type of rock, lurked the silent killer of silicosis, a malady that could strike down a healthy and robust man in a number of years.

14.10. Millstone quarry ownership and control

There is very little evidence indicative of millstone quarry ownership and control in southern Spain. In general, and through time, modest quern and millstone exploitations for local consumption probably went unnoticed or ignored by the ruling bodies. Simple surface workings along riverbeds or taluses, for example, would have been extremely difficult to control. Isolated extractions in remote areas far from the urban centres most likely remained unnoticed. Larger production centres with a high potential of profit did, however, most certainly catch the eye of the ruling bodies.

These larger sites were probably under the jurisdiction of the authorities since Antiquity. At this time, production at certain volcanic outcrops was in high gear and the owners of the land with the outcrops and/or the local political authorities surely coveted their control, as they would have with other potential sources of wealth, such as mines.

In the Middle Ages under Islamic rule, with the abandonment of volcanic exploitations, most production appears to have been undertaken at local and regional levels, exploiting different types of sedimentary rocks. It is reasonable to assume that small surface productions remained anonymous and were not awarded special attention. But larger productions, notably the vast true extractive workings, may have either been under direct control of the ruling body or obliged to pay tribute for the right to extract and commercialise millstones.

After the Christian Reconquest, authority over production centres passed, as recorded in a number of legal codices and historical documents, under the direct jurisdiction of the authorities. The *Fuero de Cuenca,* a set of laws dating to the late 12th century related to the transfer of property after the Reconquest, is unequivocal on the question of ownership of these sites:

"All the quarries, gypsum beds, ***millstone quarries****, tile works, and also the perennial springs should be common property of* ***the council****. Whoever has a* ***millstone quarry*** *or any of these things mentioned previously on a property of his should sell it to* ***the council*** *for a double-sized property, and it becomes communal. If someone occupies it against another of the council, he should pay a hundred* aurei*"* (translation by Powers 2000: 57).

There are several other examples in old written records that exemplify the control of the authorities (nobility and the clergy) over millstone production, as well as other commercial activities. The council of Loja in 1502, for example, prohibited, by means of an oral proclamation, millstone extraction without proper authorisation, under the threat of confiscation not only of the millstone but the oxen used for its transport (Loja Ordinance1502, http://www.reocities.com/tdcastros/Historyserver/Fuentes/DocMunic/DOCLOJA.6.htm). At Llerena (BA-2) in Extremadura, a similar municipal ordinance dating to 1566 prohibited "foreigners" from extracting millstones from the local quarry, unless they, like the locals, paid the official tariff (http://manuelmaldonadofernandez3.blogspot.com.es/2010_04_01_archive.html). In 1501, the dispute between the councils of Antequera and Málaga over the right to extract millstones in lands under the authority of the council of Antequera, ultimately reached the highest authorities at the time, the Catholic Monarchs (Isabella and Ferdinand) based in Granada (Fernández López 1982: 222-223; AML, Leg. 49, p. 9. 1 cuartilla + 1).

In Contemporary times, the situation changed little. The dynasty of the Duque de Medina in the Province of Cádiz maintained not only control for several centuries over the quarry of El Berrueco (CA-8) (probably one of the larger productions in the south of the Iberian

Peninsula), but control over a number of watermills in the area. It is even recorded that this family, in the early 19th century, received 18 millstones a year in exchange for a concession at the Berrueco quarry (Cruz y Bahamonde 1813: 91, note 1). This was very profitable business for the noble family because they received the best millstones in the region free of charge for the flour mills they owned.

The ownership of the Berrueco quarry by the Duque de Medina is typical of recent times. These large sites dominated the regional markets. However, smaller productions, like the intermittent workings in the mountains of Palencia described by Maestro Hernández (2011), did not disappear. These modest quarries offered an economic alternative to the more expensive quality millstones from the larger centres. However, the local workings certainly did not offer millstones that could yield the highly prized white flour necessary to make white bread.

14.11. Millstone quarry chronology

Establishing the chronology of a millstone quarry is not a simple task. It is well-established by examples in France and Switzerland that a rock with good grinding properties could have been exploited at the same outcrop over a long period of time - centuries or more. This longevity is evidenced, at times, at sites with a wide range of extraction hollow diameters, from small rotary querns to large hydraulic millstones. Sites producing only a single, standardised product were probably relatively short-lived, as could to be the case of the Roman Cerro de Limones (AL-1) and Hoya del Paraíso (AL-2) sites in Almería.

There are, nonetheless, indicators that provide clues to dating millstone quarries. The first is written sources, such as historical and notarial archives, and geographical and geological texts. As useful as these records are, they only furnish a *terminus* that reflects production at the moment the texts were penned and do not provide proof of older or younger workings. Certain sites, nonetheless, benefit from more than one written reference, spread over a wide frame of time, indicating a long, possibly continuous, life span. This is the case of El Torcal (MA-1), Loja (GR-2/3) and El Berrueco (CA-8), with citings ranging from the turn of the 15th - 16th centuries and the 19th century.

A second indicator of chronology is the extraction morphometrics, that is, the dimensions of 1) unfinished millstones abandoned at the site and/or 2) the extraction hollows. In general, large abandoned cylinders or extraction hollows tend to point to more recent quarry work, whereas small rotary quern extractions usually (although not always) point to a date from Antiquity or the Middle Ages. These measurements, in any case, are a starting point for a finer date attained by the combination of other chronological indicators. It must be noted that products abandoned at quarries, due to their unfinished nature, rarely bear typological features (rynd cuttings, handle holes, dressing) from which a finer date can be gleaned.

From measurements obtained at quarry sites and on stones in museum depositories, as well as rare citings in millstone literature, I have arrived at the following morphometric categories that serve to establish chronological tendencies of millstone quarries in southern Spain.

1- Extractions ranging in diameter from 35 to 42 cm most often correspond to Iron Age and Roman querns. Iron Age models are usually smaller and thicker than their Roman counterparts. To complicate matters, querns from the Middle Ages, although generally larger than Iron Age and Roman models, are also known in this diameter range. The difference is that Iron Age and Roman querns, on the whole, are proportionally thicker than discoidal models from the Middle Ages.

2- Thin, discoidal quern extractions 50 cm in diameter are associated, with a high degree of confidence, to the Middle Ages. Thicker, sometimes slightly larger models, correspond to animal fodder querns from Contemporary times.

3- The 60 to 90 cm diameter indicator is the most complex of the four and covers all of the chronological periods. In general, Iron Age and Roman "drum" extractions in this range are for cylindrical or ring-mills, that are often as thick as they are wide. The tendency is for the Roman models to be slightly larger than their Iron Age counterparts. Lower, discoidal models of this diameter are assumed to be either animal-driven or water-driven millstones ranging from the Middle Ages to Contemporary times. This 60 to 90 cm dimension range, independent of other indicators, is therefore *per se* not valid for dating.

4- Extractions with the diameter of 1.00 m to 1.50 m can date from at least the Late Middle Ages to Contemporary times. This wide range, although not very useful for finer dating, does permit discarding a date from earlier periods.

Rock type is also an invaluable indicator because there are certain chronological periods when specific rocks were more sought after than others. Volcanic materials (dacites, rhyolites, lamproites and basalts) are usually indicative of the Roman period, whereas, fine, white limestones are most often associated with Modern or Contemporary productions.

At a level below that of written sources, morphometrics and rock type, is a series of lesser chronological measures: 1) place names, 2) geographical proximity of a quarry to a dated feature, 3) relative chronology and vertical stratigraphy, 4) extraction techniques and 5) oral information. All of these additional indicators, applied independently, do not suffice to establish a reliable date for a quarry. Even so, they do serve to reinforce or corroborate the stronger, first- level chronological indicators.

To illustrate the chronology of the millstone quarries identified in the south of the Iberian Peninsula, I have divided them into three assemblages represented in three tables.

Chronological assemblage 1 *(see Table 11.2)*

This table represents the earliest quarries in chronological order based on their earliest phase of production. The quarries range from Late Prehistory/Protohistory through the Middle Ages. These sites saw production of saddle querns, rotary querns, cylindrical ring-mills, and small hydraulic millstones. Some, such as Rota (CA-3), Moclín (GR-1) and Castillo de Locubín (J-1), present later extraction phases certified through texts. The rock type indicator, especially that associated with Roman volcanic productions, is essential to dating some of these sites.

Chronological assemblage 2a and 2b *(see Tables 11.3a and 11 3b)*

This second assemblage is established exclusively from written sources and is illustrated in two tables separated by a hiatus of about 30 years. These sites, like those of the first table, are founded on the earliest known extractive phase. Since the chronology of these sites can be defined with a high degree of precision, this table (contrary to that of Table 11.2) is divided into columns representing centuries. Assemblage 2a (Table 11.3a) therefore corresponds broadly to the Modern period and presents the sites ranging from 1481 to 1794. It, in turn, is divided into four subgroups (a, b, c, d). Assemblage 2b, although a continuation of the first, designates the transition from Modern to Contemporary times (end of the 18th century). Most of the dating of these quarries is identified by the texts of Miñano (1826-1829) and Madoz (1845-1850). A few, from the end of the 19th or beginning of the 20th century, are dated from other written or oral sources.

Chronological assemblage 3 *(see Table 11.4)*

This last assemblage records the approximate date of the remaining millstone making sites that do not benefit from any means of dating except for the size of their extractions (all over 1.00 m in diameter). They fall anywhere between the Middle Ages and Contemporary times. Only future work, both in the field and in historical archives, will provide more precise information as to their dating.

14.12. From quarry to mill: millstone distribution through time

I have attempted in this study to dedicate a section to the question of quern and millstone distribution for each chronological period. The archaeological and written records, however, is very unbalanced from period to period. Therefore, a higher degree of precision is only applicable to the Roman period (based mostly on quern and millstone finds in museums) and to the Contemporary period by means of written sources. Distribution in the other periods is highly speculative.

In general, the traditional divisions of "local, regional and long-distance" distribution are valid for southern Spain. For the pre-Roman periods the little evidence points to local and a few regional productions. For the Roman period, based on a series of theoretical distances between potential quarry and find-spot, in particular those related to volcanic rocks, the limits are roughly established as follows: *c.* 20 km for local distribution, from *c.* 20 to *c.* 80 km for regional distribution, and beyond 100 km (sometimes attaining several hundred kilometres) for long-distance distribution.

The earliest productions were mostly local and regional. Long-distance trade of querns and millstones, a phenomenon that attained a large-scale only in Roman times, diminished toward the end of Antiquity or beginning of the Middle Ages with the abandonment of volcanic millstones. This is indicative of a "return" to local and regional productions of sedimentary rocks. Volcanic rocks were probably less apt at milling at higher speeds of rotation at this time of a great expansion of watermills. However this specific question is purely theoretical in southern Spain, because, as noted throughout this work, evidence of Roman and Early Middle Age watermills is practically nonexistent.

It appears, nonetheless, that local and regional quarries would once again dominate millstone trade for centuries, until the Contemporary period. These were, for the most part, small productions that outfitted the mills of surrounding areas. There were most likely certain quarries yielding products of higher quality that surely supplied mills farther away. Nonetheless, there is no evidence that any of these larger sites attained the commercial reach comparable to volcanic workings from Roman times.

Toward the middle of the 19th century, economic and infrastructure conditions related to the Industrial Revolution made the south of Spain ripe for the influx, once again, of long-distance imports. At this time, it was no longer volcanic material, but siliceous rocks from flourishing *meulières* in France, in particular, arriving by means of a new rail network from the Brie region in the Parisian Basin. French burrstones, although very costly, would have a profound effect on milling, reducing the time invested in maintenance and dressing, and increasing considerably the yield and quality of flour - and providing a better tasting and finer, white bread.

14.13. From stone to bread

Dust and grit in the bread that wore down people's teeth are as old as bread itself. These fine particles resulted in part from not properly cleaning the grain (Agustí 1722: 187) and, in some measure, from minute rock particles detaching themselves from millstones during grinding. The predicament was recognised in southern Spain since at least the 16th century (Gómez Ruíz 2003: 31). All millstones, to a certain degree, are culprits. Rock grit was especially prevalent directly after dressing (Agustí 1722: 187), a routine that, depending on the rock type, had to be undertaken frequently, in some cases (softer limestones) on a daily basis. There is evidence since at least the Late Middle Ages of measures taken to eradicate these "impurities".

In many old texts, there are terms or expressions associated with quarries, millstones, flour and bread that have various meanings connected to the preoccupation of a grit-free loaf of bread. *Piedra blanca*, for example, is synonymous with both white-coloured millstones and, more specifically, the upper stone (runner) of the pair. Conversely, the word for "dark" (*bazo* or *baza*, *baxo*) designates both dark-coloured millstones, as well as the lower stone of the pair. These connotations, as noted, go beyond simple definitions of stones, quarries, flour and bread and have social and economic implications. White bread (*pan blanco*), "free" of grit, was traditionally associated with the noble class and the clergy, whereas dark bread (*pan moreno, bazo, baxo*) was the bread of the lower classes.

In spite of all the skill and experience of a miller, white flour for white bread could not be attained with just any cereal or any millstone. The cereal that lent itself best to white bread, due to its starchy endosperm, was bread wheat, a grain that was readily available in the south of Spain. Furthermore, a prerequisite to obtaining white bread was to separate the white endosperm from the bran and germ. This depended greatly on the miller's experience and ability. He determined the quantity of water to add to the grains previous to grinding that facilitated separating the bran. He also controlled the tentering gear that determined the amount of "light" between the upper and lower stones so as to ensure a coarser or finer grinding. He also decided when to accelerate the millstone rotation and when to halt grinding to remove and dress the stones. Finally, the miller determined if the flour needed a second round of grinding, and the choice of the sieves to attain the desired calibre of flour.

In spite of all the miller's abilities, white bread could not be attained with any type of rock. The advantage of the bright white limestones, such as those from Moclín (GR-1) in Granada and Berruecos (CA-8) in Cádiz, and certainly that of Colmenar de Oreja (M-2) in Madrid, is that the grit released into the flour was white and fine and therefore would not have affected the colour of the bread (Belmont 2011: 15). In this sense, the advantage of grinding with bright, white rocks was not the absence of stone dust in the flour, but the perception of its absence.

Defining what is interpreted in past times as white and dark bread is not simple. There were certainly a wide range of bread types from region to region. In any case, it appears that a truly fine, white bread devoid of grit only was attained with the arrival of French burrstones.

14.14. Final comments and perspectives

The principal focus of this research is to tell the story of the procurement of querns and millstones in quarries from Late Prehistory and Protohistory to Modern times, a project that has never been undertaken in the south of Spain. The term "Modern" in the title of this work is to be interpreted in its broad sense, corresponding to the early 20th century. It is in this time frame that the Iberian millstone production industry waned, initially, as a consequence of the introduction of imported silicious French burrstones.

These *meulas francesas,* still seen decorating public spaces throughout Spain today, were a cut above any millstone produced elsewhere in Spain (and the rest of Europe). An early example of their import into Spain is narrated by Meijide Pardo in his article about the 18th-century French merchant Juan Lestache based in El Ferrol, Galicia. This pioneer of the French milling system built two watermills along the Xubia River and equipped them with *aceñas* (vertical waterwheels) and four pairs of French burrstones. Owing to the quality of the flour and the high yield, in a few years, toward the end of the 18th century, Lestache's factories became the dominant flour industry in the whole of Galicia (Meijide Pardo 1997: 39-40).

Nevertheless, up until the 20th century, most watermills were equipped with local or regional stones. The ultimate demise of the Iberian millstone industry came to pass with the proliferation of French millstones by means of the new rail network and the later introduction of the steel roller, which would inexorably precipitate the end not only of local and regional quarries, but the whole of the traditional milling industry and the milling profession.

As I have stated on a number of occasions, much research still remains to be undertaken on this subject in southern Spain and elsewhere. I have only been able to "survey" about a third of the sites (and there are certainly still more to be found). And it must be born in mind that these visits were most often of short duration, limiting the quantity and quality of the data.

Therefore, in the perspective of establishing a future chrono-typology of millstone quarries and their products, there is the need of developing coordinated research with well-established objectives and techniques, similar to those undertaken by the *Groupe Meule* in France or the "Norwegian Millstone Landscape" in Norway. These projects have combined fieldwork, work in museum depositories, research in historical archives and petrographical analyses.

One of the main contributions of this study to general millstone literature is the identification of two new Roman volcanic rock districts (the S-E Spanish volcanic district and the Campo de Calatrava volcanics). These new members of the European volcanic millstone quarry "club" are essential elements to the study of the distribution of millstones of different types in the larger European, North African and Middle Eastern framework.

Like other features of cultural heritage, millstone quarries deserve due recognition and should be recorded and mapped. Authorities should afford them the same protection as any other monument of cultural heritage. Alas, all too often these spectacular workings with their towering, tubular faces are neglected (often used as rubbish dumps), or even worse, destroyed. In fact, no millstone quarry in southern Spain has benefited from an official valorisation. That is, no site has benefited from a combination of archaeological and archive research so as to furnish the general public (by means of paths and explanatory panels) with information regarding the site's extraction techniques, products, personnel, chronology and socioeconomic context.

Hopefully, in spite of the difficult economic conditions that present society is experiencing, the new field of millstone quarry research will not be dismissed, as happened to the brilliant work on millstones initiated by Fernando de Avilés in the 1940s (Berrocal Rángel 2007). It is obvious that the subject is essential to the understanding of the history of human nutrition. It is therefore my hope that the recent international round table meeting held in Arbeca, Catalonia (organised by the Department of Archaeology of the University of Lleida with proceedings published in the journal *Revista d'Arqueologia de Ponent* 24, 2014), bringing together researchers from Spain and France to discuss Iron Age millstones and quarries, will be a starting point for future millstone research in the Iberian Peninsula.

BIBLIOGRAPHY AND SOURCES

ACUÑA CARABANTES, Lucas. *De la Gerena que venimos,* Sevilla. 2004.

AGAPAIN (collective work). *Les Meuliers: Meules et Pierres Meulièrs*, Étrépilly. 2002, 278 p.

AGÜERA MARTÍNEZ, Saturnino, INIESTA SANMARTÍN, Ángel, MARTÍNEZ ALCALDE, María. Carta Arqueológica de Mazarrón, Resultados de la Campaña de 1992-1993. *Memorias de Arqueología* 8, Murcia. 1999, p. 507-522.

AGUIRRE ANDRÉS, Antonio. Hallazgos Arqueológicos Realizados Recientemente en la Provincia de Ciudad Real, *Cuaderno de estudios Manchegos*, No. 2. 1948, p. 117-130.

AGUSTÍ, Miquel. *Libro de los Secretos de Agricultura, Casa de Campo y Pastoril*. Barcelona. 1722, 508 p.

AKERRAZ, Aomar, LENOIR, Maurice. Instruments de broyage en Maurétonie Tingtane à l'époque romaine, le cas de Volubilis. PROCOPIOU, H., TREUIL, R., (dir.), *Moudre et broyer II: Archéologie et histoire : du Paléolithique au Moyen Âge*, Paris. 2002, p. 197-207.

ALAÑON FLOX, Luis. Arqueología del Bajo Jabalón, Mapa arqueológico de la Zona de Aldea del Rey (Ciudad Real), *Cuaderno de estudios Manchegos*, 13. 1982, p. 219-237.

ALEGRÍA SUESCUN, David. El molino harinero de Zubieta. Evolución histórica. *Cuadernos de Etnología y Etnografía de Navarra*, 82. 2007, p. 5-15. http://www.vianayborgia.es/CUET-0082-0000-0005-0015.html.

ALLEPUZ MARZÀ, Xavier. *Borriol*. 2003, 335 p.

ALONSO, Natàlia. *De la llavor a la farina, Els processos agricoles protohistorics a la Catalunya Occidental. Monographies d'archéologie méditerranéenne*, 4. 1999, Lattes. 328 p.

ALONSO, Natàlia, AULINAS, MeritxelL, GARCÍA, Maria Teresa, MARTIN, Félix, PRATS, Georgina, VILA, Sílvia. Manufacturing rotary querns in the 4th c. BC fortified settlement of Els Vilars (Arbeca, Catalonia, Spain). In: WILLIAMS, David, PEACOCK, David (eds), *Bread for the People: The Archaeology of Mills and Milling, Proceedings of a Colloquium held in the British School at Rome, 4th-7th November, 2009*, University of Southampton Series in Archaeology, no. 3, Archaeopress, Oxford. 2011, p. 55-65.

ALONSO CHAVES, F.M., GARCÍA-NAVARRO, E., CAMACHO, M.A., MANTERO, E. Mª. Análisis de los Tipos de Intestabilidades del Macizo Rocoso de LaVirgen de la Peña (Huelva, España), *GEOGACETA* 44. 2008, p. 43-46.

ALONSO VILLALBOS, Carlos, GRACIA PRIETO, F. Javier, MÉNANTEAU, Loïc. Las Salinas de la Bahía de Cádiz Durante la Antigüedad: Vision Geoarqueológica de un Problema Histórico, *Revista de Prehistoria y Arqueología de la Universidad de Sevilla SPAL* 12. 2003, p. 317-332.

ALTAMIRANO GARCÍA, Manuel, ANTÓN AGUILAR, Lourdes. Una Cantera de Piedra de Molino Inédita en el Término Municipal de Córdoba, *Antiqvitas* 24. Córdoba. 2012, p. 331-339.

ÁLVAREZ MARTINEZ, José María. El Templo de Diana. *Templos Romanos de Hispania. Cuadernos de Arquitectura Romana*, Vol. 1. 1991, p. 83-93.

AMANTE, Manuel. Representaciones Iconográficas en Lucernas Romanas de la Región de Murcia, Arte y poblamiento en el SE peninsular. *Antigüedad y Cristianismo: Monografías Históricas sobre la Antigüedad Tardía (Murcia)*, No. 5. 1988, p. 213-254.

AMARÉ, María Teresa. *Lucernas Romanas en Aragón*, Institución Fernando el Católico, Volumen 1128. 1988, Zaragoza. 166 p.
AMEZCUA OGÁYAR, Juan Manuel. *Estudios Histórico-Tecnológico de los Molinos de Viento del Campo de Níjar. Aplicación al Estudio en detalle y Reconstrucción Gráfica del Molino del Collado*, Tesis Doctoral, Universidad de Jaén. 2003. 312 p.

AMOURIC, Henri. L'anille et les meules. In: GARCIA Dominique, MEEKS. Dimitri (eds), *Techniques et économie antiques et médievales. Le temps de l'innovation, Colloque international CNRS, Aix-en-Provence 21-23 mai 1996*. Paris. 1997, p. 39-47.

ANDERSON, Timothy. Three Roman Quern Quarries in Switzerland. In: BELMONT, Alain, MANGARTZ, Fritz, Les Meulières: Recherche, Protection et Valorisation d'un Patrimoine Industriel Européen. Mainz. 2006, p. 41-46.

ANDERSON, Timothy. Los Pensadores y Las Rajas (2009-2010), Poblados Prehistóricos en el Margen de la Vega de Granada, Municipio de Pinos Puente, Informe Preliminar de las Actividades Arqueológicas al Finalizar las Obras en Dos Préstamos. 2010, Unpublished Report, 27 p.

ANDERSON, Timothy. Un premier bilan sur la production de meules dans la Péninsule Ibérique depuis la protohistoire à la période médiévale, In: BUCHSENSCHUTZ, Olivier, JACCOTTEY, Luc, JODRY, Florent, BLANCHARD, Jean-Luc (eds), *Actes de la Table Ronde, Evolution Typologique et Technique des Meules du Néolithique à l'An Mille sur le Territoire Français, St. Julien-sur-Garonne, 2-4 Octobre 2009*, Supplément 23, *Aquitania*. 2011a, p. 227-238.

ANDERSON, Timothy. Los Molinos Manuales de Cebo de Priego de Córdoba de la Collección Etnográfica de Marcos Campos (abril 2011). Unpublished report. 2011b, 12 p.

ANDERSON, Timothy. Moleras en la Península Ibérica: una primera clasificación de canteras de molinos. *Revista d'Arqueologia de Ponent* 24. 2014, p. 127-144.

ANDERSON, Timothy, AGUSTONI, Clara, DUVAUCHELLE Anika, SERNEELS, Vincent, CASTELLA, Daniel. *Des Artisans à la Campagne. Carrière de Meules, Forge et Voie Gallo-Romaines à Châbles (FR), Archéologie Fribourgeoise* 19, Academic Press, Fribourg. 2003, 392 p.

ANDERSON, Timothy, CASTELLA, Daniel. *Une Ferme Gauloise à Courgeveaux (FR, SUISSE), Archéologie Fribourgeoise* 21, Academic Press, Fribourg. 2007, 214 p.

ANDERSON, Timothy, CASTELLA, Daniel, DOSWALD, Cornel, VILLET, Damien. Meules à bras et meules hydrauliques en Suisse romaine: répartition et pétrographie, *Minaria Helvetica*, 24a. 2004, p. 3-16.

ANDERSON, Timothy, GRENNE, Tor, FERNÁNDEZ SOLER, Juan Manuel. Volcanic Quern and Millstone Quarries in the Cabo de Gata (Almería) and Campos de Calatrava (Ciudad Real), Spain. In: WILLIAMS, David, PEACOCK, David (eds), *Bread for the People: The Archaeology of Mills and Milling, Proceedings of a Colloquium held in the British School at Rome, 4th-7th November, 2009*, University of Southampton Series in Archaeology, no. 3, Archaeopress, Oxford. 2011a, p. 151-167.

ANDERSON, Timothy, JACCOTTEY, Luc, SCARROW, Jane H., CAMBESES, Aitor. Premier Bilan sur les Moulins Pompéiens dans la Péninsule Ibérique. *Actes du Colloque: Archéologie des Moulins Hydrauliques, à Traction Animale et à Vent, des Origines à l'Époque Mé diévale, Lons-le-Saunier du 2 au 5 Novembre 2011*,in press.

ANDERSON, Timothy, JACCOTTEY, Luc, Vers une classification et normalisation de termes pour définir les carrières de meules, *2ème colloque sur les meules à grain du Néolithique à l'époque médiévale : technique, culture, et diffusion, Reims, 15-17 mai 2014*, in preparation.

ANDERSON, Timothy, SCARROW, Jane H. Millstone Quarries in Southern Spain: Preliminary Pinpointing of Provenance and Production - Exploiting the Internet, In: WILLIAMS, David, PEACOCK, David (eds), *Bread for the People: The Archaeology of Mills and Milling, Proceedings of a Colloquium held in the British School at Rome, 4th-7th November, 2009*, University of Southampton Series in Archaeology, no. 3, Archaeopress, Oxford. 2011, p. 259-275.

ANDERSON, Timothy, SCARROW, Jane H., CAMBESES, Aitor. Propuesta de Estudio Geoarqueológico de Molinos Rotatorios Romanos Volcánicos del Museo Arqueológico de Murcia. Unpublished report. 2011b, 53 p.

ANDERSON, Timothy, SCARROW, Jane H., CAMBESES, Aitor. Continued Characterization of Querns and Quern Quarries in Southern Spain. In: SELSING, Lotte (ed.), *Seen Through a Millstone,* AmS-Skrifter 24, University of Stavanger. 2014, p. 111-131.

ANTONELLI, Fabrizio, LAZARINI. Lorenzo. Mediterranean trade of the most widespread Roman volcanic millstones from Italy and petrochemical markers of their raw materials, *Journal of Archaeological Science*, 37. 2010, p. 2081-2092.

ASENSIO i VILARÓ, David, BELARTE FRANCO, Mari Carme, SANMARTÍ, Joan, SANTACANA, Joan. Les Meules Rotatives du Site Ibérique d'Alorda Park (Calafell, Baix Penedès, Tarragona). Pyrenae, Vol. 31-32. 2001, p. 57-73.

ARRIBAS, Antonio. *El barco del Sec (Calviá, Mallorca): estudio de los materiales*, Ayuntamiento de Calviá. 1987, 659 p.

BAATZ, Dietwulf. Die Wassermühle bei Vitruv X 5,2. Ein archäologisehet Kommentar, *Saalburg-Jahrbuch* 48. 1995, p. 5-18.

BALTANÁS, Horacio. Los Molinos de la Presa de la Aldehuela (Colmenar de Oreja): Obras de Remodelación de un Complejo Hidráulico en el Río Tajo (siglos XVI y XVII). In: BORES F., FERNÁNDEZ J., HUERTA, S., RABASA E., *Actas del Segundo Congreso Nacional de Historia de la Construcción, A Coruña, 22-24 octubre 1998*. Madrid. 1998, 529 p.

BARBERÁ MIRALLES, Benjami. La Provincia de Castellón. Molinos Harineros y Muelas. In: BARBOFF, Mouette, *et al.*, *Meules à grains: Actes du Colloque International de la Ferté-sous-Jouarre, 16-19 mai 2002*. Paris. 2003, p. 188-196.

BARBOFF, Mouette, SIGUAT, François, GRIFFIN-KREMER, Colette, KREMER, Robert (eds), *Meules à grains: Actes du Colloque International de la Ferté-sous-Jouarre, 16-19 mai 2002*. 2003, Paris, 472 p.

BARRERA, José Luis. *La Decoración Arquitectónica de los Foros de Augusta Emerita, Volumen 25 de Bibliotheca archaeologica*. 2000, 487 p.

BASTERRA ADÁN, Miguel. Convenio y Concierto Subscrito en el Siglo XVIII Entre los Concejos del Valle de Redondo, Brañosera, Celada de Roblecedo, Salcedillo y Herreruela para la Fabricación de Muelas de Molino. PIITM, 74, Palencia. 2003, p. 249-268.

BELMONT, Alain. *La Pierre à Pain. Les carrières de meules de moulins en France, du Moyen Age à la révolution industrielle, PUG*, Grenoble. 2006a, Vol. 1, 332 p.; Vol. 2, 232 p.

BELMONT, Alain. Les Meulières Médiévales: Résultats d'une Moisson Dauphinoies. In: BELMONT, Alain, MANGARTZ, Fritz, Le Meulières: Recherche, Protection et Valorisation d'un Patrimoine Industriel Européen. Mainz. 2006b, p. 81-90.

BELMONT, Alain. Why Dig a Millstone Quarry?, In: WILLIAMS, David, PEACOCK, David (eds), *Bread for the People: The Archaeology of Mills and Milling, Proceedings of a Colloquium held in the British School at Rome, 4th-7th November, 2009*, University of Southampton Series in Archaeology, no. 3, Archaeopress, Oxford. 2011, p. 1-17.

BELMONT, Alain, ANDERSON, Timothy. La meulière de Grand'Gueule, Commune de Viuz-en-Sallaz, Département de Haute-Savoie, Rapport de Fouilles Programmées. 2010, Unpublished report, 163 p.

BELMONT, Alain, in collaboration with ANDERSON, Timothy, GAILLARD, Jacques, MERCIER, Jean-Claude. Quinze Siècles d'Extraction de Meules à Grains. Les Carrières de Claix (Charente), Ve-XIEe Siècles. In: BUCHSENSCHUTZ, Olivier, JACCOTTEY, Luc, JODRY, Florent, BLANCHARD, Jean-Luc (eds), *Actes de la Table Ronde, Evolution Typologique et Technique des Meules du Néolithique à l'an Mille sur le Territoire Français St. Julien-sur-Garonne, 2-4 Octobre 2009*, Aquitania. 2011, p. 199-225.

BENEROSO SANTOS, José. Una Aproximación a la Toponimía Medieval Musulmana en el Término de San Roque, *Alameda, Nuestra Historia*, Junio. 2007, p. 13-15.

BENEROSO SANTOS, José. Guadalquitón: Un Posible Sistema Hidráulico Andalusí, *Almoraima* 42. 2011, p. 477-499. www.sanroque.es/ayuntamiento/...2008.../download [accessed November 7, 2012].

BENÍTEZ de LUGO ENRICH, Luis. El registro arqueológico en Alhambra (Ciudad Real), *Cuadernos del Instituto de Estudios Manchegos-C.S.I.C.*, nº 23-24, Ciudad Real. 2001, p. 9-25.

BENÍTEZ de LUGO ENRICH, Luis, ÁLVAREZ GARCÍA, H. J., MATA TRUJILLO, E., LÓPEZ-MENCHERO BENDICHO, V. M., MORALEDA SIERRA, J. Villae en el Municipium de Mentesa Oretana. Termas Romanas y Necrópolis Tardo-Romana en la Ontavia (Terrinches, Ciudad Real). Resultados de la Investigación y Proyecto de Musealización. *Herakleion* 4. 2011, p. 69-124.

BERMÚDEZ AZNAR, Agustín. La Organización del Concejo de Sepúlveda según el Fuero de 1305. In: ALVARADO PLANAS, Javier (ed.), Los fueros de Sepúlveda : I Symposium de Estudios Históricos de Sepúlveda, *2004*. 2005, p. 151-183.

BERNI MILLET, Piero. *Epigrafía Anfórica de la Bética. Nuevas Formas de Análisis*. Col-lecció INSTRUMENTA 29, Publicacions i Edicions, Universitat de Barcelona. 2008, 639 p.

BERROCAL RÁNGEL. Luis. El asentamiento "Céltico" del Castrejón de Capote (Higuera la Real, Badajoz), *CuPAUAM* 16. 1989, p. 245-295.

BERROCAL RÁNGEL, Luis. De la mecánica a la molienda: un ensayo sobre los molinos giratorios de la España antigua. In BLÁNQUEZ, J., ROLDÁN, L., JIMÉNEZ, H. (eds), *Augusto Fernández de Avilés. En homenaje. Universidad Autónoma de Madrid, Madrid 2006*. 2007, p. 275-297.

BESSAC, Jean-Claude. *L'outillage traditionnel de la taille de pierre : technique, chronologie, classification. Thèse de 3e cycle d'Archéologie*, Paris, E.H.E.S.S. 1982, 3 Volumes, 550 p.

BESSAC, Jean-Claude. *La Pierre en Gaule Narbonnaise et lesCarrières du Bois des Lens (Nîmes). Histoire, Archéologie, Ethnographie et Techniques* (Journal of Roman Archaeology, Supplementary Series 16). Ann Arbor. 1996, 335 p.

BESSAC, Jean-Claude. L'extraction des pierres de taille et des roches marbrières dans l'Antiquité: les principales stratégies d'exploitation. In: POUPART Laurent. *Marbres en Franche-Compté. Actes des Journées d'études, 10-12 Juin, Besançon,* Besançon. Asprodic. 2003, p. 21-34.

BISSO, José. *Crónica General de España, Crónica de la Provincia de Cádiz, Crónica General de España*, Madrid. 1868, 127 p.

BISSO, José. *Crónica General de España, Crónica de la Provincia de Málaga, Crónica General de España*, Madrid. 1869, 99 p.

BORGES, Nelson. Mós Manuais de Conimbriga. *Conimbriga (Coimbra)* XVII. 1978, p. 113-132.

BRENAN, Gerald. *South from Granada: A Sojourn in Southern Spain*. Kodansha International, London. 1957. Republished 1998, 282 p.

BRUN, Jean-Pierre. Um primeiro moinho hidraulico romano na Peninsula Ibèrica em Conimbriga. In: MOUTINHO ALARCAO, A. (ed.), *Portugal romano, exploracao dos Recursos Natura*. 1997a, p. 30-31.

BRUN, Jean-Pierre. L'introduction des moulins dans les huileries antiques. In: GARCIA D. and MEEKS. D. (eds), *Techniques et économie antiques et médievales. Le temps de l'innovation, Colloque international CNRS, Aix-en-Provence 21-23 mai 1996*. Paris. 1997b, p. 69-78.

BRUN, Jean-Pierre, BORRÉANI, Marc. Deux moulins hydrauliques du Haut-Empire romain en Narbonnaise: Villae des Mesclans à la Crau et de Saint-Pierre/Les Laurons aux Ares (Var), *Gallia: Archéologie de la France antique*, 55. 1998, p. 279-326.

BRUN, Jean-Pierre, Les moulins hydrauliques dans l'Antiquité. *Actes du Colloque: Archéologie des Moulins Hydrauliques, à Traction Animale et à Vent, des Origines à l'Époque Mé diévale, Lons-le-Saunier du 2 au 5 Novembre 2011*,in press.

BUCHSENSCHUTZ, Olivier, JACCOTTEY, Luc, JODRY, Florent, BLANCHARD, Jean-Luc (eds), *Actes de la Table Ronde, Evolution Typologique et Technique des Meules du Néolithique à l'an Mille sur le Territoire Français St. Julien-sur-Garonne, 2-4 Octobre, 2009, Aquitania*. 2011, 487 p.

BURÉS i VILASECA, Lurdes, MACÍAS, Josep María, RAMÓN, Ester. El jaciment iberic del Barranc del Prat (La Juncosa del Montmell). *Miscel•lània penedesenca*, 17. 1993, p. 114-136.

CABEZUELO, Ulysese. Le site de La Zac des "Meules" à Vic-le-Comte, (Puy-de-Dôme), In: BELMONT, Alain, MANGARTZ, Fritz, Les Meulières: Recherche, Protection et Valorisation d'un Patrimoine Industriel Européen. Mainz. 2006, p. 109-114.

CAMBESES, Aitor. *Characterization of the Volcanic Centres at Zeneta and La Aljorra, Murcia: Evidence of Minette Formation by Lamproite-Trachyte Magma Mixing*. Master thesis, University of Granada, Spain. 2010, 249 p.

CAMBESES, Aitor. Informe Petrográfico Sobre las Canteras de Molinos Cercas de Zagra, Granada. 2011, 4 p. Unpublished report.

CAPMANY y de MONTPALAU, Antonio. *Memorias Históricas Sobre la Marina Comercio y Artes de la Antigua Ciudad de Barcelona*. 1779, Madrid.

CARA BARRIONUEVO, Lorenzo, GARCÍA LÓPEZ, José Luis, LENTISCO PUCHE, José Domingo, ORTIZ SOLER, Domingo. *Los Molinos Hidráulicos Tradicionales de la Apujarra (Almería)*: Instituto de Estudios Almerienses, Almería.1999, 196 p.

CARCAVILLA, Luis, RUÍZ, Rafael, RODRÍGUEZ, Esaú. *Geomofología del Alto Tajo*, no. 4, Guía del Parque Natural del Alto Tajo. 2008, 270 p. http://es.scribd.com/doc/48886123/GUIA-P-N-ALTO-TAJO-I-carcavilla-total [accessed November 15, 2012].

CARRIÓN, Francisco, ALONSO, J. M., RULL, E., CASTILLA, J., CEPRIÁN, B., MARTÍNEZ, J. L., HARO, M., MANZANO, A. Los recursos abióticos y los sistemas de aprovisionamiento de rocas por las comunidades prehistóricas del S.E. de la Península Ibérica durante la prehistoria reciente. *Investigaciones Arqueológicas en Andalucía. 1985-1992*. Proyectos, Consejería de Cultura de la Junta de Andalucía, Sevilla. 1993, p. 295-309.

CASILLAS ANTÚNES, José Francisco. Los estudios toponomásticos en Extremadura: Estado de la cuestión. In: GORDÓN PERAL,María Dolores (coord.), *Toponomía de España: Estado Actual y Perspectivas de la Investigación, Patronimica Romanica* 24, Berlin 2010, p. 269-287

CASO AMADOR, Rafael. Molinos Hidráulicos Harineros en la Baja Extremadura a finales del Siglo XVIII, In: *Jornadas del Patrimonio de la Comarca de la Sierra, Federación de Asociasiones de la Sierra de Huelva*. 2008, p. 115-134. www.federacionsierra.es/media/documentos/doc238.pdf.

CASTAÑO FERNÁNDEZ, Santiago, LÓPEZ ROS, Joaquín, De la MORA, Julián. Itinerarios Geológicos de la Provincia de Albacete: Tobarra, Hellín, Minateda, Cancarix y La Celia, *Separatas de Al-Basit, Revista de Estudios Abacetenses*. 1985, p. 79-125. http://bibliotecahellin.blogspot.com.es/2011/04/itinerario-geologico-de-la-provincia-de.html#.Ud0B7OA27rk.

CASTEL, Carlos. *Descripcióno Física Geognóstica y Agrícola de la Provincia de Guadalajara*, Madrid. 1881, 270 p.

CASTELLA, Daniel. *Le moulin hydraulique gallo-romain d'Avenches «En Chaplix». Fouilles 1990-1991, Cahiers d'archéologie romande*, Lausanne. 1994, 156 p.

CASTELLA, Daniel, ANDERSON, Timothy. Les meules du Musée Romain d'Avenches. *Bulletin de l'association Pro Aventico* 46. 2004, p. 115-169.

CASTRO, Javier. Las Canteras Moleras de Euskal Herria, *Aranzadiana*. 2012, p. 138-141.

CHECA, Alicia, JIMENO, Alfredo, TRESSERRAS Jordi Joan, BENITO, Juan, SANZ, Alberto. Molienda y economía doméstica en Numancia. In: BURRILLO, Francisco (ed.), *IV Simposio sobre los Celtíberos, Economia, Homenaje a José Luis Argente Oliver, Daroca 1997*, Zaragoza. 1999, p. 63-68.

CISNEROS CUNCHILLOS, Miguel, LAPUENTE MERCADAL, Mª P.ilar, MAGALLÓN BOTAYA, Mª A., ORTIGA CASTILLO, M. Estudio Arqueológico-Geológico de Cerro Redondo (Pardos, Zaragoza), *TVRIASO* 6. 1985, p. 137-164.

CLEMENTSON LOPE, Julio. *Caracterización de las propiedades de la arenisca de Montoro, sus alteraciones y su aplicabilidad para la construcción*, Tesis Doctoral, Universidad de Córdoba. 2012, Servicio de Publicaciones. [http://helvia.uco.es/xmlui/handle/10396/7983].

CONSUEGRA RODRÍGUEZ, Susana, GALLEGO GARCÍA, Mª. Mar, CASTAÑEDA CLEMENTE, Nuria. Minería Neolítica de Sílex de Casa Montero, (Vicálvaro, Madrid), *Trabajos de Prehistoria,* 61, no. 2. 2004, p. 127-140. http://tp.revistas.csic.es/index.php/tp/article/viewFile/47/47.

CÓRDOBA de la LIAVE, Ricardo. Aceñas, Tahonas y Almazaras. Técnicas Industriales y Procesos Productivos del Sector Agroalimentario en la Córdoba del Siglo XV. *Hispania*. XLVIII/170.1988, p. 827-874.

CÓRDOBA de la LIAVE, Ricardo. Los Molinos Hidráulicos de la Cuenca del Guadalquivir a Fines de la Edad Media, Instrumental y Equipamiento Técnico, *Anuario de Estudios Medievales*, 33/1. 2003, p. 291-337.

CÓRDOBA de la LIAVE, Ricardo,VARELA ROMERO, Juan. *El Patrimonio Histórico Hidráulico de la Cuenca del Guadajoz: Estudio y Catalogación*, SALSVM 2, Monografías del Museo Histórico Municipal de Baena, Baena. 2011, 349 p.

CÓRDOBA DE LA LIAVE, Ricardo, *et al.* Documentación Técnica para Realizar el Proyecto de Musealización del Molino de Martos. Resumen. Informe interno. 2003, 34 p. Unpublished report.

COVARRUBIAS HORROZCO, Sebastián. *Tesoro de la Lengua Castellana, o Española*, Madrid. 1611, 1403 p.

CRAWFORD, Osbert Guy Stanhope. The quern-quarries of Mayen in the Eifel, *Antiquity* 29. 1955, p. 68-76

CRUZ y BAHAMONDE de la, Nicolás. *Viaje por España, Francia è Italia*.Vol. 13, Cádiz. 1813, 540 p.

CUEVAS RUIZ, Fernando. Breves notas sobre la Fabricación de Muela de Molino en la Montaña Palentina, *Al Socayo, revista palentina de cultura tradicional*, nº 5. 2006. http://branosera.com/piedras-de-molino.php.

CZYSZ, Wolfgang. *Steinhauer im Oberbayherishcne Inntal. Mittelalterliche Mühlsteinbrüche Zwischen Neubeuern, Oberaudorf und Brannenburg*. Bayerisches Landesamt für Denkmalpflege. Kempten. 2014, 128 p.

DAHLIN HAUKEN, Åsa, ANDERSON, Timothy J. *Collection Report: Rotary Querns in the Museum of Archaeology, University of Stavanger*, The Norwegian Millstone Landscape Project, Report 2014.002. 2014a, 161 p. http://www.ngu.no/upload/Publikasjoner/Rapporter/2014/2014_002.pdf

DAHLIN HAUKEN, Åsa, ANDERSON, Timothy J. Samlingsrapport: Manual til Registreringsskjema dreiekvern / Guide to the Rotary Quern Registration Form, The Norwegian Millstone Landscape, Report 2014.008. 2014b, 21 p. http://www.ngu.no/upload/Publikasjoner/Rapporter/2014/2014_008.pdf

DELGADO DOMÍNGUEZ, A., REGALADO ORTEGA, M.C. Rotae Urionensis las norias romanas de Rio Tinto (Huelva, España), *Ingeniería minera antigua y medieval*. 2012, p. 1-24. http://www.traianvs.net/pdfs/2011_riotinto-02.pdf.

DELGADO RAACK, Selina. *Prácticas económicas y gestión social de recursos técnicos (macro) líticos en la Prehistoria Reciente (III-I milenios AC) del Mediterráneo occidental*, Tesis Doctoral, Universitat Autònoma de Barcelona. 2008.

DÉSIRAT, Guy. Bagnols en Fôret, Var, Contribution à l'étude de la Provence orientale, 1, Millau. 1996, 192 p.

DIAGO HERNANDO, Máximo. Pastores, Carreteros y Arrieros. *Medievalismo,* No 13-14. 2004, p. 219-227.

de DIOS AYUDA, Juan. *Examen de las Aguas Medicinales de más Nombre que hay en las Andalucías*, Madrid. 1794, 348 p.

DOMÍNGUEZ-BELLA, Salvador, BERNAL CASASOLA, Darío. Fish-Based Subproducts in Late Antiquity. Archaeolmetric and Archaeological Evidence from the Fish Factories at Traducta (Algeciras, Cádiz, Spain). In: TURBANTI-MEMMI, I. *Proceedings of the 37th International Symposium of Archaeometry*, 12th - 16th May 2008, Siena, Italy, Berlin-Heidelberg. 2011, p. 453-458.

ESOJO AGUILAR, Francisco. Obras Hidráulicas en el Cauce Medio del Genil y su Aprovechamiento en la Agricultura e Industria. *Revista Singlis* 6. 2011, p. 12-21.

ESPAÑOL, Francesca. Las Manufacturas Arquitectónicas en Piedra de Girona Durante la Baja Edad Media y su Comercialización. *Anuario de Estudios Medievales (AEM)*, 39/2. 2009. p. 963-1001.

ESPEJO LARA, Juan Luis. La Arriería en Málaga en Época de los RR.CC. *Baetica. Estudios de Arte, Geografía e Historia*, 8. 1985, p. 281-300.

ESPINOSA, Manuel. Tera. In: de los RÍOS, Fernando. *Seminario Pintoresco Español, Lectura de las Familias, Enciclopedia Popular*, Madrid. 1848, p. 265-272.

EZQUERRA del BAYO, Joaquín. Ensayo de una Descripción General de la Estructura Geológica del Terreno de España en la Península, Sección V, In: *Memorias de la Real Academia de Ciencias de Madrid*, Vol. 4. 1856, p. 351-399.

FARGET, Virginie, FRONTEAU, Gilles. Les carrières de meules des Fossottes, La Salle (Vosges) In: BUCHSENSCHUTZ, Olivier, JACCOTTEY, Luc, JODRY, Florent, BLANCHARD, Jean-Luc (eds). *Actes de la Table Ronde, Evolution Typologique et Technique des Meules du Néolithique à l'an Mille sur le Territoire Français St. Julien-sur-Garonne, 2-4 Octobre 2009, Aquitania*. 2011, p. 137- 145.

FARÍAS,P., GALLASTEGUI, G., GONZÁLEZ LODEIRO, F., MARQUÍNEZ, J. MARTÍN-PARRA, L. M., MARTÍNEZ CATALÁN, J. R. de PABLO

MARCÍA, J. G., RODRÍGUEZ FERNÁNDEZ, L. R. Aportaciones al conocimiento de la litoestratigrafía y estructura de Galicia Central, *Memorias do Museo e Laboratorio Mineralogico e Geologia, Faculdade de Ciências da Unversidade do Porto* 1. 1987, p. 411-431.

FERNÁNDEZ LÓPEZ, Sebastián. El Molino Hidráulico Medieval en la Provincia de Málaga. *Acta historica et Archaeologica Mediaevalia* 3. 1982, p. 209-225.

FERNÁNDEZ OCHOA, Carmen, ZARZALEJOS PRIETO, Mar, CRISTÓBAL BURKHALTER, Thiébalt, HEVIA GÓMEZ, Patricia, ESTEBAN BORRAJO, Germán. *Arqueominería del Sector Central de Sierra Morena. Introducción al Estudio del Área Sisaponense*. ANEJOS de AEspA XXVI *Seminario Pintoresco Español, Lectura de las Familias, Enciclopedia Popular.* 2002, 128 p.

FERNÁNDEZ de los RÍOS, Ángel. *La Tierra: Descripción Geográfica y Pintoresca de las Cinco Partes del Mundo. Tomada del Libro de Victor Levasseur*, Madrid. 1849, 293 p.

FERRER, Narciso. *Reales Decretos y Órdenes de S. M. que producen resolución general en su Real Hacienda expedidos en el año de 1829.* Madrid, 1830. 443 p.

FORRER, Robert. Ein versunkener pätantiker Mühlsteintransport in Wanzenau bei Strassburg, *Anzeiger für Elsässische Altertümer* 7/8.1911, p. 131-143.

FRANKEL, Rafi. *Wine and Oil Production in Antiquity in Israel and other Mediterranean Countries*. Sheffield: Sheffield Academic Press. 1999, 230 p.

FRÖLICH, Jiri, WALDHAUSER, Jiri. Prispevkey K Ekonomice Ceskyych Keltu (Kamenictvi A Ditribuce Zernvu, *Archeologické Rozhledy XLI*, Praha 1989, p. 16-58.

GÄHWILER, Adolf, SPECK, Josef. Die Römische Wassermühle von Hagendorn bei Cham ZG: Versuch einer Rekonstruktion, *Helvetia Archaeologica*, 86. 1991, p. 33-75.

GARCÍA BLÁNQUEZ, L. A., LÓPEZ MARTÍNEZ, J. D., MARTÍNEZ SÁNCHEZ, C., RUÍZ PARRA, I. La Prospection Arqueológica de las Fincas Casa Grande y Buena Vista (General Electric) de la Aljorra, Cartagena, Murcia. *Memorias de Arqueología*. 1990, p. 628-639.

GARCÍA de la LEÑA, Cecilio. *Conversaciones Historicas Malagueñas ó Materiales para Noticias Seguras Para Formar la Historia Civil, Natural y Eclesiástica de la M. I. Ciudad de Málaga*, Málaga.1789, 279 p.

GARCÍA ROMERO, José. El Papel de la Minería y la Metalurgia en la Córdoba Romana, Universidad de Córdoba, Tesis Doctoral. 2002. http://helvia.uco.es/xmlui/bitstream/handle/10396/244/13078938.pdf?sequence=1 [accessed December 13, 2012].

GARCÍA SÁNCHEZ, Expiración. La Alimentación en la Andalucía Islámica. Estudio Histórico y Bromatológico: Cereales y Leguminosas. Añejos de Cuadernos de Historia del Islam, Granada. In: BOSCH VILA, J., HOENERBACH, W. *Andalucía Islámica, Textos y Estudios*. 1983, p. 139-177.

GARCÍA VALCÁREL, Reyes, ÉCIJA, Ana María, VALCÁREL MORENO, Soledad (Documadrid). *La Cuenca Alta de la Sierra de Manzanares*, Consejería de Educación y Cultura, Comunidad de Madrid, Madrid. 1998, 161 p.

GARRIDO PALACIOS, Manuel. Julio Caro Baroja y los molinos de Puebla de Guzman. 2001. *Revista de Folklore,* 245. 2001, p. 164-167. http://www.funjdiaz.net/folklore/07ficha.cfm?id=1945 [accessed March 10, 2011].

GASCÓN, Antonio. Ferrocarril Secundario de Zafra a Burguillos, Badajoz, *Revista de Obras Públicas*, Tomo 105. 1904, p. 378-381.

GENÍS i ARMADA, Maria Teresa. Cap a una tipologia dels molins d'època Ibèrica a Ullastret. *Faventia*, 8/2. 1986, p. 99-113.

GILES PACHECO, Francisco, VALIENTE CÁNOVAS, Santiago, GILES GUZMÁN, Francisco, GUTIÉRREZ LÓPEZ, José Mª. Avance al Estudio de una Explotación Minera Postpaleolítica de Ofitas (Doleritas) para Elementos de Molienda y Pulimentos en Peña Arpada (Alcalá de los Gazules, Cádiz), *Almoraima* 42. 2011, p. 117-141.

GLICK, Thomas, F. Hydraulic Technology in Al-Andalus. In: KADRA JAYYUISI, Salma (ed.). *The Legacy of Muslim Spain*. Leiden. 1994, p. 974-986.

GÓMEZ RUIZ, Ricardo. *Molinos en el Río Odiel. Un Estudio de Arqueología Industrial en los Límites de El Andévalo.* 2003, 138 p.

GÓMEZ ZOTANO, José. Aproximación Geomorfológica al Karst de la Sierra de La Utrera (Provincia de Málaga), *Baetica. Estudios de Arte, Geografía e Historia*, 25. 2003, p. 137-158.

GONZÁLEZ ALCANTALUD, José Antonio. *El Clientelismo Político: Perspectiva Socioantropológica,* Barcelona. 1997, 209 p.

GONZÁLEZ PERALBO, José Luis. Los Primeros Maestros Canteros que Intervienen. Blog of the Department of Social Science, Geography and History of the Secondary School IES "Antonio M" Calero" of Pozoblanco (Córdoba). http://depgeografiaehistoria.blogspot.com/2008/02/virgen-de-luna.html [accessed February 13, 2012].

GRANDAL LÓPEZ, Alfonso. Introducción Histórica y Lingüística a la Toponimia de Cartagena y su Campo. *Revista Murciana de Antropología*, Nº 11. 2004, p. 251-259.

GREEN, Chris. Hertfordshire Puddingstone Querns - Working with a Difficult Rock. In: WILLIAMS, David, PEACOCK, David (eds), *Bread for the People: The Archaeology of Mills and Milling, Proceedings of a Colloquium held in the British School at Rome, 4th-7th November, 2009*, University of Southampton Series in Archaeology, no. 3, Archaeopress, Oxford. 2011, p. 121- 130.

GRENNE, Tor, HELDAL, Tom, MEYER, Gurli B.,BLOXAM, Elizabeth. From Hyllestad to Selbu: Norwegian Millstone Quarrying through 2000 Years. In SLAGSTAD, T. (ed.), *Geology for Society, Geological Survey of Norway Special Publication*, 11, Trondheim. 2008, p. 47-66.

GUÉRIN, Pierre. *El Castellet de Bernabé, y el Horizonte Ibérico Pleno Edetano*. Servicio de Investigación Prehistórica, no. 101, Valencia. 2003, 389 p.

GUICHARD, Pierre. *Al-Andalus frente a la Conquista Cristiana*. Valencia. 2001, 781 p.

GUTIÉRREZ GARCÍA-MORENO, Anna. *Roman Quarries in the Northeast of Hispania (Modern Catalonia)*. Tarragona. 2009, 316 p.

GUTIÉRREZ LLORET, Sonia, *La Cora de Tudmir, de la Antigüedad Fardía al Mundo Islámico: Poblamiento y Cultura Material. Colección de la Casa Velázquez* 5. 1996, Madrid. 476 p.

GUTIÉRREZ LLORET, Sonia, MORET, Pierre, ROUILLARD, Pierre, SILLIÉRES, Pierre, ECHALLIER, Jean-Claude. Le Peuplement du Bas Segura de la Protohistoire au Moyen Âge (Prospections 1989-1990), *LVCENTUM*, XVII-XVIII. 1998-1999, p. 25-74.

GUTIÉRREZ SOLER, Luis María, CORPAS IGLESIAS, Francisco Antonio. Archaeometric Research in the Upper Guadalquivir in Relation to Roman Mining in the Eastern Sierra Morena Mountaines and its Precedent in the Iberian Culture. *ITÁLICA* 1, *Journal of Andalusian Classical Archaeology*. 2011, p. 13-28.

HABA QUIRÓS, Salvadora. *Medellín Romano, La Colonia Medellinensis y su Territorio. Badajoz*. 1998, 515 p.

HARB, Pierre. Roder/Klein Büel, *Archäologie und Denkmalpflege im Kanton Solothurn*, 7, Solothurn. 2002, p. 68-72.

HARMS, Eduard, MANGARTZ, Fritz. *Vom Magma zum Mühlstein, Eine Zeitreise durch die Lavaströme des Bellerberg-Vulkans (Vulkanpark-Forschungen* 5), Mainz. 2002, 107 p.

HARO NAVARRO, Martín, CARRIÓN MÉNDEZ, Franciso, GARCÍA GONZÁLEZ, David. Territorio y Georecursos en el Cabo de Gata (Níjar, Almería) Durante la Edad de Cobre. In: MARTÍNEZ FERNÁNDEZ, G., MORGADO RODRÍGUEZ, A., AFONSO MARRERO, J.A. (eds), *Sociedades Prehistóricas, Recursos Abióticos y Territorio, Actas de la III Reunión de Trabajo, Loja, 2004*. 2006, p. 315-326.

HERNANDO LUNA, Rafael. Anotaciones sobre Antiguas Obras Hidráulicas de la Provincia de Córdoba, *Brac - Boletín de la Real Academia de Córdoba*, 117. 1989, p. 273-277.

HIDALGO ESTÉVEZ, Mª C., REY ARRANS, J., DUEÑAS MOLINA, J. Visita al Área Minera de Linares - Hidrogeología del Sector. In: LÓPEZ GETA, J.A., RUBIO CAMPOS, J.C. (eds), *Presente y Futuro de las Aguas Subterráneas en la Provincia de Jaén*, Madrid. 2002, p. 309-317.

HOCKENSMITH, Charles. *The Millstone Industry: A Summary of Research on Quarries and Producers in the United States, Europe and Elsewhere*. Jefferson. 2009, 269 p.

HÖRTER, Fridolin. *Getreidereiben und Mühlsteine aus der Eifel. Ein Beitrag zur Steinbruch- und Mühlengeschichte*, Mayen. 1994, 192 p.

HURBIN, Werner. *Le pain romain, la mouture, la cuisson, recettes.* Auguster Bläter zur Römerzeit 4, Augst. 1982. 28 p.

IBÁÑEZ MARTÍNEZ, Pedro Miguel. Van Den Wyngaerde, Una Vista de Belmonte y la Campaña de Trabajo de 1563. *Archivo español de arte*, LXXVI.2003, 301, p. 71-77.

IMAMUDDIN, S. M. *Muslim Spain: 711-1492 A.D. : A Sociological Study*, Leiden. 1981, 273 p.

JACCOTTEY, Luc. Seven Thousand Years of Millstone Production in the Serre Mountain Range of the French Jura. In: WILLIAMS, David, PEACOCK, David (eds), *Bread for the People: The Archaeology of Mills and Milling, Proceedings of a Colloquium held in the British School at Rome, 4th-7th November, 2009*, University of Southampton Series in Archaeology, no. 3, Archaeopress, Oxford. 2011, p. 293-307.

JACCOTTEY, Luc, ANDERSON, Timothy, JODRY, Florent. La circulation du matériel de mouture de part et d'autre de l'Arc jurassien durant l'Antiquité. In: BÉLET-GONDA, C., MAZIMANN, J.-P., RICHARD, A., SCHIFFERDECKER, F. (eds), *Premièrs Journées Archéologiques Frontalières de l'Arc Jurassien actes, Delle (F) - Boncourt (CH), 21-22 octobre 2005*, Besançon-Porrentruy. 2007, p. 223-228.

JACCOTTEY, Luc, FARGET, Virginie, FRONTEAU Gilles, MAERTEN M., BEUCHOT, S. Le site du Portus à Collonge-en-Charollais (Saône-et-Loire), exemple d'une chaîne opératoire de fabrication de moulins rotatifs antiques. In: BUCHSENSCHUTZ, Olivier, JACCOTTEY, Luc, JODRY, Florent, BLANCHARD, Jean-Luc (eds). *Actes de la Table Ronde, Evolution Typologique et Technique des Meules du Néolithique à l'an Mille sur le Territoire Français St. Julien-sur-Garonne, 2-4 Octobre 2009, Aquitania*. 2011, p. 163-197.

JACCOTTEY, Luc, LABEAUNE, R. Découvert d'un moulin hydraulique Antique à Longvic, (Côte-d'Or), *Revue Archéologique de l'Est*, 59. 2010, p. 665-668.

JACCOTTEY, Luc, LONGEPIERRE, Samuel, *et al.* Pompeian Millstones in France. In: WILLIAMS, David, PEACOCK, David (eds), *Bread for the People: The Archaeology of Mills and Milling, Proceedings of a Colloquium held in the British School at Rome, 4th-7th November, 2009*, University of Southampton Series in Archaeology, no. 3, Archaeopress, Oxford. 2011, p. 97-116.

JAGAILLOUX, Serge. État de Santé de Meuliers, Acccidents du Travail, Affections Professionnelles, In: Agapain, *Les Meuliers: Meules et Pierres Meulièrs*, Étrépilly. 2002, p. 151-164.

JIMÉNEZ ÁVILA, Javier. Cancho Roano y los complejos monumentales post-orientalisantes del Guadiana, *Complutum* 8, 1997, p. 141-159.

JIMÉNEZ PUERTAS, Miguel. *El poblamiento del Territorio de Loja en la edad media.* Granada. 2002. 374 p.

JIMÉNEZ MARTIN, Alfonso. *La Catedral Gótica de Sevilla: Fundación y Fábrica de la Obra Nueva*, Sevilla. 2006, 433 p.

JOBEY, George. Millstones and Millstone Quarries in Northumberland, *Archaeologia Aeliana*, Series 5, Vol. 14. 1986, p. 49-80.

JODRY, Florent. Le Lexique. In: BUCHSENSCHUTZ, Olivier, JACCOTTEY, Luc, JODRY, Florent, BLANCHARD, Jean-Luc (eds). *Actes de la Table Ronde, Evolution Typologique et Technique des Meules du Néolithique à l'an Mille sur le Territoire Français St. Julien-sur-Garonne, 2-4 Octobre 2009, Aquitania*. 2011a, p. 19-33.

JODRY, Florent. First Century Querns of the Roman Army in the Light of Modern Texts, In: WILLIAMS, David, PEACOCK, David (eds), *Bread for the People: The Archaeology of Mills and Milling, Proceedings of a Colloquium held in the British School at Rome, 4th-7th November, 2009*, University of Southampton Series in Archaeology, no. 3, Archaeopress, Oxford. 2011b, p. 85-91.

JONCHERAY, Jean Pierre, SÉNAC, Philippe. Une Nouvelle Épave Sarrasine du Haut Moyen Age. *Archéologie Islamique*, 5. 1995, p. 25-34.

JOOS, Marcel. Eine permische Brekzie aus dem Südschwarzwald und ihre Verbreitung als Mühlstein im Spätlatène und in frührömischer Zeit, *Archäologisches Korrespondenzblatt 5*. 1975, p.197-199.

JORDAN MONTES, Juan. Las Salinas de Madax en Cancarix (Hellín, Albacete): Apuntes de Campo y Aspectos Históricos, Etnográficos, Arqueológicos y Geográficos. *Al-Basit: Revista de Estudios Albacetenses*, 41. 1997, p. 7-32.

JORDAN MONTES, Juan, MATILLA SEIQUER, G. Poblamiento Rural Tardoantiguo y Monasterios Visigodos en el Curso Bajo del Río Mundo (Hellín y Tobarra, Provincia de Albacete). In: NOGUERA, J. M. (ed.), *Poblamiento Rural Romano en el Sureste de Hispania: Actas de las Jornadas, Nov. 8-11, Jumilla 1993*. 1995, p. 323-338.

JORDAN MONTES, Juan, RAMALLO ASENSIO, Sebastián, SELVA INIESTA, Antonio. El Poblamiento Romano en el Valle de Minateda-Agramón, In: SANZ GAMO, Rubi (ed.). *Congreso de Historia de Albacete, I, Arqueología y Prehistoria,8-11 de Diciembrede 1983*, Albacete. 1984, p. 211-255.

LAGARDÈRE, Vincent, Moulins d'Occident Musulman au Moyen Âge (IX au XVe siècles): al-Andalus, *Al-Qantara*, 12:1. 1991, p. 59-118.

LEDUC, Mathieu. Les pistrina volubilitains, témoins majeurs du dynamisme économique municipal. In: GONZÁLEZ, Julián, RUGGERI, Paola, VISMARA, Cinzia, ZUCCA, Raimondo (eds), *L'africa romana Le ricchezze dell'Africa. Risorse, produzioni, scambi.Atti del XVII convegno di studio, Sevilla, 14-17 dicembre 2006*. Vol. 1, Sassari. 2008, p. 475-505.

LEDUC, Mathieu. L'artisanat au coeur de la ville: l'exemple des pistrina de Volubilis. In: FONTAINE, Souen, SATRE, Stéphanie, TEKKI, Amel (eds). *La ville au quotidien. Regards croisés sur l'habitat et l'artisanat antiques. Afrique du Nord, Gaule et Italie,* Actes du colloque international Maison Méditerranéenne des Sciences de l'Homme Aix-en-Provence, 23-24 novembre 2007. Aix-en-Provence. 2011, p. 181-190.

LEHMKUHL, Achim. Late La Tène Rotary Querns found in the Oppidum of Heidengraben: a New Type of Volcanic Rock and its Origin. In: WILLIAMS, David, PEACOCK, David (eds), *Bread for the People: The Archaeology of Mills and Milling, Proceedings of a Colloquium held in the British School at Rome, 4th-7th November, 2009*, University of Southampton Series in Archaeology, no. 3, Archaeopress, Oxford. 2011, p. 81-84.

LEON CÁCERES, Manuel, CARMONA BENÍTEZ, María Carmen. Las Villae Romanas de Quintana de la Serena. 2006. http://www.quintanadelaserena.org/villaes_romanas.pdf.

LERALTA, Javier. *Madrid: Cuentos, Leyendas y Anécdotas*, Vol. 1, Madrid. 2001, 331 p.

LEVEAU, Phillipe. Les Moulins de Barbegal (1986-2006), *TRAIANVS*. 2006, p. 1-21. http://www.traianvs.net/index.php.

LIZARRALDE ELBERDIN, Koldo. *Molino de Aldea de Ebro Cantabria. Madoz y su Diccionario-Geográfico-Estadístio-Histórico.* 2010, 449 p. Electronic book. http://murcia.mu/~molinosa/wp-content/uploads/2012/05/MADOZ-KOLDO.pdf.

LLORENTE PINTO, José Manuel. El Valor Indicador de los Topónimos: El Caso de la Voz Berrocal y sus Variantes. *Boletín de la Asociación de Geógrafos Españoles,*56. 2011, p. 59-77.

LONGEPIERRE, Samuel. *Les meules à grains et les meulières dans le Sud-Est de la Francedu IV siècle avant J.-C. au XII siècle après J.-C.* Thèse Universitaire, Université Aix-Marseille I. 2011. 361 p.

LÓPEZ, Juan Antonio, MARTINEZ ROBLES, Francisco. *Memorias Sobre los Montes de Toledo*. Madrid. 1821, 55 p.

LÓPEZ ANDRÉS, José, MARTÍN-CARO SAURO, Faustino. Organización, Distribución y Problemas Derivados de la Administración del Agua en Almería y su Vega en los Años Anteriores a la Reconquista. *Instituto de Estudios Almerienses Departamento de Historia, I Coloquio de Historia y Medio Físico.* 1989, p. 1019-1032.

LÓPEZ CORDERO, Juan Antonio, CABRERA ESPINOSA, Manuel. Patrimonio Histórico-Cultural de Arbuniel. *Sumuntán No.* 20, *Revista Estudios sobre Sierra Mágina*. 2004, p. 185-218. http://www.pegalajar.org/articulos/0_patrimonio_arbunielx.htm [accessed May 11, 2012].

LÓPEZ CORDERO, Juan Antonio, GONZÁLEZ CANO, Jorge. *Tipologías de Vivenda Rural en Sierra Mágina*. Torredonjimeno. 2005, 94 p. http://www.magina.org/biblioteca/publicaciones/tipologiavivienda.pdf.

LÓPEZ MARCOS, Antonio. Algunas observaciones Económico-Territoriales del Yacimiento Ibérico de Cerro Castillejo (Jódar, Jaén). *Sumuntán*, No. 5, *Revista de Estudios sobre Sierra Mágin*a. 1995, p. 29-35.

LOZANO MUÑOZ, Francisco. *Crónica de la Provincia de Jaén, Crónica General de España, Historia Descriptiva de sus Provincias*, Madrid. 1867, 82 p.

LUQUET, Armand. Blé et meunerie à Volubilis, *Bulletin d'Archéologie Marocaine*, 6. 1966, p. 301-316.

MADOZ, Pascual. *Diccionario Geográfico-Estadístico-Histórico de España y sus Posesiones de Ultramar*, Madrid, Vol. 1-16. 1845-1850.

MADRAZO, Santos, FONTANA, Josep. *El sistema de transportes en España*, 1750 – 1850, Vol. 1, Madrid. 1984, 996 p.

MAGARIÑO RUBIO, Salvador, SÁNCHEZ VELA, Rafael J. Bucear en Tarifa. Guía de la Mejores Inmersiones, *Almoraima* 37. 2008, p. 363-391.

MAESTRO HERNÁNDEZ, Juan. *La Elaboración de Muelas de Molino en la Montaña Palentina, Colección de Historia Montaña Palentina* 5. 2011, 201 p.

MALALANA UREÑA, Antonio. La Evolución de los Recintos Urbanos Amurallados Castellano-Leoneses a lo largo del Siglo XII, *Arqueología y Territorio Medieval* 16. 2009, p. 75-136.

MARQUIEGUI SOLOAGA, Ágata. Los Molinos Hidráulicos en Ibi (Alicante): Catalogación e Historia. *8º Congreso Internacional de Molinología, 28-30 de Abril de 2012, Tui (Pontevedra)*, submitted. http://pdf.depontevedra.es/ga/148/pbxUhiJrar.pdf.

MARTÍ SOLANO, Josefa, RODRÍGUEZ MARISCAL, Nuria, Problemática y Situación Actual de la Arqueología Subacuática en la Bahía de Cádiz. *Monte Buciero 9, La Conservacíon del Material Arqueológico Subacuático*. 2003, p. 398-416.

MARTÍN ALGARRA, Agustín, (ed.). Zonas Internas Béticas. In: VERA, J.A. (ed.), *Geología de España, SGE-IGME*. Madrid. 2004, p. 395-437.

MARTÍN GALINDO, José Luis. Los Chozos Extremeños: Referente Histórico y Recurso Socio-Cultural Para el Futuro, *Revista de estudios extremeños*, Vol. 62, Nº 2. 2006, p. 839-890. http://www.dip-badajoz.es/cultura/ceex/reex_digital/reex_LXII/2006/T.%20 LXII%20n.%202%202006%20mayo-ag/RV000832.pdf [accessed July 14, 2013].

MARTÍN PRIETO, Pablo. Aportación al Estudio del Molino Hidráulico en la Castilla Medieval: Los Molinos del Monasterio de Santa Clara de Alcocer, *HISPANIA. Revista Española de Historia*, Vol. LXVI, núm. 224. 2006, p. 833-850.

MARTÍNEZ y DELGADO, Francisco. *Historia de la Ciudad de Medina Sidonia*, Cádiz. 1875, 436 p.

MARTÍNEZ LÓPEZ, Francisco, ANDERSON, Timothy, GRANERO GALLEGOS, Antonio. The Quern and Millstone Quarry of the Rambla Honda, Almería, Spain. In: WILLIAMS, David, PEACOCK, David (eds), *Bread for the People: The Archaeology of Mills and Milling, Proceedings of a Colloquium held in the British School at Rome, 4th-7th November, 2009*, University of Southampton Series in Archaeology, no. 3, Archaeopress, Oxford. 2011, p. 277-283.

MARTÍNEZ LÓPEZ, Francisco, GRANERO GALLEGOS, Antonio. Cantera de piedras de moler. In: *Actas del IV Congreso Internacional de Molinología, Mallorca, 1- 3 de mayo, 2003*, Vol. 2, Palma de Mallorca. 2005, p. 125-142.

MARTÍNEZ MARTÍNEZ, María. Molinos Hidráulico en Murcia (S. III- VI), *Miscelánea Medieval Murciana XIV*. 1987-88, p. 219-250.

MARTÍNEZ PRADES, José Antonio. *Los Canteros Medievales*. 1998, Madrid, 48 p.

MARTÍNEZ TORRES, Luis M. The Typology of Ancient Quarries within the Paleocene Limestone of Álava in Northern Spain, *Geoarchaeology* 24 (1). 2008, p. 42-58.

MARZO y SÁNCHEZ, Ildefonso. *Historia de Málaga y su Provincia*, Tomo II, Málaga. 1851, 511 p.

MATILLA SÉIQUER, Gonzalo. Canteras de Piedras de Almazara en Fortuna. *Revista de Antropología*, 7. 2001, p. 267-277.

MEIJIDE PARDO, Antonio. El empresarismo industrial del negociante francés J. Lestache en el Ferrol del siglo XVIII. *Anuario Brigantino*, 20. 1997, p. 35-50.

MENA CABEZAS, Juan Eugenio. La Ribera de los Molinos. Referencias Históricas y etnográfiaca para Futuras Actuaciones. In: LORENZANA de la PUENTE, Felipe, MATEOS ASCACIBAR, Franciso (ed.). *Actas de la II Jornada de Historia de Llerena*, Llerena. 2001, p. 245-260.

MERIDETH, Craig. La Mina El Cerro de San Cristobal: a Bronze Age Tin Mine (Extremadura, Spain). *Papers from the Institute of Archaeology (PIA) University College London* 9. 1998, p. 57-69.

MIÑANO y BEDOYA, Sebastián. *Diccionario Geográfico y Estadístico de España y Portugal*. 11 volumes. 1826-1829, Madrid.

MOLINA, Eduardo, CULTRONE, Giuseppe. Litodiversidad de los Elementos Arqueológicos de Molino de la Zona Septentrional Granadina. *MACLA, Revista de la Sociedad Española de Mineralogía*,16. 2012, p. 38-39.

MONTERO BASTARRECHE, Ángel. Origen de la Piedras de Moler de Córdoba. In: GARRIDO ARANDA, J.M., MORENO PÉREZ, M. F. ROLDÁN CAÑA (eds), J. *Actas del 6º Congreso Internacional de Molinología*. Córdoba. 2008, p. 161-174.

MORITZ, Ludwig A. *Grain-Mills and Flour in Classical Antiquity*, New York. 1958. Republished 1979, 230 p.

MOTOS GUIRAO, Encarnación. *Un Poblado Medieval de Andalucía Oriental: El Yacimiento del "Cerro del Castillón" (Montefrío, Granada). Los Materiales de Excavación. Esttudio y Conclusiones.* Tesis Doctoral, Universidad de Granada. 1987. 913 p.

MUÑIZ JAÉN, Ignacio, LÓPEZ FLORES Inmaculada, LÓPEZ LÓPEZ, Gabriel. Nueva necrópolis romana de incineración en Almedinilla y redescubrimiento de una tumba hipogeo en la conocida necrópolis romana de "El Ruedo". *ANTIQVITAS*, 24. 2012, p. 163-185.

ORDOÑÉZ VERGARA, Pilar. Los Molineros de la Alpujarra. *Gazeta de Antropología* 10, no. 13. 1993. http://www.ugr.es/~pwlac/G10_13Pilar_Ordonez_Vergara.html [accessed March 10, 2013].

OTTE, Enrique. *Sevilla y sus Mercaderes a Fines de la Edad Media*, Sevilla. 1996, 339 p.

PADILLA MONJE, Aurelio. Consideraciones en Torno a la Explotación del Mármol en la Bética Durante los Siglos I-II. *Habis*, 30. 1999, p. 271-281.

PALOMO PALOMO, Juan, FERNÁNDEZ, María Pilar. Los Molinos Hidráulicos en la Antigüedad, *Espacio, Tiempo y Forma, Serie II, Historia Antigua, UNED*, 19-20. 2007. p. 499-524.

PASCUAL MAYORAL, Pilar, GARCÍA RUÍZ, Pedro. Canteras y Tecnología molinar en el Río Jubera (La Rioja)", *Revista Murciana de Antropología* 7, Murcia. 2001, p. 237-266.

PASCUAL MAYORAL, Pilar, GARCÍA RUÍZ, Pedro. Los Castillares de Villa Vieja y los Orígenes de la Villa de Jubera (La Rioja), *Kakathos* 20-21, Teruel. 2001-2002, p. 261-282.

PASCUAL MAYORAL, Pilar, GARCÍA RUÍZ, Pedro. Nuevas Canteras de Piedras de Molino y Trujal, Valle del Cidacos (Arnedillo, La Rioja), *Kalakorikos* VII, Calahorra (La Rioja). 2002, p. 209-216.

PASCUAL MAYORAL, Pilar, GARCÍA RUÍZ, Pedro. Nuevo Hallazgo de Canteras de Piedras de Molino, Río Iregua, *El Serradero* 76, Abril, Torrecilla en Cameros (La Rioja). 2003a, p. 16-18.

PASCUAL MAYORAL, Pilar, GARCÍA RUÍZ, Pedro. Canteras de Molino y Trujal. Cuenca del Río Linares: Muro de Aguas, Villarroya y Grávalos", *Graccurris* 14, La Rioja. 2003b, p. 221-241.

PASCUAL MAYORAL, Pilar, GARCÍA RUÍZ, Pedro. Las Canteras de Piedra de Molino: una Industria Riojana Desconocida", *Altza* VII, San Sebastián. 2003c, p. 135-146.

PASCUAL MAYORAL, Pilar, GARCÍA RUÍZ, Pedro. Nuevo Hallazgo de Canteras de Piedras de Molino, Río Iregua, *El Serradero* 76, La Rioja. 2003d, p. 16-18.

PASCUAL MAYORAL, Pilar, GARCÍA RUÍZ, Pedro. Quern and Millstone Quarries in the North of Spain. In: WILLIAMS, David, PEACOCK, David (eds), *Bread for the People: The Archaeology of Mills and Milling, Proceedings of a Colloquium held in the British School at Rome, 4th-7th November, 2009*, University of Southampton Series in Archaeology, no. 3, Archaeopress, Oxford. 2011, p. 285-292.

PASCUAL MAYORAL, Pilar, ANDERSON, Timothy, GARCÍA RUÍZ, Pedro. Un Molinillo de Biela – Manivela en Cubo de la Sierra (Soria). *La Voz de Trébago*. 2010, p. 9-12.

PASCUAL MAYORAL, Pilar, GARCÍA RUÍZ, Pedro, CASTRO, Javier. Canteras de Piedras de Molinos y Moleros en Navarra. CUADERNOS de Etnología y Etnografía de Navarra, Enero - Diciembre, AÑO XLIII - No 86. 2011, p. 225-255.

PAVÓN MALDONADO, Basilio. *Guadalajar Medievel: Arte y Arqueología Árabe y Mudejar*. Madrid. 1984, 518 p.

PEACOCK, David. The Roman Millstone Trade: a Petrological Sketch, *World Archaeology*, Vol. 12.1. 1980, p. 43-53.

PEACOCK, David. The Production of Roman Millstones near Orvieto, Umbria, Italy, *The Antiquaries Journal*, 66. 1. 1986, p. 45-51.

PEACOCK, David. Iron Age and Roman Quern Production at Lodsworth, West Sussex, *The Antiquaries Journal* 67.1. 1987, p. 61-85.

PEACOCK, David. The Mills of Pompeii, *Antiquity*, 63. 1989, p. 205-214.

PEACOCK, David, CUTLER, Lyn. The Earliest Rotary Querns in Southern England. In: WILLIAMS, David, PEACOCK, David (eds), *Bread for the People: The Archaeology of Mills and Milling, Proceedings of a Colloquium held in the British School at Rome, 4th-7th November, 2009*, University of Southampton Series in Archaeology, no. 3, Archaeopress, Oxford. 2011, p. 77-80.

PEACOCK, David, WILLIAMS, David,. Pompeian style mills in Jordan and Syria. . In: SELSING, Lotte (ed.). *Seen Through a Millstone*, AmS-Skrifter 24, University of Stavanger. 2014, p. 97-108.

PENCO VALENZUELA, Fernando, MORENO ALMENARA, Maudilio, GUTIERREZ DEZA, Mª Isabel. Dos Canteras Romanas en Colonia Patricia Corduba. Peñatejada y Santa Ana de la Albaida, *Anales de Arqueología Cordobesa* 15, p. 229-248. 2004. http://helvia.uco.es/xmlui/handle/10396/3598.

PEÑA CERVANTES, Yolanda. *La Producción de Vino y Aceite en España*, Institut Català d'Arqueologia Clàssica, Documenta 14. 2010, 1048 p.

PEREJIL DEREY, Antonio. El Ferrocarril del Guadalquivir-Aynalcólla-Camas. Digital publication. http://issuu.com/camasdigital/docs/el_ferrocarril_del_guadalquivir. 2005, 43 p.

PÉREZ MÍNGUEZ, Rafael. *Aspectos del Mundo Rural Romano en el Territorio Comprendido entre los Ríos Turia y Plancia*. Serie de Investigación Prehistórica 106, Valencia. 2006, 137 p.

PIOT, Auguste. *Traité Historique et Pratique sur la Meulerie et la Meunerie,* Montelimar,1860, 368 p.

PONCE CORDONES, Francisco. *Speculum Rotae: Antología de Estampas Roteñas, "La Marina Roteña del Novecientos",* Rota. 1981, 209 p.

PONSICH, Michel. *Implantation rurale antique sur le Bas-Guadalquivir*, 4 Volumes., Paris. 1974. 1979; Madrid. 1987. 1991.

PORTILLO RAMÍREZ, M. *La Mòlta i Triturat d'Aliments Vegetals Durant la Protohistòria a la Catalunya Oriental*. Tesis Doctoral, Universidat de Barcelona. 2006. 521 p. + annexes. http://diposit.ub.edu/dspace/handle/2445/42627.

POWERS, James F. (Translation and introduction). *The Code of Cuenca. Municipal Law on the Twelfth-Century Castillian Border.* Philadelphia. 2000, 243 p.

PUCHE RIART, Octavio, MAZADIEGO, Luis.Feliope. Explotaciones Históricas de Calizas en la Comunidad de Madrid, Colmenar de Oreja. *Actas IV Sesión Científica de la SEDPGYM: Patrimonio Geológico y Minero. Bélmez, 28-30 Octubre 1999*. Tomo I. 1999, p. 91-104.

PUERTAS, Rafael. *Excavaciones Arqueológicas en Lacipo (Casares, Málaga): Campañas de 1975 y 1976. Excavaciones Arqueológicas en España*, Vol. 125, Ministerio de Cultura, Dirección General de Bellas Artes y Archivos, Subdirección General de Arqueología y Etnografía, Madrid. 1982.

PY, Michel. Meules d'époque protohistorique et romaine provenant de Lattes, Recherches sur l'économie vivrière des Lattarenses, *Lattara* 5, Lattes. 1992, p. 183-232.

QUESADA SANZ, Fernando, KAVANAGH de PRADO, Eduardo, MORALEJO ORDAZ, J. El Asentamiento de Época Ibérica en el Cerro de la Cruz. In: MUÑIZ JAÉN, Ignacio, QUESADA SANZ, Fernando. (eds), *Oikos, 2, Un Drama en Tres Actos, Dos Milenios de Ocupación Humana en el Cerro de la Cruz, Almedinilla, Córdoba, Cuadernos Monográficos del Ecomuseo del Río Caicena*. 2010, p. 75-95.

QUESADA SANZ, Fernando, KAVANAGH, Eduardo, LANZ, Mercedes. Los molinos del yacimiento del Cerro de la Cruz (Almedinilla, Córdoba): clasificación y análisis de los ejemplares de época ibérica y emiral, *SPAL, Revista de Prehistoria y Arqueología de la Universidad de Sevilla*. 2014, p. 83-118.

de la RADA y DELGADO, Juan de Dios. *Crónica de la Provincia de Granada, Crónica General de España, Historia Descriptiva de sus Provincias*, Madrid. 1869, 192 p.

RADLEY, Jeffrey. A Millstone Maker's Smithy on Gardom's Edge, Barlow, *Derbyshire Archaeological Society*, 84. 1964, p. 123-127.

RAMALLO ASENSIO, Sebastián. *Canteras Romanas de Carthago Nova y Alrededores (Hispania Citerior)*, Universidad de Murcia, Servicio de Publicaciones, Murcia. 1987, 163 p.

RAMOS FOLQUES, Alejandro. Mapa Arqueológico del Termino Municipal de Elche (Alicante), Consejo Superior de Investigaciones Científicas, Instituto de Arqueología y Prehistoria "Rodrigo Caro", Madrid. 1953, p. 323-354.

RAMOS MILLÁN, Antonio, RIESCO GARCÍA, J. C. La Villa Romana de Las Canteras (Chipiona, Cádiz): Procesos Formativos y Transformativos del Registro Arrqueológico e Inferencias Preliminares del Asentamiento. *Cuadernos de Prehistoria y Arqueología de la Universidad de Granada*, 8. 1983, p. 375-416.

REYES MESA, José Miguel. *Los Molinos de la Ciudad de Granada*, Granada. 2000a, 154 p.

REYES MESA, José Miguel. Tecnología y Arquitectura Popular. Los Molinos Hidráulicos en la Provincia de Granada, *Gazeta de Antropología*, Nº 16, Artículo 21. 2000b, p. 1-13. http://www.ugr.es/~pwlac/G16_21JoseMiguel_Reyes_Mesa.html [accessed June 8, 2008].

REYES MESA, José Miguel. *Los Molinos Hidráulicos Harineros de la Provincia de Granada*. 2006, 249 p.

RISCH, Rorberto. *Recursos Naturales y Sistemas de Producción en el Sudeste de la Península Ibérica entre 3000 y 1000 ANE*. Tesis Doctoral, Universidad Autónoma de Barcelona. 1995, 610 p. http://www.tdx.cat/handle/10803/5524.

RODRÍGUEZ MONTEOLIVA, Faustino. Los molinos de Harina en la Alpujarra de Granada, Durante los Diglos XVI al XVIII. Léxico, Etnografía e Historia. In: *Coloquio de Historia y Medio Físico, I, Instituto de Estudios Almerienses, Departamento de Historia*. 1989, p. 683-709.

RODRÍGUEZ RODRÍGUEZ, Amelia, CRUZ, Valentín Barroso. Labrar la piedra para moler grano: La explotación prehistórica de las canteras de molinos de toba en la isla de Gran Canaria. *El Pajar: Cuaderno de Etnografía Canaria*, Nº. 10. 2001, p. 4-9.

ROJAS SOLA, José Ignacio, AMEZCUA-OGÁYAR, Juan Manuel. Origen y Expansión de los Molinos de Viento en España. *Revista de Ciencia y Tecnología de América*, Vol. 30, Nº. 6. 2005, p. 316-325.

ROMÁN LOZANO, Andrés. La piedra o losa de Tarifa. *ALJARANDA* 65. 2007, p. 18-24.

RÖDER, Josef. Die antiken Tuffsteinbrüche des Pellenz, Zur Steinbruchgeschichte des Pellenz- und Brohltaltuffs, *Bonner Jahrbücher*, 157. 1957, p. 214-271.

RUNNELS, Curtis. *A Diachronic Study and Economic Analysis of MIlstones from the Argolid, Greece*. Doctoral thesis, University of Indiana, Bloomington. 1981, 402 p.

SAINZ ALBERO, Mª Inés. Etnografía Histórica de Mendavia.Panadería, Molino y Tiendas. *Cuadernos de Etnología y Etnografía de Navarra*, no. 82. 2007, p. 163-199.

SÁNCHEZ LUQUE, María. Consideraciones sobre la imagen urbana de Coín en los siglos XVII Y XVIII, *Isla de Arriarán, Revista cultural y científica* 23-24. 2004, p. 103-126.

SÁNCHEZ NAVARRO, Joaquin. Estudi de les pedres de molins manuals i de les seves zones d'extracció a Menorca, Treballs de la Secció d'Estudis, *Publicacions des Born 10*, Ciutadella. 2001, p. 49-179.

SÁNCHEZ NAVARRO, Joaquin. Estudi de les pedres de molins manuals i des les zones d'extracció a Menorca, In: BARCELO, M. and RETAMERO, F. (eds), *Els barrancs trancats. L'ordre pagès al sud de Menorca en època andalusina (segles X-XIII)*, Barcelona. 2005, p. 235-267.

SÁNCHEZ NAVARRO, Joaquin. Estat actual de investigacions sobre les pedres de molins manuals i de les seves zones d'extracció a Menorca, In: *I jornades de recerca històrica de Menorca. La Manurqa de Sa'id Ibn Hakam, Un País Islàmic a Occident, Publicacions des Born 15-16, Ciutadella*. 2006, p. 154-196.

SÁNCHEZ NAVARRO, Joaquin. Les Meulières de l'Ile de Minorque. In: WILLIAMS, David, PEACOCK, David (eds), *Bread for the People: The Archaeology of Mills and Milling, Proceedings of a Colloquium held in the British School at Rome, 4th-7th November, 2009*, University of Southampton Series in Archaeology, no. 3, Archaeopress, Oxford. 2011, p. 193-204.

SANCHO i PLANAS, Marta. Agua, Rocas y Metales: Arqueología y Explotación de Recursos Minerales en la Edad Media. *Acta Historica et Archaeologica Mediaevalia*, 29. 2008, p. 519-545.

SANZ ELORZA, Mario. Toponomástica Molinológica de la Provincia de Burgos. *Revista de Folklore*, Tomo 32, no. 361. 2012, p. 4-33. http://www.funjdiaz.net/folklore/07ficha.cfm?id=2699.

SCARROW, Jane, CAMBESES, Aitor, ANDERSON, Timothy. Roman Volcanic Querns and Mills from the Region of Murcia, *Tilting at Mills, The 5th International Colloquium on the Archaeology and Geology of Mills and Millstones, Almería, 5-8 March, 2014*, in preparation.

SERRANO RUANO, Delfina. Minas en colecciones de fetuas y casos jurídicos del Occidente islámico (ss.XII-XVId.C.): el problema de la propiedad de los yacimientos mineros, *Espacio,Tiempo y Forma*, Serie III, *H.a Medieval*, t.23. 2010, p. 185-203.

SILLIÈRES, Pierre, FINCKER, Myria, LABARTHE, Jean-Michel. *Baelo Claudia: Una ciudad romana de la Bética*. 1997, 237 p.

SOTO, Juan Ignacio. Petrographical Report of a Conglomerate Sample from the from the Caniles Millstone Quarry. 2010, Unpublished report.

SOTO JIMÉNEZ, Luis. Descubrimiento de la Cilniana en Estepona. *Jábega* 17. 1977, p. 59-65.

SWINBURNE, Henry. *Travels Through Spain in the Years 1775-1776*, Vol. 1, London. 1787, 373 p.

TIANA FERRER, Alejandro. Educación obligatoria, asistencia escolar y trabajo infantil en España en el primer tercio del siglo XX, *Historia de la educación*. 1987, Vol. 6., p. 43-59. http://gredos.usal.es/jspui/handle/10366/79434.

TORRES ABRIL, Francisco Luis, GARCÍA DÍAZ, Margarita, GÓMEZ ARROQUIA, María Isabel, MARISCAL RIVERA, Domingo. Aprovisionamiento de Materias Primas Líticas en el Territorio del Campo de Gibraltar Durante la Prehistoria, *Almoraima* 29. 2003, p. 59-69.

TORRES MONTES, Francisco. Los Antiguos Molinos de Viento en el Campo de Níjar. Estudio Etnográfico-Lingüístico. In: ESCOBEDO RODRÍGUEZ, Antonio (ed.), *Homenaje a la Profesora Elena Pezzi*. 1993, p. 265-279.

TUCKER, Gordon T. Millstone Making in the Peak District of Derbyshire: The Quarries and Technology. *Industrial Archaeology Review*, Vol. 8, 1. 1985, p. 42-59.

TURRIANO, Juanele. *Veintiun libros de los ingenios y máquinas de Juanelo Turriano* - Transcripcion al español - Tomo III by LAÍN ENTRALGO, Pedro. Madrid. 1996. http://issuu.com/juaneloturriano/docs/21librosingeniosymaquinastomo_iii [accessed October 11, 2013].

UTANDA MORENO, Luisa. La Evolución Histórica de la Industria en la Comarca de Las Vegas (1750-1950). In: *Historia, clima y paisaje: estudios Geográficos en memoria del profesor Antonio López Gómez*, Valencia. 2004, p. 233-252.

VALLEJO, José Mariano. *Tratado sobre el Movimiento y Aplicaciones de las Aguas*, Vol. 2, Madrid. 1833, 533 p.

VARGAS GARCÍA, Enrique. BERNAL CASASOLA, Darío. Roma y la Producción de Garum y Salsamenta en la Costa Meridional de Hispania. Estado actual de la Investigación. In: BERNAL CASASOLA, D. (ed.), *Arqueología de la Pesca en el Estrecho de Gibraltar, De la Prehistoria al Fin del Mundo Antiguo*, Cádiz. 2009, p. 133-181.

VERA, Juan Antonio. (ed.). *Geología de España. Edited by the Sociedad Geológica de España (SGE)/Instituto Geológico y Minero de España (IGME)*. Facultad de Ciencias, Universidad de Salamanca. 2004, 885 p.

VERDEJO PAEZ, Franciso, *Guía Práctico de Agrimensores y Labradores*, Madrid. 1822, 248 p.

VIDAL TERUEL, Nuria de la O. *El Modelo de Implantación en el Extremo Occidental de la Baetica. Doctrina y Praxis en la Occupación del Territorio Onubense*. Tesis Doctoral, Universidad de Huelva. 2001. http://rabida.uhu.es/dspace/handle/10272/2630.

VILLAREAL de Berriz, Pedro. *Máquinas Hydraulicas de Molinos, y Herrerías*, Madrid. 1739, 168 p.

VIÑAS MEY, Carmelo, PAZ, Ramón (ed.). *Relaciones histórico-geográfico-estadísticas de los pueblos de España hechas por iniciativa de Felipe II: Reino de Toledo. Madrid. 1951* Available at the website of the University of Castilla-La Mancha. http://www.uclm.es/ceclm/b_virtual/libros/Relaciones_Toledo/index.htm.

VIVAR, Gustavo. El transporte de redistribución en el nordeste de la Península Ibérica en época tardorepublicana: el ejemplo de los Yacimientos de Illa Pedrosa y Cala Galladera,. In: GALLINA ZEVI, Anna TURCHETTI, Rita (eds), *Méditerranée Occidentale Antique: Le Échanges, III, Seminario, Marseille, 14-15 Mai, 2004.* 2004, 229 p.

WALTON, Geoffrey. *Secure and Sustainable Final Slopes for SME Agrregate Qarries, A Handbook.* 2004, http://www.gwp.uk.com/secureslopesdl.html [accessed November 12, 2012].

WHITE, Donald. Survey of Millstones from Morgantina. *American Journal of Archaeology* 67. 1963, p. 199-206.

WILLIAMS, David, PEACOCK, David (eds), *Bread for the People: The Archaeology of Mills and Milling, Proceedings of a Colloquium held in the British School at Rome, 4th-7th November, 2009*, University of Southampton Series in Archaeology, no. 3, Archaeopress, Oxford. 2011a, 359 p.

WILLIAMS, David, PEACOCK, David. Pompeian Style Mills in Britain. In: WILLIAMS, David, PEACOCK, David (eds). *Bread for the People: The Archaeology of Mills and Milling, Proceedings held in the British School at Rome, Nov. 4-7, 2009, University of Southampton Series in Archaeology, no. 3, Archaeopress*, Oxford. 2011b, p. 117- 122.

WILLIAMS-THORPE, Olwen. Provenancing and Archaeology of Roman Millstones from the Mediterranean Area, *Journal of Archaeological Science*, 15.3. 1988, p. 253-305.

WILLIAMS-THORPE, Olwen, THORPE, Richard. Els Origens Geologics dels Molins Romans de Pedra del Nord-Est de Catalunya. *Vitrina*. 1987, p. 50-58.

WILLIAMS-THORPE, Olwen, THORPE, Richard. Millstone Provenancing used in Tracing the Route of a Fourth-Century BC Greek Merchant Ship. *Archaeomertry* 32. 1990, p. 115-137.

XIMÉNEZ de GÚZMAN, Antonio. *Historia de Coín*, Coín. Edición de Manuscrito Original de 1796. Compilación, Adaptación y Realisación Digital de José Manuel GARCÍA AGÜERA, Coín 2002. cited in: SÁNCHEZ LUQUE, María. Consideraciones sobre la Imagen Urbana de Coín en los Siglos XVII y XVIII, *Isla de Arriarán*, XXIII-XXIV. 2004, p. 103-126.

Divers dictionaries, historical sources, catalogues, databases

Archimedes Project. Database of Machine Drawings. [http://dmd.mpiwg-berlin.mpg.de/author/dmd/database/].
A
El Catastro del Marqués de Ensenada en el Antiguo Reino de Granada, Instrumentos de Descripción, Archivos 16, Consejería de Cultura, Archivo Histórico Provincial de Granada. Edited by JIMÉNEZ SERRANO, Bárbara, SERRANO GUTIÉRREZ, Juan. Compact Disk.

COMISION RÉGIA DE ESPAÑA. *Exposición Universal de 1867, Catálogo General de la Sección Española* . Paris.1867, 505 p.

Diccionario de la Lengua Española, 20th Edition, Real Academia Española [http://lema.rae.es/drae/].

Diccionario Geográfico Universal, por una Sociedad de Literatos, S.B.M.F.C.L.D., Tomo I-X, Barcelona, José Torner. 1831-1834. [http://books.google.es/books].

Encyclopedia Metódica, Geografía Moderna, translated by ARRIBAS y SORIA and de VELASCO, Julian, Tomo 1, Madrid. 1792.

Gazeta de Madrid. España. Ministerio de la Gobernación 1829 (Boletín Oficial del Estado), [http://books.google.es/books?id=jJofhorWCnEC&pg=PA347&dq=%22saca+piedras+de+moler%22&hl=fr&sa=X&ei=Tvy5UL3rL9D3sgawm4CoCw&ved=0CDAQ6AEwAQ#v=onepage&q=%22saca%20piedras%20de%20moler%22&f=false].

Instituto Geológico y Minero de España (Geological Survey of Spain) [http://www.igme.es].

Millstone quarries.e.u.atlas - Meulières.eu, Atlas des carrières de meules de moulins en Europe / Millstonequarries.eu European millstone quarries: a database. [http://meuliere.ish-lyon.cnrs.fr/.

Relaciones Topográficas for Guadalajara [http://www.uclm.es/ceclm/b_virtual/libros/relaciones_gu/index.htm].

Relaciones Topográficas de los Pueblos de España de Felipe II, 1574-1578.

Respuestas Generales del Catastro del Marqués de la Ensenada (Zenón de Somodevilla y Bengoechea), 1750-1754 [http://pares.mcu.es/Catastro/servlets/ServletController?accion=2&opcion=10].

Revista de Obras Públicas, Boletín, Canteras Españolas, Caliza de Colmenar, Vol. 12. 1896, p. 145-147.

Captions and credits of illustrations and photographs from other sources

Despite our efforts, I did not always succeed in identifying the possessors of rights of certain illustration. Therefore, I kindly ask those concerned to address me via timanderson.granada@gmail.com where corrections or licenses are required.

Chap. 1:

Fig. 1.4: Photograph from c. 1915 of the quarry of Montjuïc (Barcelona) from the blog of Ana Ma Ferrin, http://amf2010blog.blogspot.com.es/2012/03/gaudi-y-la-piedra-un-magico-acuerdo.html) [accessed March 10, 2013].

Fig. 1.5: Montjuïc quarry: http://barcelonaimatgesambhistoria.blogspot.com.es/2012_01_01_archive.html [accessed March 9, 2013]

Fig. 1.6: Photograph dating to March 1933 of millstone makers (moleros) *at the quarry of Brañosera in the mountains of Palencia (from Cuevas Ruiz 2006).*

Fig. 1.8: Reconstruction of a Moorish donkey-driven olive oil mill dating to the 15th century, Nigüelas, Granada. Photograph by Eve Andersson.

Fig. 1.9: Rotary sharpening stone (1808-1812). Painting by Francisco de Goya. Displayed in the Budapest Museum of Fine Arts.

Fig. 1.11: Abandoned granite quern extraction or property boundary near Ávila (Ávila) (from the blog of M. Serna Martínez; http://terraeantiqvae.com/profiles/blogs/el-medievo-a-las-puertas-de-vila#.UV0wl45FfuE).

Chap. 2:

Fig. 2.1: Cover pages of the geographical and historical dictionaries known as "the Madoz" and "the Miñano": a) Diccionario Geográfico-Estadístico-Histórico de España y sus Posesiones de Ultramar, 16 vol., 1845-1850; b) Diccionario Geográfico y Estadístico de España y Portugal, 11 vol., 1826-1829.

Fig. 2.2: Extract and detail of the Padul (Granada) entry to the Census of the Marqués de Ensenada. El Catastro del Marqués de Ensenada en el Antiguo Reino de Granada, Instrumentos de Descripción, Archivos 16, A.H.P.G (in CD format).

Chap. 3:

Fig. 3.2: Examples of saddle querns of various lithologies. a) Late Neolithic calcareous sandstone quern from the site of Los Pensadores (Granada), excavation T. Anderson, unpublished; b) Chalcolithic garnet mica schist from Las Rajas (Granada) excavation T. Anderson, unpublished; c) Early Bronze Age garnet mica schist querns from Fuente Álamo (Murcia); d) Iron Age sandstone and conglomerate querns from Numantia (Soria); e) granite querns from the Early Iron Age site of Cancho Roano (Badajoz) (photographs a-b by author; photograph c from http://www.elargar.com/caracterizacion/Artefactos/Instrumentos/; photograph e from http://extremosdelduero.blogspot.com.es/2011/12/edificio-protohistorico-de-la-mata.html).

Fig. 3.4: Examples of rotary handquerns attributed to the Iron Age. a-b) Upper stone (ø: 32 cm) and lower stone (ø: 34 cm) from Ventorros de San José (Granada), possibly a pair; b) upper stone with lugs with vertical slits (Alcalá la Real, Jaén); upper stone (ø: 38 cm) with inverted "keyhole" handle cuttings (Jódar, Jaén); and e) top and side view of an upper stone with lug and inverted "keyhole" cutting (Cerro de la Cruz, Almedinilla, Córdoba). Petrography: a-c) porous limestone; d) probably conglomerate; e) granite. I thank E. Kavanagh for the photograph of the Cerro de la Cruz quern.

Fig. 3.5: 1) Roman quern upper stone (ø: 39 cm) (photograph by the Museo Arqueológico de Linares, Jaén; 2) Example of a "sombrero" type lower stone (ø: 36 cm), Museo Municipal de Úbeda. Both are of volcanic rock.

Fig. 3.9: Rotary querns from the 1981 excavation of the medieval settlement of El Castillejo, Montefrío (Granada) dating from the 9th to the 12th century (Motos Guirao 1987: 480). a) Upper stone with handle hole and lower rynd cutting, ø c. 38 cm; b) upper stone with no rynd cutting; c) lower stone; and d) upper stone with handle and rynd cutting of a very small hand quern (photographs by T. Anderson; the sections from the original drawings from Motos Guirao 1987, plates I-III).

Fig. 3.11: Schema of the Contemporary (20th century) handmill from Cubo de la Sierra (Soria) (drawing by T. Anderson, from Pascual et al. *2010, 10, fig. 5).*

Fig. 3.14: Iron Age "travertine" mill in situ *at the Cerro de la Cruz (photo a from http://www.rutasconhistoria.es/loc/poblado-ibero-el-cerro-de-la-cruz).*

Fig. 3.16: Volcanic ring catillus *from the Roman city of* Castulo *(Jaén). The type of rock of the lower stone (not the original pair) has not been determined. ø: 64 cm (photograph, Domus, Linares museum, inv. no. CE01430_R).*

Fig. 3.18: Photograph c, Domus http://www.juntadeandalucia.es/culturaydeporte/WEBDomus/fichaCompleta.do?acron=MAEGR&musid=9&ninv=CE14119&volver=portal&k=CE14119&tipoBusqueda=simple.

Fig. 3.20: Representations of Pompeian millstones on Roman lamps. a-b) Bilbilis, *Zaragoza (from Amaré 1988, p. 83-84, fig. 152-153; c) Murcia (from Amante 1988, p. 227, pl. IV, 72).*

Fig. 3.21: Pompeian mill from Mazarrón (Murcia). The lower stone and podium are reconstructed (photograph Museo Arqueológico de Murcia).

Fig. 3.23: Extract from the Codex of pseudo Juanele Turriano (approx. 1595) depicting a tahona *(animal-driven grain mill) (from Archimedes Project, http://dmd.mpiwg-berlin.mpg.de/author/dmd/database/).*

*Fig. 3.24: View of part of the animal-driven flour mill (*tahona*) in the castle of Calatrava la Nueva (Aldea del Rey, Ciudad Real) (photograph left http://patrindustrialquitectonico.blogspot.com.es/2011/08/un-molino-del-siglo-xii-es-las-nueva.html; photograph right by T. Anderson).*

Fig. 3.27-1: Engraving from 1842 entitled Molinos Árabes de la Mina en Alcalá de Guadaira by Genaro Pérez de Villaamil. The scene is the mill house of a subterranean watermill under the town of Alcalá de Guadaira (Seville) (from http://sevillalegendaria.blogspot.com.es/2013/05/los-canos-de-carmona-agua-e-historia.html).

Fig. 3.28: Extract from the Codex of pseudo Juanele Turriano (approx. 1595) depicting an aceña *(vertical wheel grain mill) (from Archimedes Project, http://dmd.mpiwg-berlin.mpg.de/author/dmd/database/).*

Fig. 3.29: Extract from an engraving by the Flemish artist Van den Wyngaerde (1567) with a view of the city of Córdoba. The arrows point to the waterwheels of the aceñas *in the Guadalquivir River (original illustration in the Victoria and Albert Museum of London).*

Fig. 3.30: Extract from the Codex of pseudo Juanele Turriano (approx. 1595) depicting a rodezno *(horizontal wheel grain mill) (from Archimedes Project, http://dmd.mpiwg-berlin.mpg.de/author/dmd/database/).*

Fig. 3.31: Drawings from the Catastro del Marqués de Ensenada (1750-1754) of the towns of Nivar and Viznar (Province of Granada) with the location of the rodezno *watermills. These drawings supplement the answers of a questionnaire and illustrate the principal features of the towns such as mills, churches, houses, surrounding farms, irrigation ditches (*acequias*) and landmarks (crosses). El Catastro del Marqués de Ensenada en el Antiguo Reino de Granada, Instrumentos de Descripción, Archivos 16, A.H.P.G. (in CD format).*

Fig. 3.33: Drawing of the windmills of Belmonte (Cuenca) by Van de Wyngaerde (1563). The work is in the Victoria and Albert Museum, London (from Ibáñez 2003: 75).

Fig. 3.34: Photograph by Jean Dableda of the town of Torrevieja, Alicante in the late 19th century (photograph from http://franciscorebollo.blogspot.com.es/2011_09_01_archive.html).

Fig. 3.35: Examples of windmills (Cartagena, Murcia and Cabo de Gata, Almería) in southeastern Spain driven by vanes with triangular sails (photograph left by Torres Ros, http://www.forocartagena.com/t800-molinos-de-viento-y-arquitectura-del-campo-de-cartagena; photograph right Cabo de Gata, Nijar, Almería by T. Anderson).

Chap. 4:

Fig. 4.1. Map of the principal geological zones of Iberia (after Farias et al. 1987). A black and white version of this map appeared in Anderson & Scarrow 2011: 261.

Fig. 4.2: Detailed map of the principal geological formations of the Bétic range (after Martín Algarra 2004). A black and white version of this map appeared in Anderson & Scarrow 2011: 261.

Fig. 4.5: Detail of the SE Spain volcanic district with the location of the reputed Roman quern and millstone quarries (adapted from Cambeses 2010: 8, fig. 1.3).

Chap. 5

Fig. 5.5: Photograph dating to 1933 of moleros *(millstone makers) at the millstone quarry of Brañosera (Palencia) in northern Spain (photograph from Cuevas Ruiz 2006).*

Fig. 5.6: Examples of tools of direct percussion. Left: Engraving from Antiquity of a quarryman with his pick, Kruft, Kr. Mayen (Germany) (photograph from Röder 1957, Plate 21, 1). Right: Scene of a quarryman cutting or shaping a block on a capital of a column of the western entrance to the church of San Miguel de Biota (Zaragoza), 12th century (photograph by Ray Escámez Rivero, flicker, http://www.flickr.com/photos/adfinem/6849827592/).

Fig. 5.14: Examples of vertical extractions at the site of El Campillo Viejo (HU-1) (photographs by A. García Veiga).

Fig. 5.23: Schema of the technique of splitting a quern cylinder directly with a pick, avoiding the use of wedges (from Anderson et al. 2003: 49, fig. 44, drawing by A. Pulido).

Fig. 5.24: Examples of tool marks on the quarry floor of the quern quarry of Châbles, Fribourg (Switzerland) (from Anderson et al. *2003: 51, fig. 52). We thank Jean-Claude Bessac for pointing out this significant detail.*

Fig. 5.26: Reconstruction of the technique of detaching an angular block from the quarry floor for millstone production (drawing by T. Anderson based on a photograph from http://burgess-shale.rom.on.ca/en/transcripts/slideshow_1998.html).

Fig. 5.33: Example of a "chaîne opératoire" *of quern fashioning from angular blocks based on observations at the quarry site of Portus (Collonge-en-Charollais, Saône-et-Loire (from Jaccottey* et al. *2011: 194, fig. 39).*

Fig. 5.38: Cylindrical roughout from the true extractive quarry of Châbles, Fribourg, Switzerland. The pecks (no. 3) on the upper surface are from single pointed chisel strikes, whereas the thin vertical lines on the edge (no. 4) are multiple contiguous chisel strikes. The "paunch" (bourrelet) indicating work on one half at a time is visible to the bottom left of each representation. The larger diagonal lines, no. 1, are the original extraction pick marks and no. 2 are the splitting cavities (from Anderson et al. *2003: 55, fig. 61, drawing by T. Anderson).*

Fig. 5.39: Example of fashioning a sandstone (grès coquillier) *rotary quern with hammer and chisel at the site of Châbles, Fribourg, Switzerland. One half was fashioned before turning the cylinder around to fashion the second half, resulting in a "paunch" along the base (from Anderson* et al. *2003: 54, fig. 60, drawing by A. Pulido).*

Fig. 5.46: Schema of the two types of composite millstones proposed in the pseudo Juanele Turriano Codex. The schema to the left with five stones is more solid, according to the author, than the model to the right (from pseudo Juanelo Turriano - 1996,Vol. III: 351, fig. 204 1996. http://issuu.com/juaneloturriano/docs/21librosingeniosymaquinastomo_iii.

Chap. 6:

Fig. 6.18: Photograph of Miraflores de la Sierra (M-5) by Fernando Colmenarviejo.

Fig. 6.19: Photographs of El Campillo Viejo a and b by Alonso García Veiga;

Chap. 8:

Fig. 8.1: Plan of the Gallo-Roman site of Châbles-Les Saux dating to the 1st-2nd century AD (Canton of Fribourg, Switzerland). The a) millstone quarry and the b) smithy are circled. The study of this excavation revealed a tight link between the smithy and the quarry (from Anderson et al. *2003: 34, fig. 23). The other noteworthy features are the c) road and d) house interpreted as the residence of the quarryman.*

*Fig. 8.9: Plan of part of the Gallo-Roman site of Châbles-Les Saux dating to the 1st-2nd century AD (Canton of Fribourg, Switzerland). The rectangular house (*bâtiment est/east house*), built with wood, is between the road and the quarry. This modest structure is interpreted as the residence of the quarryman (from Anderson* et al. *2003: 196, fig. 241, drawing by T. Anderson).*

Fig. 8.13: Drawing from the pseudo Juanelo Turriano codex (c.1595) of a quarry cart driven by mules (from Archimedes Project, http://dmd.mpiwg-berlin.mpg.de/author/dmd/database/).

Fig. 8.14: Modern flatbed quarry cart from Extremadura (from Martín Galindo 2006: 849).

Fig. 8.15: Reconstruction of an oxen-drawn cart for the transport of millstones and other heavy production, based on information from the region of Navarra (from Pascual et al. 2011: 250, fig. 25, drawing by J. Castro).

Fig. 8.16: Examples of recent "V" shaped rastra *sledges designed to transport heavy loads (photo a from http://cuentoquenoescuento.blogspot.com.es/2012_02_01_archive.html; photo b from http://espanolinternacional.blogspot.com.es/2008/03/el-transporte.html).*

Fig. 8.17: Example of a narria *(from Pascual Mayoral et al. 2011: 249, fig. 23; drawing by Javier Castro).*

Fig. 8.18: Reconstruction of a scene of moving a millstone by means of the "rolling stone" technique (from Gómez Ruíz 2003: 85, fig. 7).

Fig. 8.19: Example of a rutted road cut through a conglomerate embankment. This road is near the quern and millstone quarry of the Puerto de la Cadena (MU-1) (from http://senderosdecartagena.wordpress.com/2010/12/16/cabezo-del-puerto-de-la-cadena/).

Fig. 8.20: Scene at the loading platform of the train station of Gerena (Seville), where granite blocks were loaded onto wagons. Oil rollers and millstones probably also benefitted from this means of transport (photograph from http://lafactoria-cuencaminera.blogspot.com.es/2010_12_01_archive.html).

Fig. 8.21: Huelva train station towards the end of the 19th century with five millstones in transit. Behind the upright example (to the left), are three others. The stone to the right, under the sitting official, appears to have a fracture along its lower rim. The absence of iron girth bands reduces the likelihood that they were composite French Burr imports. The Huelva-Zafra rail line was built between 1867 and 1900. The station is by the port, as seen by the boats in the background to the right (photograph from http://lafactoria-cuencaminera.blogspot.com.es/2011/06/casas-espanolas-al-borde-de-un.html).

Chap. 9:

Fig. 9.1: Photograph dating to 1933 of moleros *(millstone makers) at the quarry of Brañosera (Palencia) in northern Spain (photograph from Cuevas Ruiz 2006).*

Fig. 9.2: Photographs of the working crews of the quarries of Montjuïc (Barcelona) and Gerena, Seville (SE-7). (Montjuïc from http://amf2010blog.blogspot.com.es/2012/03/gaudi-y-la-piedra-un-magico-acuerdo.html); photograph of Gerena from Acuña Carabantes 2004.

Fig. 9.3: Scene of workers from the 1920s from a granite quarry (building blocks or paving stones) in the area of Gerena, Seville, a district known to have produced millstones. In the background, to the left, is what appears to be a child (from Acuña Carabantes 2004).

Chap. 10:

Fig. 10.5: View of the alignment of 255 millstones in the Fuentelárbol (Quintana Redonda, Soria) (photograph by P. Pascual and P. García).

Chap. 11:

Fig. 11.4: View of taking the measurements of the extraction hollows of Rambla Honda (AL-3a) (photograph by. F. Martínez).

Fig. 11.5: Line graph of the diameters of 151 quern and millstone hollows measured in cm (from Martínez et al. *2011). Four diameter groups stand out at this site: a) from 40 to 50; b) from 80 to 95; c) from 105 to 120; d) a final group between 130 and 140 is more modest.*

Fig. 11.10: Cerrp de la Cruz: Photograph left: http://www.rutasconhistoria.es/loc/poblado-ibero-el-cerro-de-la-cruz

Fig. 11.19: Example of a small unfinished quern extraction at Ibi (A-2). The quern is difficult to date. The proportion of its diameter to its width suggests a Roman or Iron Age date (photograph by Ágata Marquiegui).

Fig. 11.23: Quern extraction in the heart of an abandoned millstone extraction at the quarry of Puerto de la Cadena (MU-1) (from : http://senderosdecartagena.wordpress.com/2010/12/16/cabezo-del-puerto-de-la-cadena/).

Fig. 11.24: View of a tubular face at the quarry of Montesa (V-2) (photograph by J. García Cerdá).

Chap. 12:

Fig. 12.5: Quern and millstone from Jódar, Jaén. The quern to the left is a conglomerate (ø: 40 cm); the stone to the right is a granite (ø: 52 cm) (photograph by Ildefonso Alcalá, Museum of Jódar).

Fig 12.15: Volcanic Pompeian hourglass catilli from the Museo de Bellas Artes de Córdoba and from the Museo Arqueológico de Murcia (photograph by the Museo de Murcia).

Fig 12.16: Pompeian hourglass catilli *from the Museum of Mellilla (photograph by Museo de Melilla).*

*Fig 12.18: Example of a cylindrical ring-*catillus *from the Roman city of* Castulo *(Jaén), ø: 64 cm (photograph Linares museum, inv. no. CE01430_R).*

Chap. 13:

Fig. 13.1: Extract of a sermon of the Dominican Friar Domingo Baltanás from Seville.

Digital sources (historical and hiking itinerary websites)

ANDALUSIA

Province of Granada

GR-2/3: Loja Ordinance: Pregón 1502, enero, 10. Loja, Pregón de la ordenanzas de Loja sobre no sacar piedras de molino. AML, Leg. 49, p. 9. 1 cuartilla + 1: http://www.reocities.com/tdcastros/Historyserver/Fuentes/DocMunic/DOCLOJA.6.htm [accessed February 9, 2013].

GR-7: Padul Municipal web site: Cantera de Piedras de Molino: http://www.adurcal.com/enlaces/cultura/zona/historia/padul/cantemolin/index.htm [accessed November 7, 2012].

GR-8: Los Guájares Municipal website "Ruta Avices, Pie Moro y Minchal" http://www.losguajares.es/turismorural.html [accessed in 2008, no longer available].

GR- 14: GAN QUESADA, Rafael, "La Comarca de Ugíjar: Tierra de Contrastes": http://sp.ideal.es/municipios/rutas-com.php?id_ruta=4&id_comarca=3&pageNum_rutas=8 [accessed February 2, 2012].

Province of Jaén

J-2: Jimena, El Lanchar, Hiking itinerary "Los Caracoles": http://www.turjaen.com/dondeiryquever/senderismofamiliar/sierramagina/ruta-caracoles.php [accessed November 9, 2012].

Province of Córdoba

CO-7: BORJA, Jose Manuel, Canteras de Santa Albaida: http://www.eltiempo.es/fotos/en-provincia-cordoba/canteras-de-santa-ana-de-la-albaida.html [accessed November 1, 2012].

CO-7: MADRIGAL, Francisco. Canteras de Santa Ana de la Albaida, Blog: http://pacomadrigal-cordoba.blogspot.com.es/2012/07/canteras-de-santa-ana-de-la-albaida.html [accessed November 1, 2012].

CO-8: Vertice photos: http://www.flickr.com/photos/vertice1/2561950036/in/photostream/ [accessed November 12, 2012].

CO-10: GONZÁLEZ PERALBO, José Luis. Blogs of the Department of Social Science, Geography and History of the Secondary School IES "Antonio M" Calero" of Pozoblanco (Córdoba). Los Primeros Maestros Canteros que Intervienen: http://depgeografiaehistoria.blogspot.com/2008/02/virgen-de-luna.html [accessed November 2, 2012].

CO-10: ELVIAR, Manu: http://www.flickr.com/photos/23969797@N04/3777759408/in/photostream [accessed November 13, 2012].

CO-10: Website of the 30 Grados Sport Club: http://www.30grados.es/Web/fotosbelmez.html [accessed November 13, 2012].

CO-14: del PINO CUTILLAS, María Teresa, *et al*. *Carta Arqueológica de Montoro*. http://www.juntadeandalucia.es/cultura/publico/BBCC/Carta_arqueologica_Montoro.pdf [accessed June 6, 2013].

CO-15: Almedinilla: General information and photograph of the site: http://www.rutasconhistoria.es/loc/poblado-ibero-el-cerro-de-la-cruz).

Province of Cádiz

CA-2a: Photograph of the quarry from blog of Manuel L.: http://www.rutasyfotos.com/2011/09/duna-de-bolonia-punta-camarinal-cabo-de.html [accessed October 20, 2012].

CA-2b: Hiking itinerary: "Canteras Romanas de Paloma Alta": http://dcaminata.wordpress.com/2013/01/08/canteras-romanas-de-paloma-alta/ [accessed December 28, 2012].

CA-3; CA-5: Prudente ARJONA LOBATO, Historias Populares de Rota: Molinos, Tahonas y Tahoneros (II), Cosas de Andalucía: http://www.cosasdeandalucia.com/web/index.php/memoria-historica/nuestros-ayeres/1658.html [accessed November 25, 2012].

CA-8: Archive of Vejer de la Frontera: http://www.patrimoniojandalitoral.es/f_vai_8.htm [accessed February 12, 2013].

CA-11: Manuel CABELLO and Esperanza IZQUIERDO, Un recorrido por la "Trompa del Elefante", January 4, 2012: http://manuelcabelloyesperanzaizquierdo.blogspot.com.es/2012/01/un-recorrido-por-la-trompa-del-elefante.html [accessed November 6, 2012].

CA-11: Jesús ORTIZ, Rutas Serranas de la Provincia de Cádiz: http://elsenderodejesus.blogspot.com.es/2013/02/7-travesia-guia-de-la-naturalezaruinas.html [accessed February 23, 2013].

CA-12: Leandro CABELLO, ¿Como se puede trasladar una piedra de molino? in the blog of M. CABELLO and E. IZQUIERDO, October 27, 2010: http://manuelcabelloyesperanzaizquierdo.blogspot.com.es/2010/10/como-se-puede-trasladar-una-piedra-de.html [accessed November 6, 2012].

Province of Seville

SE-1: Television interview with Miguel Ángel VARGAS: http://www.canalsur.es/portal_rtva/web/noticia/id/112752/portada/hallada_una_cantera_de_epoca_romana_en_almaden [accessed November 12, 2011].

SE-4: http://rutadeltempranillo.org/index.php?page=86 [accessed Octorber 14, 2012].

SE-4: Francisco ESTEPA: http://historiadecasariche.blogspot.com.es/p/23-cerro-bellido.html [accessed Octorber 14, 2012].

SE-7: Old photographs of Gerena quarries from ACUÑA CARABANTES, 2004.

SE-7: Gonzalo HERDUGO: http://gerenadiario.blogspot.com.es/2009/03/blog-post.html [accessed December 2, 2013].

SE-7: El Ferrocarril De Aznalcóllar - Guadalquivir: http://www.portalaznalcollar.com/ferrocarril/ferrocarril.htm [accessed October 15, 2012].

SE-7: Hiking itinerary: http://montesysenderos.wordpress.com/2012/05/23/sendero-de-los-molinos-de-gerena/ [accessed October 15, 2012].

Province of Huelva

HU-1; HU-2; HU-3; HU-4: Alonso GARCÍA VEIGA. Canteras Medievales de El Campillo, Canteras Medievales de Piedras de Molino. Las Piedras Molares: http://alongarvi.blogspot.com.es/2009/11/canteras-medievales-de-el-campillo.html [accessed November 12, 2012].

HU-5: Hiking itinerary: http://montesysenderos.wordpress.com/2012/04/18/minas-de-tharsis-y-ermita-de-la-virgen-de-la-pena/ [accessed November 16, 2012].

HU-5: Anonymous black and white photograph: http://www.pueblos-espana.org/andalucia/huelva/puebla+de+guzman/galeria-fotografica/ [accessed January 7, 2013).

HU-8: Archaeological inventory of the Province of Huelva: Patrimonio Inmueble de Andalucía, Denominación: Molares, Código: 01210040006, Caracterización: Arqueológica, Provincia: Huelva, Municipio: Almonaster la Real; http://www.iaph.es/patrimonio-inmueble-andalucia/resumen.do?id=i14744 [accessed November 20, 2012].

MURCIA

MU-1: Hiking itinerary: http://senderosdecartagena.wordpress.com/2010/12/16/cabezo-del-puerto-de-la-cadena/ [accessed October 18, 2012].

MU-3: Brochure: Donde Vive el Buho. Finca Municipal los Porches. Programa de custodia del territorio entre el Ayuntamienteo de Murcia y la Asociación Vecinal para el dessarrollo sostenible de Garruchal (ADESGA): http://www.magrama.gob.es/es/ceneam/programas-de-educacion-ambiental/programas-de-otras-entidades/murcia.aspx [accessed November 13, 2012].

CASTILLA LA MANCHA

Province of Ciudad Real

CR-5: Juan Jesús DONOSO website "Molino de Viento": http://granatula.com/del-pueblo/granatula-y-sus-molinos/molino-de-viento [accessed November 12, 2012].

CR-6: Website of the research group "Geomorfología, Territorio y Paisaje en Regiones Volcánicas", University of Castilla La Mancha, Columba, El Cabezuelo, las Cuevas: http://www.uclm.es/profesorado/egcardenas/columba.htm [accessed May 23, 2012].

Province of Albacete

AB-1: *Respuestas Generales del Catastro del Marqués de la Ensenada, 1750-1754:* [http://pares.mcu.es/Catastro/servlets/ServletController?accion=2&opcion=10 [accessed 15 of October, 2012].

AB-1 Coat of Arms of Abengibre: from http://es.wikipedia.org/wiki/Abengibre [accessed October 15, 2012].

AB-2: Serge TISSRON: Una Ventana a la Naturaleza, : http://jumillanatural.blogspot.com.es/2013/01/volcan-de-cancarix.html [accessed February 9, 2013].

AB-2: Guía de Geología Albacete, Geología Día 10, El Volcán de Cancarix, 25 abril 2010: http://www.sociedadgeologica.es/archivos_pdf/gdia10_albacete_triptico.pdf [accessed February 9, 2013].

AB-2: CASTAÑO FERNÁNDEZ, Santiago, LÓPEZ ROS, Joaquín, De la MORA, Julián. Itinerarios Geológicos de la Provincia de Albacete: Tobarra, Hellín, Minateda, Cancarix y La Celia, *Separatas de Al-Basit, Revista de Estudios Abacetenses*. 1985, p. 79-125. http://bibliotecahellin.blogspot.com.es/2011/04/itinerario-geologico-de-la-provincia-de.html#.Ud0B7OA27rk: [accessed February 9, 2013].

Province of Toledo

TO-1: Responses to the questionnaire of Archbiship Lorenza of Toledo (1782) for the town of Torrecillas de la Jara were consulted at: http://www.torrecilladelajara.com/historia.htm [accessed October 15, 2012].

TO-2: Photograph of the Cerro del Águila: https://ssl.panoramio.com/photo/55252564 [accessed November 12, 2012].

Province of Cuenca

CU-1: http://www.senderosdecuenca.org/Portals/111/senderos_pdf/PR_30_sierraAlta.pdf [accessed November 2, 2012].

CU-1: Photographs from "CampiSierra": http://www.campisierra.org/portal/lang_es-ES/p28615_2/tabid_11980/default.aspx [accessed November 2, 2012].

Province of Guadalajara

GU-3: Oscar QUIRÓS. "Ruguila: El Aroma del Silencio": http://loscuadernosdeoscarquiros.blogspot.com.es/2007_12_01_archive.html [accessed November 12, 2012].

GU-4: Histgüeb, "Paseando entre antiguas minas": http://www.histgueb.net/minas_petroleo/lazona.htm [accessed November 2, 2012].

GU-5: http://www.pueblos-espana.org/castilla+la+mancha/guadalajara/tobes/ [accessed November 12, 2012].

GU-5: Photographs, Faustino CALDERÓN: http://lospueblosdeshabitados.blogspot.com.es/2010/05/tobes-guadalajara.html [accessed November 12, 2012].

GU-8: *Relaciones Topográficas de los Pueblos de España de Felipe II*, 1574-1578. Montarrón: http://www.uclm.es/ceclm/b_virtual/libros/relaciones_gu/MONTARRÓN.htm [accessed November 17, 2012].

VALENCIA

Province of Valencia

V-1: Website of the Municipality of Canals: http://www.canals.es/canals/index.php/ciudad/paisaje.html [accessed November 22, 2012].

V-2: "Montesa, Enclave Megalítico": http://viajesmagicosporiberia.blogspot.com.es/ [accessed January 7, 2013].

Province of Castillón

CS-1: Hiking itinerary: http://www.soneja.es/node/249 [accessed November 2, 2012].

CS-1: Hiking itinerary "Andamontes": http://andamontesenblog.blogspot.com.es/2013/03/laguna-de-la-dehesa-paraje-natural.html, [accessed April 21, 2013].

CS-1: Short video: http://www.youtube.com/watch?v=yFdj83OQTe0 [accessed November 2, 2012].

CS-1: Biking itinerary"BikepedalValencia": http://bikepedalvalencia.blogspot.com.es/2013/07/alfara-de-la-baronia-sot-de-ferrer.html.

EXTREMADURA

Province of Badajoz

BA-1: Website of the town of Alconera with the Royal Census of 1791: http://www.nuevoportal.com/andando/pueblos/extrema/badajoz/alconera.html [accessed October 25, 2012].

BA-1: Website El patrimonio histórico y cultural de la comarca: http://www.nodo50.org/asierrasuroeste/patrimonio.htm [accessed October 25, 2012].

BA-2: 1566 Ordinance blog by MALDONADO FERNÁNDEZ, Manuel: http://manuelmaldonadofernandez3.blogspot.com.es/2010_04_01_archive.html [accessed October 25, 2012].

BA-2: El Molar blog by MENA CABEZAS, Juan Eugenio: http://juaneugeniomena.blogspot.com.es/2010/12/etnologia-para-mejorar-la-oferta.html [accessed October 25, 2012].

BA-5: Forum of the Comarca de Jerez de los Caballeros: http://www.foros.net/search.php?mforum=jerezhabla [accessed July 3, 2012].

Province of Cáceres

CC-1: Montaña DOMÍNGUEZ CARRRO: http://blogs.hoy.es/geografiainterior/2012/10/01/una-cantera-unica-en-espana-en-pleno-barrio-del-pilar [accessed 25 December, 2012].

CC-3: Forum Arqueólogos: http://www.historiayarqueologia.com/photo/cazoletas-en-villar-de-7 [accessed February 11, 2013].

CC-4: Javier PÉREZ ROSS "Cosas de un Extremeño": http://extremadurense.blogspot.com.es/2012/06/molino-gualija-peraleda-de-san-roman.html [accessed February 15, 2013].

CC-5: Internet photographs: http://www.pueblos-espana.org/extremadura/caceres/guijo+de+galisteo/galeria-fotografica/ [accessed February 13, 2012].

CC-6: Geological website by Juan GIL MONTES: http://jugimo.blogspot.com.es/2010/12/las-villuercas-lugares-de-especial.html [accessed October 25, 2012].

MADRID

M-1: Website of Gloria GÓMEZ TRAPERO, http://prejubiladasinfronteras-glo.blogspot.com.es/2012/10/el-berrueco-madrid.html.

M-1: DÍAS MARTÍNEZ, Enrique, RODRÍGUEZ ARANDA, Juan Pablo. *Itinerario geológico por el norte de la Comunidad de Madrid. Paseos por la Geología Madrileña*. 1 Instituto Geológico y Minero de España. Archivo PDF para la página web de la Comunidad de Madrid, 2007 [accessed October 25, 2012].

M-2: Canteras Históricas de la Comunidad de Madrid website: http://www.madrimasd.org/cienciaysociedad/patrimonio/rutas/geomonumentales/rutas/canteras/calizas.asp?pest=3 [accessed October 26, 2012].

M-2: Carlos Osorio blog "Caminando por Madrid": http://caminandopormadrid.blogspot.com.es/2012/02/dos-ruedas-de-molino-en-una-corrala-de.html [accessed October 26, 2012].

M-2: Community of Madrid website: http://www.colmenardeoreja.es/colmenardeoreja/opencms/site/web/conoce_el_pueblo/colmenar_de_oreja/lugares_de_interes/canteras_de_piedra/ [accessed October 26, 2012].

M-3: Canteras Históricas de la Comunidad de Madrid website: http://www.madrimasd.org/cienciaysociedad/patrimonio/rutas/geomonumentales/rutas/canteras/calizas.asp?pest=3 [accessed 22 December 2012].

M-3: Tesoros de ayer website: http://www.tesorosdelayer.com/esp/subseccion.php?id=31 [accessed 22 December 2012].

M-3: Todo coleccion: http://www.todocoleccion.net/colmenar-viejo-madrid-vista-una-cantera-microgranito~x21497345 [accessed 22 December 2012].

Annexes

1. List of millstone quarries
2. Spanish-English millstone quarry glossary
3. Extracts from the dictionaries of Sebastián Miñano and Pascual Madoz
4. Pinilla de Jadraque: estimated distances from quarry to mill

Annex 1

List of millstone quarries

Province	Municipality	Name	Code	Field confirmed	Confirmed by photography	Indicated in text, not confirmed in field	Supposed
Granada	Moclín	Canteras	GR-1a	X	-	-	-
Granada	Moclín	Las Pedrizas	GR-1b	X	-	-	-
Granada	Loja	Cerro de Fuensanta	GR-2	X	-	-	-
Granada	Loja	Camino del Calvario	GR-3	X	-	-	-
Granada	Loja	La Merced 1	GR-4a	X	-	-	-
Granada	Loja	La Merced 2	GR-4b	X	-	-	-
Granada	Loja	La Merced 3	GR-4c	X	-	-	-
Granada	Zagra	La Atalayuela	GR-5	X	-	-	-
Granada	Alhama de Granada	Fuente de los Morales	GR-6	X	-	-	-
Granada	Padul	Barranco de los Guillares	GR-7	X	-	-	-
Granada	Los Guájares	Los Mochos	GR-8	X	-	-	-
Granada	Motril	Playa de Carchuna	GR-9	X	-	-	-
Granada	Vélez de Benaudalia	Barranco de la Piedras	GR-10	X	-	-	-
Granada	Caniles	Rambla de las Canteras	GR-11	X	-	-	-
Granada	Pinos Puente	Cantera de Zujaira	GR-12	-	-	-	X
Granada	Otívar		GR-13	-	-	X	-
Granada	Ugíjar	Las Canteras	GR-14	X	-	-	-
Almería	Níjar	Cerro de Limones	AL-1	X	-	-	-
Almería	Níjar	La Hoya del Paraíso	AL-2	X	-	-	-
Almería	Albox	Cantera de la Rambla Honda	AL-3a	X	-	-	-
Almería	Albox	Los Leonardos	AL-3b	X	-	-	-
Almería	El Ejido	Guardias Viejas	AL-4	X	-	-	-
Almería	Alcolea	Barranco Baena-Pedreros	AL-5	-	-	X	-
Almería	Bayárcal	Barranco de Palancón	AL-6	-	-	X	-
Almería	Sorbas	Los Loberos	AL-7	-	-	X	-
Almería	Adra	Cerro el Chispas	AL-8	X	-	-	-
Almería	Vera		AL-9	-	-	X	
Almería	Níjar	El Barronal	AL-10	-	-	X	
Almería	Níjar	Rodalquilar	AL-11	-	-	-	X
Jaén	Castillo de Locubín	Las Canteras	J-1	X	-	-	-
Jaén	Jimena	El Lachar	J-2	X	-	-	-
Jaén	Huelma-Solera	Las Canteras	J-3	-	-	X	-
Jaén	Cambil	Arbuniel – Los Batanes	J-4	-	X	-	-
Jaén	Andújar	Morales	J-5	-	-	X	-
Jaén	Andújar	El Pedroso	J-6	-	-	X	-
Jaén	Linares		J-7	-	-	X	-

Province	Municipality	Name	Code	Field confirmed	Confirmed by photography	Indicated in text, not confirmed in field	Supposed
Córdoba	Cabra	Cantera de los Frailes	CO-1	X	-	-	-
Córdoba	Cabra	Los Cortaores	CO-2	X	-	-	-
Córdoba	Carcabuey	Cudillas	CO-3	X	-	-	-
Córdoba	Priego de Córdoba	Vega de los Morales	CO-4	-	X	-	-
Córdoba	Baena	Molino de la Piedra	CO-5	X	-	-	-
Córdoba	Baena	Monte de Iscar (Izcar)	CO-6	X	-	-	-
Córdoba	Córdoba	Santa Ana de Albaida	CO-7	X	-	-	-
Córdoba	Posadas	Cantera Honda	CO-8	-	X	-	-
Córdoba	Córdoba	Los Arenales	CO-9	-	-	X	-
Córdoba	Bélmez	La Pedrera, Arroyo Albardado	CO-10	-	X	-	-
Córdoba	Villanueva de Córdoba	Piedras Moleras	CO-11	-	-	-	X
Córdoba	Córdoba	El Partiarca	CO-12	X	-	-	-
Córdoba	Hornachuelos		CO-13	-	-	X	-
Córdoba	Montoro		CO-14	-	-	X	-
Córdoba	Almendinilla		CO-15	-	-	-	X
Málaga	Anterquera	El Torcal Bajo	MA-1	X	-	-	-
Málaga	Teba	El Tajo	MA-2	X	-	-	-
Málaga	Alhaurín el Grande	Las Canteras	MA-3	-	-	X	-
Málaga	Coín	Sierra Gorda	MA-4	-	-	X	-
Málaga	Guaro		MA-5	-	-	X	-
Málaga	Alozaina	Mulera	MA-6	-	-	X	-
Málaga	Málaga	El Jabonero	MA-7	-	-	-	X
Málaga	Casares	Sierra de Utrera Karst	MA-8	-	X	-	-
Cádiz	Barbate	Bahía de Trafalgar	CA-1	X	-	-	-
Cádiz	Tarifa	Punta Camarinal	CA-2	-	-	-	X
Cádiz	Rota	Bahía de Cádiz	CA-3	X	-	-	-
Cádiz	Rota	Playa de Aguaduce	CA-4	-	-	-	X
Cádiz	Rota	Roa Martín	CA-5	-	-	X	-
Cádiz	Chipiona	Playa de las Canteras	CA-6	X	-	-	-
Cádiz	Tarifa	Isla de la Paloma	CA-7	-	-	X	-
Cádiz	Medina Sidonia	El Berrueco	CA-8	X	-	-	-
Cádiz	Medina Sidonia	Pila de Casares	CA-9	X	-	-	-
Cádiz	Alcalá de los Gazules	Peña Arpada	CA-10	X	-	-	-
Cádiz	Ubrique	Salto de la Mora	CA-11	-	-	X	-
Cádiz	Benaocaz	El Esparragal	CA-12	-	-	X	-
Cádiz	San Roque	Las Canteras – Guadalquitón	CA-13	-	-	X	-
Seville	Almadén de la Plata	Arroyo de la Calzadilla 1	SE-1a	X	-	-	-
Seville	Almadén de la Plata	Arroyo de la Calzadilla 2	SE-1b	X	-	-	-
Seville	El Pedroso	El Castillejo	SE-2	X	-	-	-
Seville	Lora de Estepa	El Hacho	SE-3	X	-	-	-
Seville	Casariche	Cerro Bellido	SE-4	X	-	-	-
Seville	Alanís		SE-5	-	-	X	-
Seville	Villanueva de San Juan		SE-6	-	-	X	-
Seville	Gerena		SE-7	-	-	-	X
Seville	Estepa	Tajo de Gilena	SE-8	X	-	-	-

Province	Municipality	Name	Code	Field confirmed	Confirmed by photography	Indicated in text, not confirmed in field	Supposed
Huelva	El Campillo	Umbría del Hornillo	HU-1	-	X	-	-
Huelva	Linares de la Sierra	El Prao de Abad I-II	HU-2	-	X	-	-
Huelva	Linares de la Sierra	Las Malenas I-II	HU-3	-	X	-	-
Huelva	Aracena	La Obra Pía	HU-4	-	X	-	-
Huelva	Puebla de Guzmán	Cerro del Águila	HU-5	-	-	X	-
Huelva	Zalamea la Real		HU-6	-	-	X	-
Huelva	Aroche	Fuente de la Aliseda	HU-7	-	-	X	-
Huelva	Almonaster la Real	Los Molares	HU-8	-	-	X	-
Murcia	Murcia	Puerto de la Cadena	MU-1	X	-	-	-
Murcia	Mazarrón	Cabezo de la Oliva	MU-2	-	-	X	-
Murcia	Murcia	Cantera de los Porches	MU-3	-	X	-	-
Murcia	Fortuna	Sierra de los Baños	MU-4	-	X	-	-
Ciudad Real	Almodóvar del Campo	Sisapo-Bienvenida	CR-1	X	-	-	-
Ciudad Real	Bolaños de Calatrava	Cantera de las Herrerías	CR-2	X	-	-	-
Ciudad Real	Chillón		CR-3	-	-	X	-
Ciudad Real	Alcázar de San Juan	Pedrizas de Piédrola	CR-4	-	X	-	-
Ciudad Real	Granátula de Calatrava	Las Canteras	CR-5	-	X	-	-
Albacete	Fuentealbilla	El Molar	AB-1	-	-	X	-
Albacete	Hellín	Pitón de Cancarix	AB-2	-	-	-	X
Toledo	Torrecilla de la Jara		TO-1	-	-	X	-
Toledo	Las Ventas con Peña Aguilera		TO-2	-	-	X	-
Cuenca	Portilla	Los Molares	CU-1	-	X	-	-
Guadalajara	Brihuega		GU-1	-	-	-	X
Guadalajara	Pinilla de Jadraque	Monasterio de San Salvador	GU-2	-	-	-	X
Guadalajara	Cifuentes	Ruguilla	GU-3	-	-	-	X
Guadalajara	Sigüenza	La Cuerda	GU-4	-	X	-	-
Guadalajara	Sienes	Tobes	GU-5	-	-	-	X
Guadalajara	Corduente		GU-6	-	-	-	X
Guadalajara	Castilnuevo		GU-7	-	-	-	X
Guadalajara	Montarrón	Los Morales (Los Molares)	GU-8	-	X	-	-
Guadalajara	Cobeta	Barranco de Arrandilla	GU-9	-	X	-	-
Alicante	San Fulgencio	Sierra del Molar	A-1	-	-	X	-
Alicante	Ibi	Barranco de los Molinos	A-2	-	X	-	-
Valencia	Canals	Les Moles	V-1	-	X	-	-
Valencia	Montesa	La Mola	V-2	-	X	-	-
Castellón	Soneja	Los Arenales	CS-1	-	X	-	-
Castellón	Nules, Judicial District		CS-2	-	-	X	-
Badajoz	Alconera	Las Pedreras	BA-1	-	-	X	-
Badajoz	Llerena	El Molar	BA-2	-	-	X	-
Badajoz	Llera		BA-3	-	-	X	-
Badajoz	Salvaleón		BA-4	-	-	X	-
Badajoz	Jerez de los Caballeros		BA-5	-	X	-	-

Province	Municipality	Name	Code	Field confirmed	Confirmed by photography	Indicated in text, not confirmed in field	Supposed
Badajoz	Zafra (Judicial District)		BA-6	-	-	X	-
Badajoz	Mérida district		BA-7	-	-		X
Cáceres	Villar de Plasencia		CC-2	-	-	X	-
Cáceres	Villar de Plasencia		CC-3	-	X	-	-
Cáceres	Bohonal de Ibor	Molino Gualija	CC-4	-	X	-	-
Cáceres	Guijo de Galisteo	Dehesa Boyal	CC-5	-	X	-	-
Cáceres	Logrosán	Sierra de San Cristóbal	CC-6	-	-	X	-
Madrid	El Berrueco		M-1	-	X	-	-
Madrid	Colmenar de Oreja	Las Canteras	M-2	-	-	X	-
Madrid	Colmenar Viejo		M-3	-	-	X	-
Madrid	Chapinería		M-4	-	-	X	-
Madrid	Miraflores de la Sierra		M-5	X	X	-	-

Annex 2

Spanish-English millstone quarry glossary

Molino: mill

Molino vaivén: saddle quern

Molino rotatorio manual: rotary quern

Tahona: animal-driven flour mill or, probably more often, a bakery (originally from Arabic)

Aceña: watermill with vertical waterwheel (originally from Arabic)

Rodenzo: watermill with horizontal waterwheel

Almazara: olive oil mill (originally from Arabic)

Muela: millstone

Muela: quern or millstone. This term (like the French *meule*) covers all types of millstones, from the earliest saddle quern to the most recent French burrstone import. Not to be confused with *muela* meaning tooth (molar), *muela* meaning hill with a flat top or m*uela (almorta)*, an indigenous legume (*Lathyrus sativus*), growing especially in the north of Spain and consumed by humans in the 19th century

Francesas: literally "French", refers to siliceous burrstones imported from quarries in France (such as La Ferté sous Jouarre in the Paris Basin). Used with *muelas* and *piedras*

Grano: millstone specifically for *tahonas* (animal-driven mills). Possibly a local term used at the Berrueco quarry site (CA-8). No other reference to this term has been identified with millstones in any dictionaries

Volandera: upper stone

Piedra blanca: upper stone

Corredera: upper stone

Piedra móvil: upper stone

Solera: lower stone

Yusera: lower stone

Bermeja: lower stone

Piedra baza: lower stone

Cantera: quarry

Pedrera: quarry

Tajo: quarry, mine

Molera: millstone quarry (term from the *fueros* or codices)

Amoladera: sharpening stone (whetstone) quarry

Cantero: quarryman

Picapedrero: quarryman

Pedrero: quarryman

Molero: millstone maker or millstone merchant

Extractive terminology

Sacar: to extract

Extraer: to extract

Senos: hollows

Nidos: hollows

Rocks types

Almendrilla (piedra): conglomerate (puddingstone) with "almond"-sized clasts

Berroqueña (piedra): term usually designates granite

Blanca (piedra): term at times designates white limestone rock or white (biotite) granite. It also designates the upper stone of a mill

Bravía (piedra): coarse stone

Jabaluna (piedra): a type of limestone (not for millstones)

Lancha: large slab of rock (natural formation)

Molar (piedra): silicieous millstone rocks (at times a reference to burrstone)

Molar (piedra): in the Palencia mountains, refers to conglomerates with small rounded clasts

Molinaza (piedra): fine, reddish sandstone (*Bundsandstein*)

Panalizo (piedra): rock that is soft when extracted but becomes very hard when dry

Piedra basta: literally coarse stone; it probably refers to conglomerates or puddingstones with large clasts

Piedra de grano: in the Palencia mountains, this term refers to conglomerates or puddingstones with pebble inclusions (Basterra 2003: 249). In certain contexts corresponds to granite (presumably white biotite granite)

Sal y pez (piedra): designates a white and black biotite granite. *Sal* (salt) is white and *pez* (tar, not fish) is black

Toba or "*tova*" (*piedra calcarea*): calcareous or limestone tufa

Toba volcánica: volcanic tuff

Tosca (piedra): porous limestone

Travertino: used to designate limestone tufa

Rock formations

Berrueco, barrueco, berrocal: terms from the same root that designate granite landscapes. They are often made up of rounded boulders of different sizes sculpted by erosion. The term can refer to a small elevation or to a vast region

Bolo, bola: rounded granite surface boulders of different sizes sculpted by erosion. Typical of *Berroqueño* landscapes

Piedra caballera: pedestel rock, naturally sculpted boulder perched in equilibrium on a small base. This feature is known in *berroqueño* and limestone karstic landscapes

Canchal or *canchales* : accumulation of rocks, such as talus, usually with angular shapes

Lanchar: terrain with "*lanchas*" (slabs)

Peña: crag

Peñasco: crag

Pedregal: terrain with loose rocks

Dwellings

Covachas: hovel

Caracolas: hovel

Flour and bread

Blanco: white

Harina: flour

Pan: bread

Prieto: dark

Moreno: dark

Bazo or *Baxo*: dark; at times interpreted as *bajo* meaning low, as in unleavened bread

Millstone transport

Arriero: transporter of goods by mule

Carretero: transporter of goods by cart

Barquero: transporter of goods by boat

Corza: y-shapped wooden "sled" dragged by animals used to transport millstones

Narria: simple wooden sled

Carro: cart

Caballeria: equine (horse, mule, donkey)

Mula: mule

Buey: ox

Ownership, laws

Cabildo: ecclesiastic body that make up a ruling council; at times refers to the council

Consejo: council

Fuero: legal document or codex regulating the laws in land occupied after the Islamic domination

Apeo: legal document, related to the transfer of ownership of property

Pregón: ordinance transmitted orally

Annex 3

Extracts from the Madoz dictionary (1845-1850)

Site		Year	Vol.	Page	Reference
GR-1	Moclín	1847	8	480	*"...y las de Moclin y Velez de Benaudalla escellentes piedras de molino."*
GR-2	Loja	1847	10	360	*"... las hay de panalizo blanco y rosado en el Cojin de Loja, para piedras de molinos harineros."*
GR-6	Fuente de los Morales	1847	8	216	*"Es notable por sacarse de su terreno las mejores muelas de molino para pan blanco, de que se surten hasta los pueblos de la prov. de Malaga"*
GR-10	Velez de Benaudalla	1847	8	480	*"...y las de Moclin y Velez de Benaudalla escellentes piedras de molino."*
GR-11	Caniles	1846	5	461	*"Al E., y dist. de 1 leg. de pueblo, hay una cantera de piedras de molino para pan moreno..."*
AL-8	Adra,Cerro el Chispa	1845	1	88	*"... Hácia el E. á dist. de 1/2 cuarto hora existe una cantera de piedras de molino."*
AL-9	Vera	1850	15	670	*"... contiene canteras de yeso y para baldosas y de ruedas de molino."*
J-1	Catillo de Locubín	1845	1	382	*"A la falda del O. de la Acamuña hay una elevacion que se denomina las Canteras, pues de alli re sacan las piedras de molino."*
J-3	Huelma	1847	9	260	*"Hay 3 canteras de piedra, una de ellas en el punto llamado la* Cantera*, de la cual se sacan piedras para molino y otros usos..."*
J-5	Andújar, Los Morales	1845	2	305	*"Las canteras de granito, que solo se emplean en molinos de pan y de aceite, puede decirse que son innumerables; pero las que estan en uso son las del Pedroso y Morales."*
J-6	Andújar, El Pedroso	1845	2	305	*"Las canteras de granito, que solo se emplean en molinos de pan y de aceite, puede decirse que son innumerables; pero las que estan en uso son las del Pedroso y Morales."*
J-7	Linares district	1847	10	290	*"... adundan las canteras de arenisca, escelente para la construccion; de granito para molinos harineros y de aceite ..."*
CO-9	Los Arenales	1849	15	136	*"...y en el sitio que llaman los Arenales una cantera para piedras de molino."*
CO-10	El Albardado	1846	4	131	*"En el arroyo Albardado se encuentra una cantera de pierra basta de molino, de donde se proveen los pueblos inmediatos y aun algunos mas distantes ..."*
CO-10	El Albardado	1847	9	39	*"...y por la izq. á otro 1/4 de leg. de dist. el Albardado, notable por las diferentes canteras para piedras de molino que hay á sus inmediaciones ..."*

Code	Site	Year	Vol.	Page	Reference
CO-16	Minas de Espiel district	1849	14	382	*"Corre hácia el SO., y se compone de capas de piedra arenisca, mas ó menos basta, que prinicpalmente suele usarse para construir piedras de molino ..."*
CO-17	Fuente Obejuna	1847	8	230	*"... á las inmediaciones del mismo camino y orillas de los r. Guadiato y Suja, se encuentran 2 abundantes canteras muy útiles para piedras de molino y tahonas ..."*
MA-2	Teba	1949	14	752	*" ... estraccion de piedras de molino para molinos de agua y tahonas, que sacan de 10 canteras de jaspe blanco y encarnado, ..."*
MA-3	Alhaurín el Grande	1845	1	604	*" ... viéndose igualemente al S. del pueblo otras varias canteras de granito, propria para ruedas de molino, pilones, umbrales ..."*
MA-5	Guaro	1847	9	56	*"... comprende un monte llamado del Señor, poblado de encinas, alcornoques y matas bajas, y una cantera apropósito para piedras de molino harinero."*
MA-6	Alozaina	1845	2	186	*"En el de Alozayna se encuentran canteras para piedras de molino de mérito particular, ..."*
CA-8	El Berrueco	1846	4	290	*"BERRUECO (EL): monte en la prov. de Cádiz, part. jud. y térm. jurisd. de Medina-Sidonia... La piedra de que está compuesto es blanquísima y dura, famosa en toda la prov., por sacarse de ella las mejores muelas de molino: hay 5 pedreras en las que se ocupan 23 hombres, que viven alli mismo, con privilegio de uso de armas, no solo por la obligacion que tienen de trabajar graciosamente en allanar algunas eminencias de piedra en Medina, sino porque persiguen á los rateros que aparecen á veces por aquellos cotornos. Sacan el año que mas, 64 piedras para molinos de agua ó viento, y 480 granos ó sean piedras para moler en las tahonas, redondeándolas y poniéndolas en estado de servir para el objeto: de los desperdicios sacan la mejor cal...*
CA-8	El Berrueco	1846	5	140	*"Piedras de amolar y de molino: las primeras se labran en Ubrique ... las segundas en el cerro del Berrueco de Medina.*
CA-8	El Berrueco	1848	11	343	*"En Medina hay fáb. de alfarería; otra de tejidos de lienzo; otra de piedras de molinos en el Berrueco."*
CA-10	Peña Harpada	1845	1	376	*"... hay una cantera de piedra para ruedas de molinos harineros en* Peña-Jarpa, *..."*
SE-3	Lora de Estepa	1847	7	690	*"... en la sierra existen tres canteras, una de jaspe aceitoso y dos de cipia; en la nombrada el Hacho otra de piedra bravia, blanca y apropósito para molinos de harina ..."*

	Site	Year	Vol.	Page	Reference
SE-5	Alanís	1845	1	190	*"... una cantera de piedra azul, á una leg. al SE., escelente para afilar las navajas de afeitar, y otras de piedras para molinos ..."*
SE-6	Villanueva de San Juan	1850	16	207	*"... y una cantera de molino que no se beneficia por falta de caminos."*
SE-7	Gerena	1847	8	348	*"... muchas canteras de piedra que se emplea bastante para los molinos de aceite."*
HU-6	Zalamea la Real	1850	16	450	*"El TERRENO es casi todo de sierra y secano... Encuéntranse diferentes canteras para piedras de molinos."*
CR-3	Chillón	1847	7	327	*"... en él se encuentra 2 canteras, una de piedra de granito para molinos ..."*
GU-6	Corduente	1847	7	9	*"... hay canteras de buenas piedras de molino ..."*
GU-7	Castilnuevo	1847	6	172	*"... varias canteras de piedra tova, de las cuales se surten los molinos harineros de la circunferencia."*
AB-1	Fuentealbilla	1845	1	51	*"... IND.: la agricultura, ganadería, fáb. de alpargatas, laboreo de piedras de molino que se estraen del sitio llamado el Molar, en el térm. de Fuentealbilla ..."*
AB-1	Fuentealbilla	1845	1	256	*"... y en le Molar, térm. de Fuente-albilla, del mismo part., la de piedras de molínos, de grano poco compacto, por lo que se aprecia menos que la barcelonesa."*
BA-4	Salvaleón	1849	13	711	*"Hay otros riscos mas pequeños; dos canteras de piedra de grano; una de piedra blanca para las ruedas de molino ..."*
BA-5	Jerez de los Caballeros	1847	9	627	*"... hay tambien canteras de piedras morena y blanca para molinos harineros ..."*
CC-4	Bohonal de Ibor	1846	4	376	*"El TERRENO participa de monte y llano, bastante quebrado, y con muchos berrocales y canteras de piedra muy dura, de donde se estraen buenas ruedas de molino ..."*
CC-6	Logrosán	1847	10	355	*"Al S. de la cab. del part. hay otra sierra bastante elevada, llamada de San Cristobal; ... hay en ella canteras de piedra de grano y se estraen muchas de molino y de lagar ..."*
M-2	Colmenar de Oreja	1847	6	525	*"... á la parte del N. se ven canteras de piedra blanca llamada de Colmenar, de donde se estrajo la que se necesitó par los palacios de Madrid y Aranjuez y otros muchos edificios; en el dia hay poca extraccion, estando limitada á algun baño y piedra para las tahonas de Madrid."*
M-3	Colmenar Viejo	1847	6	530	*"... muchas canteras de piedra granito, de las que se sacan piedras para molinos harineros ..."*

Annex 4

Pinilla de Jadraque: Estimated distances from quarry to flour mill

List of the towns and hamlets (21) and the number of their flour mills in a radius of 20 km from the millstone quarry of Pinilla de Jadraque (GU-2). Pinilla, according to Miñano had "excellent millstones ... extracted from its precious quarry for other towns" (Miñano 1827, Vol. 7: 24). The list, based on data gleaned from Madoz (1845-1850) two decades after the Miñano text, suggests that the number of flour mills in the radius of 20 kilometres (total 14), would not have furnished enough demand for the Pinilla de Jadraque quarry. The quarry presumably exported its millstones to mills beyond the 20 km sphere.

Town	Distance from Pinilla quarry (km)	No. of flour mills	Bibliography
Pinilla de Jadraque	1	1	Madoz 1849, Vol. 13: 36
Pálmaces	4	1	Madoz 1849, Vol. 12: 610
Congostrina	4	0	Madoz 1847, Vol. 6: 564
Torremocha de Jadraque	5	0	Madoz 1849, Vol. 15: 94
La Toba	5	0	Madoz 1849, Vol. 14: 767
Medranda	6	2	Madoz 1848, Vol. 11: 350
San Andrés de Congosto	8	0	Madoz 1844, Vol. 2: 298
Zarzuela de Jadraque	8	0	Madoz 1850, Vol. 16: 663
Hiedelencina	9	0	Madoz 1847, Vol. 9: 143
Cendejas de Enmedio	9	0	Madoz 1847, Vol. 6: 307
La Bodera	11	1	Madoz 1846, Vol. 4: 371
Jadraque	13	2	Madoz 1847, Vol. 9: 492
Gascueña de Bornova	13	0	Madoz 1847, Vol. 8: 328
Matillas	15	1	Madoz 1848, Vol. 11: 312
Semillas	15	1	Madoz 1849, Vol. 14: 168
Riofrío del Llano	15	0	Madoz 1849, Vol. 13: 489
Baides	15	0	Madoz 1846, Vol. 3: 301
Bustares	15	1	Madoz 1846, Vol. 4: 677
Cogolludo	16	3	Madoz 1847, Vol. 6: 513
Espinosa de Henares	17	1	Madoz 1847, Vol. 7: 574
Villanueva de Argecilla	18	0	Madoz 1850, Vol. 16: 200
Total flour mills		**14**	